Lecture Notes in Computer Science 9943

Commenced Publication in 1973
Founding and Former Series Editors:
Gerhard Goos, Juris Hartmanis, and Jan van Leeuwen

More information about this series at http://www.springer.com/series/7407

Jin Akiyama · Hiro Ito
Toshinori Sakai · Yushi Uno (Eds.)

Discrete and Computational Geometry and Graphs

18th Japan Conference, JCDCGG 2015
Kyoto, Japan, September 14–16, 2015
Revised Selected Papers

Springer

Editors
Jin Akiyama
Tokyo University of Science
Tokyo
Japan

Hiro Ito
The University of Electro-Communications
Tokyo
Japan

Toshinori Sakai
Tokai University
Tokyo
Japan

Yushi Uno
Osaka Prefecture University
Sakai
Japan

The original version of the cover and the front matter pages III–V were revised: The sequence of the editor names was incorrect. Yushi Uno was not listed as volume editor. The erratum to the cover and the front matter pages III–V is available at DOI: 10.1007/978-3-319-48532-4_26

ISSN 0302-9743   ISSN 1611-3349   (electronic)
Lecture Notes in Computer Science
ISBN 978-3-319-48531-7   ISBN 978-3-319-48532-4   (eBook)
DOI 10.1007/978-3-319-48532-4

Library of Congress Control Number: 2016956484

LNCS Sublibrary: SL1 – Theoretical Computer Science and General Issues

Printed on acid-free paper

This Springer imprint is published by Springer Nature
The registered company is Springer International Publishing AG
The registered company address is: Gewerbestrasse 11, 6330 Cham, Switzerland

Preface

This volume of proceedings contains the peer-reviewed papers presented during the 18th Japan Conference on Discrete and Computational Geometry and Graphs (JCDCG^2 2015). JCDCG^2 2015 was held at Kyoto University, Kyoto, Japan, during September 14–16, 2015.

This volume is dedicated to Prof. Naoki Katoh (Fig. 1) for his retirement from Kyoto University and for his new journey to Kwansei Gakuin University. Professor Katoh obtained his Bachelor, Master, and Doctor of Engineering degrees from Kyoto University, in 1973, 1975, and 1981, respectively. He was first employed by Kobe University of Commerce as a lecturer in 1981, and then became Associate Professor and (Full) Professor before he went back to his alma mater in 1997. At Kyoto University, he entered and worked as a professor at the Department of Architecture and Architectural Engineering, Faculty of Engineering. After 18 years of dedication, he retired from Kyoto University and moved to the Department of Informatics, Faculty of Science and Technology at Kwansei Gakuin University where he is now a professor. Professor Katoh has written 137 refereed academic papers in highly competitive journals, such as *J. ACM, SIAM J. Comput., SIAM J. Discrete Math., Discrete & Comput. Geometry*, and *Combinatorica*. Moreover, he has 94 peer-reviewed papers in various international conference proceedings, which includes STOC, FOCS, SODA, and SoCG. His diverse research output includes operations research, combinatorial optimization, discrete algorithms, computational geometry, etc. His contributions to the aforementioned areas are undoubtedly outstanding and relevant. Thus, it is with great pleasure that we celebrate his move by dedicating these proceedings to him.

Fig. 1. Professor Naoki Katoh

The previous JCDCG^2 conferences were held in Tokyo as JCDCG (Japan Conference on Discrete and Computational Geometry) in 1997, 1998, 1999, 2000, 2002, 2004, 2006, and 2011 and as JCDCG^2 (Japan Conference on Discrete and Computational

Geometry and Graphs) in 2013 and 2014; in Kyoto as KyotoCGGT (Kyoto International Conference on Computational Geometry and Graph Theory) in 2007; and in Kanazawa as JCCGG (Japan Conference on Computational Geometry and Graphs) in 2009. Other conferences in this series were also held in the Philippines (2001), Indonesia (2003), China (2005, 2010), and Thailand (2012). The proceedings of these conferences were published by Springer as a part of the LNCS series in volumes 1763, 2098, 2866, 3330, 3742, 4381, 4535, 7033, 8296, and 8845. The proceedings of the fifth and 12th conferences (2001 and 2009) were published by Springer as special issues of the journal *Graphs and Combinatorics*, Vol. 18, No. 4, 2002 and Vol. 27, No. 3, 2011.

JCDCG^2 2015 received 64 proposals for oral presentations, 60 of which were accepted, and only 25 of these papers are published in these proceedings after careful scrutiny and strict review procedures.

The organizers of JCDCG^2 2015 express deep appreciation to the invited speakers, John Iacono, Toshihide Ibaraki, János Pach, David Rappaport, and Takeshi Tokuyama, for their invaluable contributed talks. It is also appropriate to record the organizers' gratitude to the conference secretariat, to all the speakers, and to each participant. This conference was a success because of all of you.

July 2016

Jin Akiyama
Hiro Ito
Toshinori Sakai
Yushi Uno

Organization

Conference Chair

Naoki Katoh   Kwansei Gakuin University, Japan

Program Committee

David Avis   Kyoto University, Japan, and McGill University, Canada
Erik D. Demaine   MIT, USA
Rudolf Fleischer   Fudan University, China, and GUtech, Oman
Takashi Horiyama   Saitama University, Japan
Hiro Ito   University of Electro-Communications, Japan (Chair)
David Kirkpatrick   University of British Columbia, Canada
Matias Korman   National Institute of Informatics, Japan
Marc van Kreveld   Utrecht University, The Netherland
Stefan Langerman   Université Libre de Bruxelles, Belgium
Yasuko Matsui   Tokai University, Japan
Chie Nara   Meiji University, Japan
Yoshio Okamoto   University of Electro-Communications, Japan
Toshinori Sakai   Tokai University, Japan
Yushi Uno   Osaka Prefecture University, Japan
Jorge Urrutia   Universidad Nacional Autónoma de México, Mexico

Organizing Committee

Shuichi Miyazaki   Kyoto University, Japan
Yota Otachi   JAIST, Japan
Toshinori Sakai   Tokai University, Japan
Kenjiro Takazawa   Kyoto University, Japan
Atsushi Takizawa   Osaka City University, Japan
Shin-ichi Tanigawa   Kyoto University, Japan
Yushi Uno   Osaka Prefecture University, Japan (Chair)
Liang Zhao   Kyoto University, Japan

Steering Committee

Jin Akiyama   Tokyo University of Science, Japan (Chair)
Erik D. Demaine   MIT, USA
Hiro Ito   University of Electro-Communications, Japan
Mikio Kano   Ibaraki University, Japan

Naoki Katoh   Kwansei Gakuin University, Japan
Stefan Langerman   Université Libre de Bruxelles, Belgium
Toshinori Sakai   Tokai University, Japan
Jorge Urrutia   Universidad Nacional Autónoma de México, Mexico

Contents

A Note on the Number of General 4-holes in (Perturbed) Grids

O. Aichholzer[1], T. Hackl[1], P. Valtr[2], and B. Vogtenhuber[1($\boxtimes$)]

[1] Institute for Software Technology, Graz University of Technology, Graz, Austria
bvogt@ist.tugraz.at
[2] Department of Applied Mathematics and Institute for Computer Science (ITI),
Charles University, Prague, Czech Republic

Abstract. Considering a variation of the classical Erdős-Szekeres type problems, we count the number of general 4-holes (not necessarily convex, empty 4-gons) in squared Horton sets of size $\sqrt{n} \times \sqrt{n}$. Improving on previous upper and lower bounds we show that this number is $\Theta(n^2 \log n)$, which constitutes the currently best upper bound on minimizing the number of general 4-holes for any set of n points in the plane.

To obtain the improved bounds, we prove a result of independent interest. We show that $\sum_{d=1}^{n} \frac{\varphi(d)}{d^2} = \Theta(\log n)$, where $\varphi(d)$ is Euler's phi-function, the number of positive integers less than d which are relatively prime to d. This arithmetic function is also called Euler's totient function and plays a role in number theory and cryptography.

1 Introduction

Let S be a set of n points in the plane. A *k-gon* in S is a simple polygon spanned by k points of S, and a *k-hole* in S is a k-gon which does not contain any points of S in its interior.

In a classical 1935 paper by Erdős and Szekeres [9] it was shown that for any k there is a smallest integer $g(k)$ such that any set of $g(k)$ points in general position (no three points on a common line) contains at least one convex k-gon, and the question of determining the value of $g(k)$ was raised. Since then, a lot of effort has been put into settling this question and a number of its variations. One family of these questions deals with (convex and non-convex) k-holes instead of k-gons. For a more detailed introduction into this area as well as a comprehensive overview of the latest results on k-gons and k-holes we refer to the recent publication [3].

In 1983, Horton used a specially constructed family of point sets to show that there exist arbitrarily large sets containing no convex 7-hole [12], by this solving a question of Erdős [8]. Valtr [13] generalized Horton's construction to

This work is partially supported by FWF projects I648-N18 and P23629-N18, by the OEAD project CZ 18/2015, and by the project CE-ITI no. P202/12/G061 of the Czech Science Foundation GAČR, and by the project no. 7AMB15A T023 of the Ministry of Education of the Czech Republic.

© Springer International Publishing AG 2016
J. Akiyama et al. (Eds.): JCDCGG 2015, LNCS 9943, pp. 1–12, 2016.
DOI: 10.1007/978-3-319-48532-4_1

a family of sets that he called *Horton sets*. This family has since then become well-known under this name and plays a special role for deriving good bounds when minimizing the number of k-holes for small k.

In this work we concentrate on the number of general (i.e., convex and non-convex) 4-holes in perturbed grids of size $\sqrt{n} \times \sqrt{n}$, especially in the so-called *squared Horton sets*. Those were first considered by Valtr in [13] (as the set A in Sect. 4); some interesting properties of them can also be found in [7]. Roughly speaking, a squared Horton set is a grid which is perturbed such that every set of originally collinear points forms a Horton set. See Sect. 1.1 for a definition of Horton sets and squared Horton sets.

It was shown in [1] that there cannot be less than $\frac{5}{2}n^2 - O(n)$ general 4-holes in any set of n points in general position. For random point sets, Fabila-Monroy et al. [10] proved that the expected number of general 4-holes in sets of n points distributed uniformly and independently in the unit square is $12n^2 \log n + o(n^2 \log n)$. In [3] (Sect. 5) an upper bound of $O(n^{\frac{5}{2}} \log n)$ general 4-holes in any squared Horton set of n points was given. In the same paper it was also stated that every ε-perturbation $p_\varepsilon(G)$ of an integer grid G of size $\sqrt{n} \times \sqrt{n}$ contains $\Omega(n^2 \log n / \log \log n)$ general 4-holes (proof in [2]). Especially, squared Horton sets are ε-perturbations of integer grids.

In this work we close the gap between the latter two bounds by showing that any ε-perturbation of an integer grid of size $\sqrt{n} \times \sqrt{n}$ contains $\Omega(n^2 \log n)$ general 4-holes (Sect. 2, Corollary 1) and that any $\sqrt{n} \times \sqrt{n}$ squared Horton set contains $\Theta(n^2 \log n)$ general 4-holes (Sect. 3, Theorem 2). To obtain the improved lower bound we show in Theorem 1 that $\sum_{d=1}^{n} \frac{\varphi(d)}{d^2} = \Theta(\log n)$, where $\varphi(d)$ is Euler's phi-function (also called Euler's totient function), the number of positive integers less than d that are relatively prime to d. As Euler's phi-function plays a central role in number theory and is also used in cryptography, this result might be of independent interest.

1.1 Definitions and Notation

Consider an integer grid G of size $\sqrt{n} \times \sqrt{n}$. As distance measure we use the L_∞-norm, i.e., the distance of two points $p, q \in G$ is the maximum of the differences of their x- and y-coordinates. The length of a segment e spanned by two points of G is defined by the (L_∞-)distance of its endpoints. A lattice line is a line containing at least two (and hence infinitely many) grid points in the whole plane.

We denote an edge spanned by two points of G which does not have any points of G in its relative interior as *prime segment*. Likewise, we denote a k-gon in G where all edges are prime segments of G as *prime k-gon*. We remark that a prime k-gon might have collinear edges (but cannot have overlapping ones). For example, consider two parallel lattice lines which do not have any points of G between them. Then three consecutive points of G on one line together with one point of G on the second line form a prime 4-hole in G. Omitting the middle one of the tree collinear points as vertex, the same region could also be interpreted as non-prime triangle in G.

An ε-*perturbation* $p_\varepsilon(G)$ of G is a perturbation of G where every point of G is replaced by a point at distance at most ε. Observe that if ε is small enough, then for every triple of points non-collinear in G, their orientation in $p_\varepsilon(G)$ is the same as in G.

Horton Sets. Two disjoint planar point sets are *mutually avoiding* if the convex hull of each of them is intersected by no line passing trough two points of the other set. A finite planar point set is a *Horton set* if it is Horton according to repeated applications of the following two rules:

1. Every set of at most two points with distinct x-coordinates is Horton.
2. If $h_1, h_2, \ldots, h_t$ are points with distinct and increasing x-coordinates and the sets $h_1, h_3, h_5, \ldots$ and $h_2, h_4, h_6, \ldots$ are Horton and mutually avoiding then the set $h_1, h_2, h_3, \ldots, h_t$ is Horton.

Note that if $h_1, h_2, h_3, \ldots, h_t$ is a Horton set, then any subset of the form $h_i, h_{i+1}, \ldots, h_j$, $i < j$ is a Horton set as well.

Squared Horton Sets. A *squared Horton set* S of size $\sqrt{n} \times \sqrt{n}$ is a specific ε-perturbation of the integer grid G of size $\sqrt{n} \times \sqrt{n}$ such that triples of non-collinear points in G keep their orientations in S and such that points along each non-vertical line in G are perturbed to points forming a Horton set in S (and points along each vertical line are perturbed to points forming a rotated copy of a Horton set in S). For more details see [7,13].

2 A New Lower Bound for the Number of General 4-holes in any Slightly Perturbed Grid

In [3] a lower bound of $\Omega(n^2 \log n / \log \log n)$ for the number of general 4-holes in any ε-perturbed integer grid of size $\sqrt{n} \times \sqrt{n}$ was stated. The proof of this statement, which only appeared in the arXiv version [2] of that work, first introduces a relation (below restated in Lemma 1) between the number of prime 4-holes in an integer grid and Euler's phi-function and then presents an estimate $\sum_{d=1}^{n} \frac{\varphi(d)}{d^2} = \Omega(\log n / \log \log n)$. In Theorem 1 we tighten this estimate to $\sum_{d=1}^{n} \frac{\varphi(d)}{d^2} = \Theta(\log n)$ and thereby improve the lower bound on the number of prime 4-holes in an integer grid (Corollary 2).

Lemma 1 ([2], **proof of Theorem 13**). *The integer grid of size $\sqrt{n} \times \sqrt{n}$ contains*

$$\Omega\left(n^2 \cdot \sum_{d=1}^{\lfloor \sqrt{n}/3 \rfloor - 1} \frac{\varphi(d)}{d^2} \right)$$

prime 4-holes.

It is well known that for any $s > 2$, it holds that $\sum_{d=1}^{\infty} \frac{\varphi(d)}{d^s} = \sum_{d=1}^{\infty} \frac{1}{d^{s-1}} / \sum_{d=1}^{\infty} \frac{1}{d^s}$, which is finite; see for example [11] (p. 255). Although $\sum_{d=1}^{\infty} \frac{\varphi(d)}{d^s} = \infty$ for $s = 2$, we show in the following theorem that asymptotically it still holds that $\sum_{d=1}^{n} \frac{\varphi(d)}{d^2} = \Theta(\sum_{d=1}^{n} \frac{1}{n} / \sum_{d=1}^{n} \frac{1}{n^2}) = \Theta(\log n)$.

Theorem 1. $\sum_{d=1}^{n} \frac{\varphi(d)}{d^2} = \Theta(\log n)$.

Proof. For the upper bound, note that $\varphi(d) \leq d$. Thus we have

$$\sum_{d=1}^{n} \frac{\varphi(d)}{d^2} \leq \sum_{d=1}^{n} \frac{d}{d^2} = \sum_{d=1}^{n} \frac{1}{d} = O(\log n).$$

For the lower bound, it is known that

$$\sum_{d=1}^{N} \varphi(d) = \frac{3}{\pi^2} \cdot N^2 + O(N \log N);$$

see for example [11] (p. 268). Using this approximation with $N = 2^{i+1}$ and $N = 2^i$ and subtracting the latter from the former, we obtain

$$\sum_{d=2^i+1}^{2^{i+1}} \varphi(d) = \frac{3}{\pi^2} \left(2^{2i+2} - 2^{2i}\right) + O\left(2^{i+1} \log 2^{i+1}\right) - O\left(2^i \log 2^i\right)$$

$$= \frac{3}{\pi^2} \left(4 \cdot 2^{2i} - 2^{2i}\right) + O\left(2^{i+1} \cdot (i+1)\right) - O\left(2^i \cdot i\right)$$

$$\geq \frac{3}{\pi^2} \left(3 \cdot 2^{2i}\right) - O\left(2^i \cdot i\right) = \frac{9}{\pi^2} \cdot 2^{2i} - O\left(2^i \cdot i\right).$$

Next, consider $\sum_{d=2^i+1}^{2^{i+1}} \frac{\varphi(d)}{d^2}$. Increasing the d^2 in the denominator by its maximum value we can apply the above bound for $\sum_{d=2^i+1}^{2^{i+1}} \varphi(d)$, which gives

$$\sum_{d=2^i+1}^{2^{i+1}} \frac{\varphi(d)}{d^2} \geq \sum_{d=2^i+1}^{2^{i+1}} \frac{\varphi(d)}{2^{2i+2}} \geq \frac{1}{2^{2i+2}} \cdot \frac{9}{\pi^2} \cdot 2^{2i} - O\left(\frac{1}{2^{2i+2}} \cdot 2^i \cdot i\right) = \frac{9}{4\pi^2} - O\left(\frac{i}{2^i}\right).$$

Finally, splitting the sum $\sum_{d=1}^{n} \frac{\varphi(d)}{d^2}$ accordingly and bounding each part from below we obtain the desired result:

$$\sum_{d=1}^{n} \frac{\varphi(d)}{d^2} \geq \sum_{i=0}^{\lfloor \log n \rfloor - 1} \sum_{d=2^i+1}^{2^{i+1}} \frac{\varphi(d)}{d^2}$$

$$\geq \sum_{i=0}^{\lfloor \log n \rfloor - 1} \left(\frac{9}{4\pi^2} - O\left(\frac{i}{2^i}\right)\right)$$

$$= \frac{9}{4\pi^2} \cdot (\lfloor \log n \rfloor - 1) - O\left(\sum_{i=0}^{\lfloor \log n \rfloor - 1} \frac{i}{2^i}\right)$$

$$= \frac{9}{4\pi^2} \cdot (\lfloor \log n \rfloor - 1) - O(1) = \Omega(\log n) \qquad \square$$

Combining Theorem 1 with Lemma 1, we obtain an improved lower bound for the number of prime 4-holes in the grid.

Corollary 1. *The integer grid G of size $\sqrt{n} \times \sqrt{n}$ contains $\Omega(n^2 \log n)$ prime 4-holes.*

Note that a non-prime k-hole in an integer grid G, i.e., a k-hole having points of G on its boundary that are not vertices of the k-hole, might be non-empty and therefore not a k-hole in a perturbed grid G'. This can happen even if the perturbation is arbitrarily small. To the contrary, for any ε-perturbation $p_\varepsilon(G)$ of G (see [3]), every prime k-hole in G corresponds to a k-hole in the perturbed grid $p_\varepsilon(G)$. Thus, we obtain the following more general statement.

Corollary 2. *For any n there is an $\epsilon > 0$ such that any ε-perturbation $p_\varepsilon(G)$ of an integer grid G of size $\sqrt{n} \times \sqrt{n}$ contains $\Omega(n^2 \log n)$ 4-holes.*

3 An Upper Bound on the Number of General 4-holes in the Squared Horton Set

Let S be a squared Horton set S of size $\sqrt{n} \times \sqrt{n}$. In the following we are frequently switching between considering the squared Horton set S and the regular integer grid G of the same size. We identify the points of S with their corresponding non-perturbed versions in G.

Note that each 3-hole in S is either *degenerate* in G, i.e., corresponds to three collinear points of G, or its three vertices form an *interior-empty triangle* in G, i.e., a triangle with no points of G in its interior. For each 4-hole in S we fix a *diagonal* that triangulates the 4-hole into two 3-holes lying on different sides of the diagonal. Note that for convex 4-holes we have two possibilities which diagonal to choose, whilst for non-convex 4-holes we use the unique diagonal going through the interior of the 4-hole. Further, each of the two 3-holes of a 4-hole is either degenerate or interior-empty in G. We distinguish two types of 4-holes in S:

– 4-holes having a prime segment as diagonal in G, and
– 4-holes having a non-prime segment as diagonal in G.

In Sects. 3.1 and 3.2, we show that for both types of 4-holes, an upper bound of $O(n^2 \log n)$ holds. Together with the results from the previous section this implies the following theorem.

Theorem 2. *The number of general 4-holes in any squared Horton set of size $\sqrt{n} \times \sqrt{n}$ is $\Theta(n^2 \log n)$.*

Further, from the arguments in the proof of the upper bound (Sects. 3.1 and 3.2), we obtain the following proposition as a side result.

Proposition 1. *In any squared Horton set of size $\sqrt{n} \times \sqrt{n}$, the maximum number of empty triangles incident to a fixed edge is $\Theta(\sqrt{n})$.*

Bárány conjectured that every point set contains a segment that spans a super-constant number of 3-holes (see [5]). Recently, Bárány et al. [6] showed that for random point sets, the expected maximum number of triangles incident to some edge is $\Omega(n/\log n)$. Hence, Proposition 1 on the one hand confirms Bárány's conjecture for Horton sets, and on the other hand shows that for that conjecture, Horton sets are quite different from random point sets.

Combining Theorem 2 and Proposition 1 with the argumentation in the proof of Theorem 6 of [3] gives

Corollary 3. *For every constant $k \geq 4$, the number of k-holes in any squared Horton set of size $\sqrt{n} \times \sqrt{n}$ is at most $O(n^{\frac{k}{2}}(\log n))$.*

3.1 4-holes Having a Prime Segment as Diagonal in G

If $0 < a < d$ are two integers then the segment $I = (0,0)(d,a)$ is a prime segment (i.e., contains no other lattice points) if and only if a and d are coprime. Moreover, for any k, $0 < k < d$, the vertical lattice line $x = k$ contains a lattice point below I at vertical distance $(ka \mod d)/d$ from I. Thus, if I is a prime segment then the unique closest lattice point below I has vertical distance $1/d$ to I and its x-coordinate is the unique integer k, $0 < k < d$, with the property $ka \equiv 1 \mod d$.

It follows that for a fixed $d > 1$, if we consider all prime segments $\mathcal{I} = \{I = (0,0)(d,a), 0 < a < d\}$, the x-coordinates of the closest lattice points below segments of $\mathcal{I}$ are exactly those integers k, $0 < k < d$, which are coprime with d. Further, each such k appears in this role exactly once - namely for the unique segment $I = (0,0)(d,a)$ for which $ka \equiv 1 \mod d$ holds[1].

Let G be an integer grid of size $\sqrt{n} \times \sqrt{n}$ that contains $(0,0)$. For a fixed d and a fixed coprime a, $0 < a < d$, consider the closest grid point c below $I = (0,0)(d,a)$ and its x-coordinate $0 < k < d$, cf. Fig. 1. Further, consider the supporting line ℓ_0 of I and the lattice line ℓ parallel to ℓ_0 that passes through c. Let $\mathcal{L}_1$ be the set of lattice lines passing through $(0,0)$ and a grid point on ℓ with x-coordinate at least k, as well as the according parallel lattice lines passing through (d,a). Analogously, let $\mathcal{L}_2$ be the set of lattice lines passing through (d,a) and a grid point on ℓ with x-coordinate at most k, as well as the according parallel lattice lines passing through $(0,0)$. Note that the stripe formed by two parallel lines in $\mathcal{L}_1$ or $\mathcal{L}_2$ does not contain any points of G in its interior. Further, any grid point r below ℓ_0 that forms a non-degenerate interior-empty triangle Δ with I in G must be on a line $\ell' \in \mathcal{L}_1 \cup \mathcal{L}_2$: In order for Δ to not contain any point of ℓ in its interior, r must be either on ℓ or pass through one stripe formed by two parallel lines in $\mathcal{L}_1$ or $\mathcal{L}_2$. No such stripe contains any points of G in its interior, and all points on the boundary of a stripe (as well as all points on ℓ) are on a line in $\mathcal{L}_1 \cup \mathcal{L}_2$.

Now consider a line $\ell_1 \in \mathcal{L}_1$ passing through $(0,0)$ and a grid point on ℓ with x-coordinate $x_\ell > 0$. Then $x_\ell = id + k$ for some integer $0 \leq i < \frac{\sqrt{n}-k}{d}$,

---

[1] If $0 < p < q$ are two coprime integers then there is a unique r, $0 < r < q$, such that $pr \equiv 1 \mod q$.

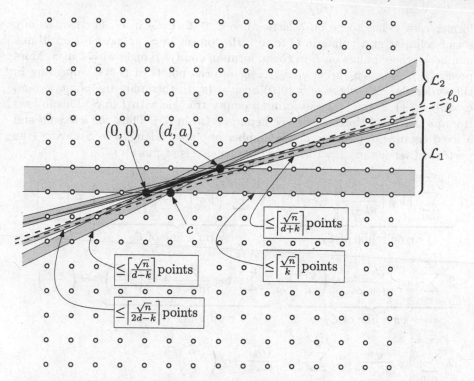

Fig. 1. Illustration for the definition of ℓ_0, ℓ, and sets $\mathcal{L}_1$ and $\mathcal{L}_2$, for $\sqrt{n} = 15$, $a = 1$, $d = 3$, and $k = 1$. Stripes formed by parallel lines of $\mathcal{L}_1$ and $\mathcal{L}_2$ are drawn shaded. For better visibility, not all lines of $\mathcal{L}_1$ and $\mathcal{L}_2$ are drawn.

and ℓ_1 contains at most $\left\lceil \frac{\sqrt{n}}{id+k} \right\rceil$ grid points. See again Fig. 1, where $d = 3$ and $k = 1$. Of course, this bound also holds for the lattice line parallel to ℓ_1 and going through (d, a). Similarly, let $\ell_2 \in \mathcal{L}_2$ be a line that passes through (d, a) and a grid point on ℓ with x-coordinate $x_\ell < d$. Then $x_\ell = d - (id - k) = k - (i - 1)d$ for some integer $1 \leq i < \frac{\sqrt{n} + k}{d}$ and ℓ_2 contains at most $\left\lceil \frac{\sqrt{n}}{id-k} \right\rceil$ grid points and the same bound holds for the lattice line parallel to ℓ_2 and through $(0, 0)$. Thus, the number of non-degenerate interior-empty triangles in G having I as one side is at most

$$2 \cdot \underbrace{\sum_{i=0}^{\left\lfloor \frac{\sqrt{n}-k}{d} \right\rfloor - 1} \left\lceil \frac{\sqrt{n}}{id+k} \right\rceil}_{r \text{ on a line of } \mathcal{L}_1} + 2 \cdot \underbrace{\sum_{i=1}^{\left\lfloor \frac{\sqrt{n}+k}{d} \right\rfloor - 1} \left\lceil \frac{\sqrt{n}}{id-k} \right\rceil}_{r \text{ on a line of } \mathcal{L}_2}.$$

If instead of the regular grid G we consider the squared Horton set S of same size, then for any point $r \in S$ that forms an empty triangle with I in S, the according triangle in G is either interior-empty or degenerate. In the latter case, r needs to be one of the at most $\frac{\sqrt{n}}{d}$ grid points on the lattice line ℓ_0 in G. In the former case, r has to be on a lattice line $\ell' \in \mathcal{L}_1 \cup \mathcal{L}_2$ in G. Further, as any set of collinear grid points in G form a Horton set in S, at most a logarithmic number of these points on ℓ' in G can form an empty triangle with I in S. More precisely, if $p_s, \ldots, p_2, p_1, w, q_1, q_2, \ldots, q_t$ are the points of $G \cap \ell'$ appearing in this order along ℓ', where $w = (0,0)$ or $w = (d,a)$, then only the points p_i and q_i with $i \in \{1, 2, 4, 8, \ldots\}$ can form an empty triangle with I in S. This follows from properties of Horton sets (see Lemma 5 (ii) in [7]). Thus, for a fixed d and a fixed coprime a, $0 < a < d$, the number of empty triangles in S having (the perturbed version of) $I = (0,0)(d,a)$ as one side is at most

$$\underbrace{2 \cdot \sum_{i=0}^{\left\lfloor \frac{\sqrt{n}-k}{d} \right\rfloor} \left(\log \left\lceil \frac{\sqrt{n}}{id+k} \right\rceil + \Theta(1) \right)}_{r \text{ on a line of } \mathcal{L}_1} + \underbrace{2 \cdot \sum_{i=1}^{\left\lfloor \frac{\sqrt{n}+k}{d} \right\rfloor} \left(\log \left\lceil \frac{\sqrt{n}}{id-k} \right\rceil + \Theta(1) \right)}_{r \text{ on a line of } \mathcal{L}_2} + \underbrace{\left\lceil \frac{\sqrt{n}}{d} \right\rceil}_{r \text{ on } \ell_0}$$

$$\leq \underbrace{2 \cdot \sum_{i=0}^{\left\lfloor \frac{\sqrt{n}-k}{d} \right\rfloor} \log \frac{\sqrt{n}}{id+k} + \Theta\left(\frac{\sqrt{n}}{d} \right)}_{r \text{ on a line of } \mathcal{L}_1} + \underbrace{2 \cdot \sum_{i=1}^{\left\lfloor \frac{\sqrt{n}+k}{d} \right\rfloor} \log \frac{\sqrt{n}}{id-k} + \Theta\left(\frac{\sqrt{n}}{d} \right)}_{r \text{ on a line of } \mathcal{L}_2} + \Theta\left(\frac{\sqrt{n}}{d} \right)$$

$$\leq \underbrace{2 \cdot \sum_{i=0}^{\left\lfloor \frac{\sqrt{n}-k}{d} \right\rfloor} \log \frac{\sqrt{n}}{id+k}}_{\text{first sum}} + \underbrace{2 \cdot \sum_{i=1}^{\left\lfloor \frac{\sqrt{n}+k}{d} \right\rfloor} \log \frac{\sqrt{n}}{id-k}}_{\text{second sum}} + \Theta\left(\frac{\sqrt{n}}{d} \right)$$

$$= \underbrace{2 \cdot \log \frac{\sqrt{n}}{k} + 2 \cdot \sum_{i=1}^{\left\lfloor \frac{\sqrt{n}-k}{d} \right\rfloor} \log \frac{\sqrt{n}}{id+k}}_{\text{first sum}} + \underbrace{2 \cdot \log \frac{\sqrt{n}}{d-k} + 2 \cdot \sum_{i=2}^{\left\lfloor \frac{\sqrt{n}+k}{d} \right\rfloor} \log \frac{\sqrt{n}}{id-k}}_{\text{second sum}} + \Theta\left(\frac{\sqrt{n}}{d} \right)$$

$$< 2 \cdot \sum_{i=1}^{\left\lfloor \frac{\sqrt{n}-k}{d} \right\rfloor} \log \frac{\sqrt{n}}{id} + 2 \cdot \sum_{i=1}^{\left\lfloor \frac{\sqrt{n}+k-d}{d} \right\rfloor} \log \frac{\sqrt{n}}{id} + 2 \cdot \left(\log \frac{\sqrt{n}}{k} + \log \frac{\sqrt{n}}{d-k} \right) + \Theta\left(\frac{\sqrt{n}}{d} \right)$$

$$< 2 \cdot \sum_{i=1}^{\left\lfloor \frac{\sqrt{n}}{d} \right\rfloor} \log \frac{\sqrt{n}}{id} + 2 \cdot \sum_{i=1}^{\left\lfloor \frac{\sqrt{n}}{d} \right\rfloor} \log \frac{\sqrt{n}}{id} + 2 \cdot \left(\log \frac{\sqrt{n}}{k} + \log \frac{\sqrt{n}}{d-k} \right) + \Theta\left(\frac{\sqrt{n}}{d} \right)$$

$$= 4 \cdot \sum_{i=1}^{\left\lfloor \frac{\sqrt{n}}{d} \right\rfloor} \log \frac{\sqrt{n}}{id} + 2 \cdot \left(\log \frac{\sqrt{n}}{k} + \log \frac{\sqrt{n}}{d-k} \right) + \Theta\left(\frac{\sqrt{n}}{d} \right).$$

Reformulating the sum of logarithms as logarithm of a product and using that by Stirling's formula $\log(x^x / \lfloor x \rfloor!) = \Theta(x)$, we can continue this calculation with

$$= 4 \cdot \log \left(\prod_{i=1}^{\left\lfloor \frac{\sqrt{n}}{d} \right\rfloor} \frac{\sqrt{n}}{id} \right) + 2 \cdot \left(\log \frac{\sqrt{n}}{k} + \log \frac{\sqrt{n}}{d-k} \right) + \Theta\left(\frac{\sqrt{n}}{d} \right)$$

$$\leq 4 \cdot \log \left(\frac{\left(\frac{\sqrt{n}}{d} \right)^{\frac{\sqrt{n}}{d}}}{\left\lfloor \frac{\sqrt{n}}{d} \right\rfloor !} \right) + 2 \cdot \left(\log \frac{\sqrt{n}}{k} + \log \frac{\sqrt{n}}{d-k} \right) + \Theta \left(\frac{\sqrt{n}}{d} \right)$$

$$= 2 \cdot \left(\log \frac{\sqrt{n}}{k} + \log \frac{\sqrt{n}}{d-k} \right) + \Theta \left(\frac{\sqrt{n}}{d} \right).$$

Hence, for the number of empty 4-gons in S having I as a diagonal, we obtain an upper bound of

$$\left(2 \cdot \left(\log \frac{\sqrt{n}}{k} + \log \frac{\sqrt{n}}{d-k} \right) + O \left(\frac{\sqrt{n}}{d} \right) \right)^2.$$

Next, we let a go through all values $0 < a < d$ where a is coprime with d. Then k also goes through $0 < k < d$ where k is coprime with d (just in a different order). Further, we skip the restriction of $0 < a < d$ and instead consider the set of all prime segments $\mathcal{I}$ with fixed length $|d| > 1$ that have a common endpoint (this gives a multiplicative factor of 8). By this, we obtain the following upper bound for the number of 4-holes in S whose diagonal is a prime segment incident to a fixed point and of fixed length $|d| > 1$:

$$8 \cdot \sum_{k \in \{0, \dots, d\}} \left(2 \cdot \left(\log \frac{\sqrt{n}}{k} + \log \frac{\sqrt{n}}{d-k} \right) + O \left(\frac{\sqrt{n}}{d} \right) \right)^2$$

$$< 8 \cdot \sum_{k \in \{0, \dots, d\}} \left(4 \cdot \max \left\{ \log \frac{\sqrt{n}}{k}, \log \frac{\sqrt{n}}{d-k} \right\} + O \left(\frac{\sqrt{n}}{d} \right) \right)^2$$

$$\leq 8 \cdot 2 \cdot \sum_{k \in \{0 \dots \lfloor \frac{d}{2} \rfloor\}} \left(4 \cdot \log \frac{\sqrt{n}}{k} + O \left(\frac{\sqrt{n}}{d} \right) \right)^2,$$

where we always sum only over values of k such that k, d are coprime. This is bounded from above by the same sum where we sum over all $k \in \{0 \dots \lfloor \frac{d}{2} \rfloor\}$. Using $(s + t)^2 \leq 2s^2 + 2t^2$, we get the upper bound

$$\sum_{k=1}^{\lfloor \frac{d}{2} \rfloor} \left(4 \cdot \log \frac{\sqrt{n}}{k} + O \left(\frac{\sqrt{n}}{d} \right) \right)^2 = O \left(\sum_{k=1}^{\lfloor \frac{d}{2} \rfloor} \left(\log^2 \left(\frac{\sqrt{n}}{k} \right) + \left(\frac{\sqrt{n}}{d} \right)^2 \right) \right)$$

$$= O \left(\sum_{k=1}^{\lfloor \frac{d}{2} \rfloor} \log^2 \left(\frac{\sqrt{n}}{k} \right) \right) + O \left(\sum_{k=1}^{\lfloor \frac{d}{2} \rfloor} \left(\frac{\sqrt{n}}{d} \right)^2 \right).$$

The second sum is obviously $O(n/d)$. The first sum can be bounded from above by

$$\sum_{k=1}^{\lceil \frac{d}{2} \rceil} \log^2 \left(\frac{\sqrt{n}}{k} \right) < \sum_{k=1}^{\lceil \frac{d}{2} \rceil} \frac{\sqrt{n}}{k} < \sum_{k=1}^{\sqrt{n}} \frac{\sqrt{n}}{k} = O(\sqrt{n} \log n),$$

implying a total upper bound of $O(\sqrt{n} \log n + n/d)$ for $d > 1$.

For the case where $0 \leq a \leq d = 1$, the two closest grid points below $I = (0,0)(1,a)$ in G are $(0,-1)$ and $(1, a - 1)$. Let again ℓ be the supporting lattice line of these points. With $k := 1$ and $\mathcal{L}_1$ and $\mathcal{L}_2$ being defined according to the case where $d > 1$, we obtain an upper bound

$$\underbrace{2 \cdot \sum_{i=0}^{\sqrt{n}-2} \frac{\sqrt{n}}{i+1}}_{r \text{ on a line of } \mathcal{L}_1} + \underbrace{2 \cdot \sqrt{n} + 2 \cdot \sum_{i=2}^{\sqrt{n}} \frac{\sqrt{n}}{i-1}}_{r \text{ on a line of } \mathcal{L}_2}$$

for the number of non-degenerate interior-empty triangles in G having I as one side (the extra term for $\mathcal{L}_2$ comes from the vertical lattice lines through the endpoints of I). With similar arguments as above, we obtain $O(\sqrt{n} \log n + n) = O(n)$ as an upper bound for the number of 4-holes in S whose diagonals are prime segments incident to a fixed point and of fixed length $|d| = 1$.

Finally, summing up over all possible distances $|d|$ and multiplying by the $O(n)$ possible starting points, we get the desired upper bound for the number of 4-holes in S having a prime segment as diagonal in G:

$$O(n) \cdot \sum_{d=1}^{\sqrt{n}-1} O\left(\sqrt{n} \log n + n/d\right) = O\left(n^2 \log n\right).$$

3.2   4-holes Having a Non-prime Segment as Diagonal in G

Consider a non-convex 4-hole Q in S having a non-prime segment as diagonal I. Then at least one of the two triangles Q consists of has to be degenerate, i.e., the according third point r is on the same lattice line ℓ_0 as I in G. The second triangle Δ can either be degenerate as well or it can be non-degenerate.

Case 1. The second triangle Δ is degenerate as well, i.e., Q is completely contained in one lattice line.

Consider a line ℓ_0 of slope a/d with $0 \leq a \leq d$, $d \geq 1$, a coprime with d. As the closest points below ℓ_0 have vertical distance $1/d$ from ℓ_0, there are at most $2\sqrt{n} \cdot d = O(\sqrt{n} \cdot d)$ lines parallel to ℓ_0 spanned by points of G. Further, summing up over all values of a, we obtain $\varphi(d)$ different directions and thus $O(\varphi(d) \cdot \sqrt{n} \cdot d)$ lattice lines where two adjacent grid points have distance d. Any such line contains $O(\sqrt{n}/d)$ points of G corresponding to a Horton set in S (see [13]). Thus, it contains $O((\sqrt{n}/d)^2)$ empty triangles and convex 4-holes [4,14] and $O((\sqrt{n}/d)^3)$ non-convex 4-holes [1]. in S. Summing up over all distances $1 \leq d \leq \sqrt{n}$, and adding also the 4-holes for the $\sqrt{n}$ vertical lattice lines we obtain

$$\sqrt{n} \cdot O\left(\sqrt{n}^3\right) + \sum_{d=1}^{\sqrt{n}} 4 \cdot O(\varphi(d) \cdot \sqrt{n} \cdot d) \cdot O\left(\left(\frac{\sqrt{n}}{d}\right)^3\right)$$

$$= O\left(n^2 + n^2 \cdot \sum_{d=1}^{\sqrt{n}} \frac{\varphi(d)}{d^2}\right) = O(n^2 \log n)$$

as an upper bound for the total number of 4-holes consisting of 2 degenerate 3-holes in S.

Case 2. The second triangle Δ is non-degenerate, i.e., the third point r' of Δ does not lie on the supporting line of I.

Assume that r lies above I in S and thus r' lies below I in S (recall that I is an inner diagonal of Q in S). I contains at least one grid point in its relative interior and has at least two closest points below it. Consider the lattice line ℓ parallel to I through these closest points, and let ℓ' be the line parallel to ℓ and through the closest points below ℓ. Then r' has to lie on ℓ or ℓ' for Δ to be interior-empty: If the third point r' of the second triangle lies below ℓ', consider the triangle Δ formed by I and r' and the segment $s = \ell \cap \Delta$. As the Euclidean length of s is at least two thirds of the Euclidean length of I, s and thus Δ would contain at least one point of G in its interior. Similarly, if r' is below I, then r needs to be on one of the two closest lines above I and parallel to I.

Now let ℓ_0 be of slope a/d with $0 \leq a \leq d$, $d \geq 1$. Then there are $O(\sqrt{n}/d)$ points on ℓ_0 and its four closest parallel lattice lines. Thus, there are $O\left((\sqrt{n}/d)^2\right)$ empty triangles on ℓ_0 in S, and $O(\sqrt{n}/d)$ possible candidates for r'. Again, adding the same counting for the $\sqrt{n}$ vertical lattice lines we obtain the same upper bound:

$$\sqrt{n} \cdot O\left(\sqrt{n}^2\right) \cdot O(\sqrt{n}) + \sum_{d=1}^{\sqrt{n}} 4 \cdot O(\varphi(d)\sqrt{n} \cdot d) \cdot O\left(\left(\frac{\sqrt{n}}{d}\right)^2\right) \cdot O\left(\frac{\sqrt{n}}{d}\right)$$

$$= O\left(n^2 + n^2 \cdot \sum_{d=1}^{\sqrt{n}} \frac{\varphi(d)}{d^2}\right) = O\left(n^2 \log n\right).$$

Hence, the number of 4-holes in S having a non-prime segment as diagonal in G, as well as the total number of 4-holes in S is $O(n^2 \log n)$, which completes the proof of the upper bound in Theorem 2.

References

1. Aichholzer, O., Fabila-Monroy, R., González-Aguilar, H., Hackl, T., Heredia, M.A., Huemer, C., Urrutia, J., Valtr, P., Vogtenhuber, B.: 4-holes in point sets. CGTA **47**(6), 644–650 (2014)
2. Aichholzer, O., Fabila-Monroy, R., González-Aguilar, H., Hackl, T., Heredia, M.A., Huemer, C., Urrutia, J., Valtr, P., Vogtenhuber, B.: On k-gons and k-holes in point sets (2014). arXiv:1409.0081
3. Aichholzer, O., Fabila-Monroy, R., González-Aguilar, H., Hackl, T., Heredia, M.A., Huemer, C., Urrutia, J., Valtr, P., Vogtenhuber, B.: On k-gons and k-holes in point sets. CGTA **48**(7), 528–537 (2015)
4. Bárány, I., Füredi, Z.: Empty simplices in Euclidean space. Can. Math. Bull. **30**, 436–445 (1987)
5. Bárány, I., Károlyi, G.: Problems and results around the Erdös-Szekeres convex polygon theorem. In: Akiyama, J., Kano, M., Urabe, M. (eds.) JCDCG 2000. LNCS, vol. 2098, pp. 91–105. Springer, Heidelberg (2001). doi:10.1007/3-540-47738-1_7

6. Bárány, I., Marckert, J.-F., Reitzner, M.: Many empty triangles have a common edge. Discrete Comput. Geom. **50**(1), 244–252 (2013)
7. Bárány, I., Valtr, P.: Planar point sets with a small number of empty convex polygons. Studia Sci. Math. Hungar. **41**(2), 243–266 (2004)
8. Erdős, P.: Some more problems on elementary geometry. Aust. Math. Soc. Gaz. **5**, 52–54 (1978)
9. Erdős, P., Szekeres, G.: A combinatorial problem in geometry. Compos. Math. **2**, 463–470 (1935)
10. Fabila-Monroy, R., Huemer, C., Mitsche, D.: Empty non-convex and convex fourgons in random point sets. Studia Sci. Math. Hungar. **52**(1), 52–64 (2015)
11. Hardy, G.H., Wright, E.M.: An Introduction to the Theory of Numbers, 5th edn. Oxford University Press, London (1979)
12. Horton, J.: Sets with no empty convex 7-gons. Can. Math. Bull. **26**(4), 482–484 (1983)
13. Valtr, P.: Convex independent sets and 7-holes in restricted planar point sets. Disc. Comp. Geom. **7**, 135–152 (1992)
14. Valtr, P.: On the minimum number of empty polygons in planar point sets. Stud. Sci. Math. Hung. **30**, 155–163 (1995)

Reversible Nets of Polyhedra

Jin Akiyama[1], Stefan Langerman[2(✉)], and Kiyoko Matsunaga[1]

[1] Tokyo University of Science, 1-3 Kagurazaka, Shinjuku, Tokyo 162-8601, Japan
ja@jin-akiyama.com, matsunaga@mathlab-jp.com
[2] Université Libre de Bruxelles, Brussels, Belgium
stefan.langerman@ulb.ac.be

Abstract. An example of reversible (or hinge inside-out transformable) figures is the Dudeney's Haberdasher's puzzle in which an equilateral triangle is dissected into four pieces, then hinged like a chain, and then is transformed into a square by rotating the hinged pieces. Furthermore, the entire boundary of each figure goes into the inside of the other figure and becomes the dissection lines of the other figure. Many intriguing results on reversibilities of figures have been found in prior research, but most of them are results on polygons. This paper generalizes those results to a wider range of general connected figures. It is shown that two nets obtained by cutting the surface of an arbitrary convex polyhedron along non-intersecting dissection trees are reversible. Moreover, a condition for two nets of an isotetrahedron to be both reversible and tessellative is given.

1 Introduction

A pair of hinged figures P and Q (see Fig. 1) is said to be *reversible* (or *hinge inside-out transformable*) if P and Q satisfy the following conditions:

1. There exists a dissection of P into a finite number of pieces, $P_1, P_2, P_3, \ldots, P_n$. A set of dissection lines or curves forms a tree. Such a tree is called a *dissection tree*.
2. Pieces $P_1, P_2, P_3, \ldots, P_n$ can be joined by $n-1$ hinges located on the perimeter of P like a chain.
3. If one of the end-pieces of the chain is fixed and rotated, then the remaining pieces form Q when rotated clockwise and P when rotated counterclockwise.
4. The entire boundary of P goes into the inside of Q and the entire boundary of Q is composed exactly of the edges of the dissection tree of P.

The theory of hinged dissections and reversibilities of figures has a long history and the book by Frederickson [11] contains many interesting results. On the other hand, Abbott et al. [1] proved that every pair of polygons P and Q with the same area is hinge transformable if we don't require the reversible condition. When imposing the reversible condition, hinge transformable figures have some remarkable properties which were studied in [3–6,9,12].

S. Langerman—Directeur de Recherches du F.R.S.-FNRS.

J. Akiyama et al. (Eds.): JCDCGG 2015, LNCS 9943, pp. 13–23, 2016.
DOI: 10.1007/978-3-319-48532-4_2

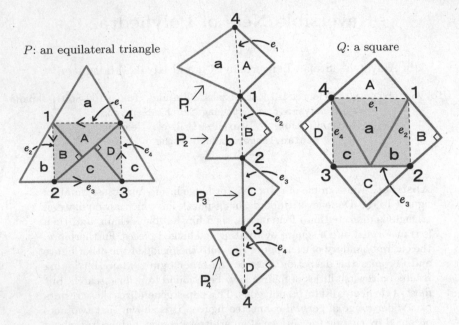

Fig. 1. Reversible transformation between P and Q.

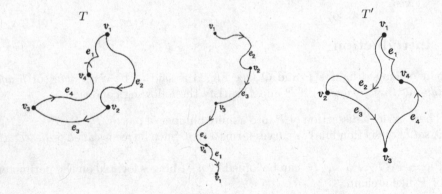

Fig. 2. T and one of its conjugate regions.

Let T be a closed plane region whose perimeter consists of n curved (or straight line) *segments* $e_1, e_2, \ldots, e_n$ and let these lines be labeled in clockwise order. Let T' be a closed region surrounded by the same segments $e_1, e_2, \ldots, e_n$ but in counterclockwise order. We then say that T' is a *conjugate region* of T (Fig. 2).

Let P be a plane figure. A region T with n vertices $v_1, \ldots, v_n$ and with n perimeter parts $e_1, \ldots, e_n$ is called an *inscribed region* of P if all vertices v_i ($i = 1, \ldots, n$) are located on the perimeter of P and $T \subseteq P$.

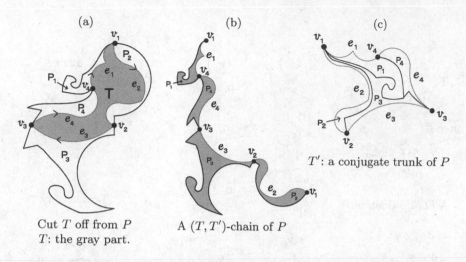

(a)

Cut T off from P
T: the gray part.

(b)

A (T, T')-chain of P

(c)

T': a conjugate trunk of P

Fig. 3. A trunk T of P, a (T, T')-chain of P and a conjugate trunk T' of P.

A *trunk* of P is a special kind of inscribed region T of P. First, cut out an inscribed region T from P (Fig. 3(a)). Let e_i ($i = 1, \ldots, n$) be the perimeter part of T joining two vertices v_{i-1} and v_i of T, where $v_0 = v_n$. Denote by P_i the piece located outside of T that contains the perimeter part e_i. Some P_i may be empty (or just a part e_i). Then, hinge each pair of pieces P_i and P_{i+1} at their common vertex v_i ($1 \le i \le n - 1$); this results in a chain of pieces P_i ($i = 1, 2, \ldots, n$) of P (Fig. 3(b)). The chain and T are called (T, T')-chain of P, and *trunk of P*, respectively, if an appropriate rotation of the chain forms T' which is one of the conjugate regions of T with all pieces P_i packed inside T' without overlaps or gaps. The chain T' is called a *conjugate trunk* of P (Fig. 3(c)).

Suppose that a figure P has a trunk T and a conjugate trunk T'; and a figure Q has a trunk T' and a conjugate trunk T. We then have two chains, a (T, T')-chain of P and a (T', T)-chain of Q (Fig. 4).

Combine a (T, T')-chain of P with a (T', T)-chain of Q such that each segment of the perimeter, e_i, has a piece P'_i of P on one side (right side) and a piece Q_i of Q on the other side (left side). The chain obtained in this manner is called a *double chain of (P, Q)* (Fig. 5).

We say that a piece of a double chain is *empty* if that piece consists of only a perimeter part e_i. If a double chain has an empty piece, then we distinguish one side of that edge from the other side so that it satisfies the conditions for reversibility. If one of the end-pieces (Say P_1 and Q_1 in Fig. 5) of the double chain of (P, Q) is fixed and the remaining pieces are rotated clockwise or counterclockwise, then figure P and figure Q are obtained respectively (Fig. 5). The following result is obtained from [3].

Theorem 1 (Reversible Transformations Between Figures). *Let P be a figure with trunk T and conjugate trunk T', and let Q have trunk T' and conjugate trunk T. Then P is reversible to Q.*

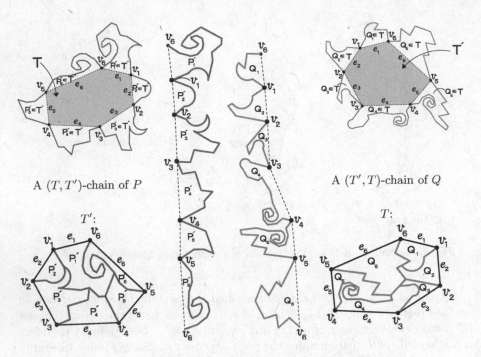

Fig. 4. A (T, T')-chain of P and a (T', T)-chain of Q.

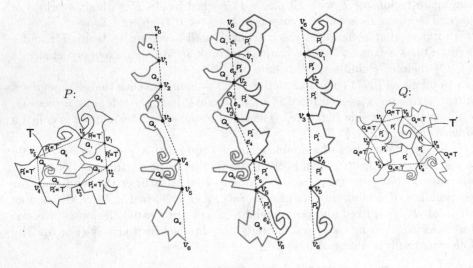

Fig. 5. A double chain of (P, Q).

Remarks

1. In Theorem 1, figure P which is the union of T and n pieces P_i' of the conjugate trunk T' reversibly transforms into figure Q which is the union of T' and n pieces of T.
2. Harberdasher's puzzle by H. Dudeney is also one such reversible pair. In this puzzle, the figures P and Q are an equilateral triangle and a square, respectively. The trunk T and conjugate trunk T' are the identical parallelogram T (the gray part in Fig. 1).

2   Reversible Nets of Polyhedra

A *dissection tree* D of a polyhedron P is a tree drawn on the surface of P that spans all vertices of P. Cutting the surface of P along D results in a *net* of P. Notice that nets of some polyhedron P may have self-overlapping parts (Fig. 6). We allow such cases when discussing reversible transformation of nets.

Theorem 2. *Let P be a polyhedron with n vertices $v_1, \ldots, v_n$ and let D_i ($i = 1, 2$) be dissection trees on the surface of P. Denote by N_i ($i = 1, 2$) the nets of P obtained by cutting P along D_i ($i = 1, 2$), respectively. If D_1 and D_2 don't properly cross, then the pair of nets N_1 and N_2 is reversible, and has a double chain composed of n pieces.*

Proof. Suppose that dissection trees D_1 (the red tree) and D_2 (the green tree) on the surface of P do not properly cross (Fig. 7(a)). Then there exists a closed Jordan curve on the surface of P, which separates the surface of P into two pieces, one containing D_1, the other containing D_2. Let C be an arbitrary such curve (Fig. 7(b)). We call C a separating cycle. The net N_1, obtained by cutting P along D_1, contains an inscribed closed region T whose boundary is C (Fig. 8(a)). On the other hand, a net N_2 which is obtained by cutting P along D_2 contains an inscribed conjugate region T' whose boundary is the opposite side of C (Fig. 8(c)). Hence, a net N_1 has a trunk T and a conjugate trunk T', and a net N_2 has a trunk T' and a conjugate trunk T. By Theorem 1 this pair of N_1 and N_2 is reversible (Fig. 8(b)).   □

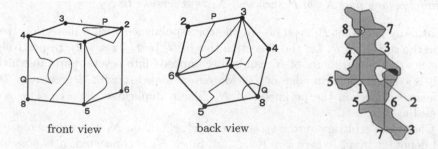

front view   back view

Fig. 6. A net of a cube with self-overlapping part (the overlap is the black part).

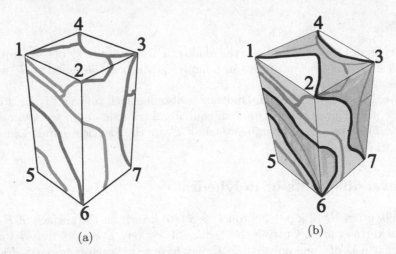

(a)   (b)

Fig. 7. A polyhedron P with dissection trees D_1 (red tree) and D_2 (green tree), a separating cycle C (black cycle). (Color figure online)

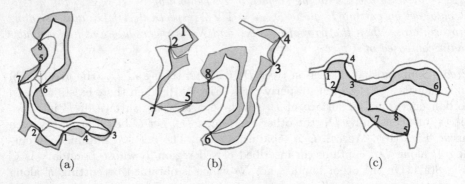

(a)   (b)   (c)

Fig. 8. Nets N_1 and N_2 obtained by cutting the surface of P along D_1 and D_2, respectively.

Theorem 3. *For any net N_1 of a polyhedron P with n vertices, there exist infinitely many nets N_2 of P such that N_1 is reversible to N_2.*

Proof. Any net N of P has a one-to-one correspondence with a dissection tree D on the surface of P. Let the dissection tree of N_i be D_i ($i = 1, 2$), respectively (Fig. 9(a)). The perimeter of N_i can be decomposed into several parts in which each is congruent to an edge of D_i. Moreover, a vertex with degree k on D_i appears k times on the perimeter of N_i. These duplicated vertices of v_i are labeled as $v_i', v_i'', \ldots$.

Choose an arbitrary vertex v_k among $v_k, v_k', v_k'', \ldots$ on N_1 as a representative and denote it by v_k^*, where $k = 1, 2, \ldots, n$. Since N_1 is connected, it is possible to draw infinitely many arbitrary spanning trees D_2, each of which connects v_k^*

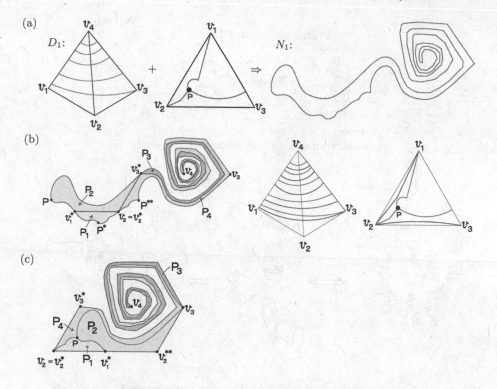

Fig. 9. A swirl net of a regular tetrahedron.

$(k = 1, 2, \ldots, n)$ inside N_1 (Fig. 9(b)). Then, any such D_2 doesn't intersect D_1.
(Fig. 9(c)). As in Theorem 2, dissect N_1 along D_2 into n pieces $P_1, \ldots, P_n$, and
then connect them in sequence using $n-1$ hinges on the perimeter of N_1 to form
a chain. Fix one of the end-pieces of the chain and rotate the remaining pieces
then forming net N_2 which is obtained by cutting P along D_2 (Fig. 9 (d)).   □

Corollary 1 (Envelope magic [7]). *Let E be an arbitrary doubly covered polygon (dihedron) and let D_1 and D_2, be dissection trees of E. If dissection tree D_1
doesn't properly cross dissection tree D_2, then a pair of nets N_1 and N_2 obtained
by cutting the surface of E along D_1 and D_2 is reversible (Fig. 10).*

The previous two theorems show that it is always possible to dissect any
polyhedron P into two nets that are reversible, however, as mentioned in the
beginning of this section, those nets may sometimes self-overlap when embedded
in the plane. One may then ask whether a convex polyhedron P always has a
pair of reversible non self-overlapping nets. The following theorem answers in
the positive.

Theorem 4. *For any convex polyhedron P, there exists an infinity of pairs of
non self-overlapping nets of P that are reversible.*

Proof. Choose an arbitrary point s on the surface of P, but not on a vertex.
The *cut locus* of s is the set of all points t on the surface of P such that the

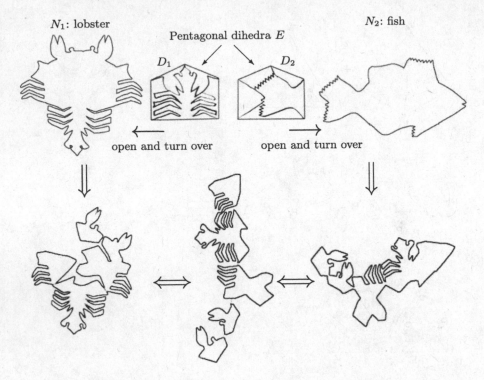

Fig. 10. A lobster transforms into a fish; The separating cycle C is the hem of a pentagonal dihedron.

shortest path from s to t is not unique. It is well known that the cut locus of s is a tree that spans all vertices of P. Cutting P along the cut locus produces the *source unfolding*, which does not overlap [10]. Let D_1 be the cut locus from s, and N_1 the corresponding non self-overlapping net. The net N_1 is a star-shaped polygon, and the shortest path from s to any point t in P unfolds to a straight line segment contained in N_1. The dissection tree D_2 is constructed by cutting P along the shortest path from s to every vertex of P. The net N_2 thus produced is a *star unfolding* and also does not overlap [8]. Note also that the shortest path from s to any vertex of P, when cutting the source tree D_1, unfolds to a straight line segment from s to the corresponding vertex on N_1. Therefore D_1 and D_2 do not properly intersect (In fact D_1 and D_2 may coincide but not properly cross. In order to avoid this, it suffices to choose s not on the cut locus of any vertex of P.) By Theorem 2, N_1 and N_2 are reversible. $\qquad\square$

3 Reversibility and Tessellability for Nets of an Isotetrahedron

A tetrahedron T is called an *isotetrahedron* if all faces of T are congruent. Note that there are infinitely many non-similar isotetrahedra. Every net of an

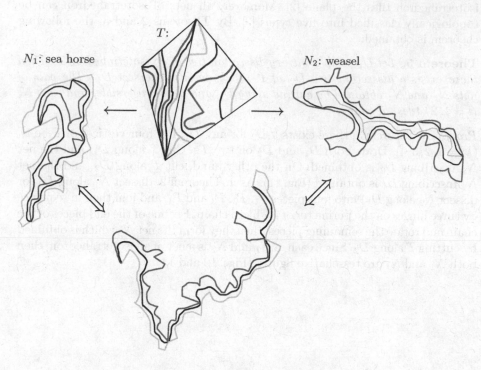

Fig. 11. sea horse ⇔ weasel

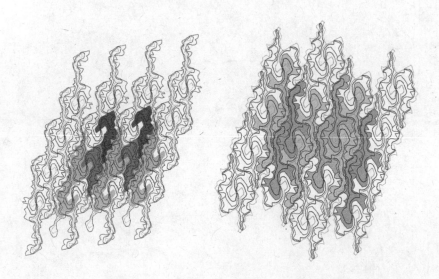

Fig. 12. Tiling by sea horse and weasel

isotetrahedron tiles the plane [2]. Moreover, all nets of isotetrahedron can be topologically classified into five types [3]. By Theorems 2 and 3, the following theorem is obtained:

Theorem 5. *Let D_1 be an arbitrary dissection tree of an isotetrahedron T. Then there exists a dissection tree D_2 of T which doesn't intersect D_1. The pair of nets N_1 and N_2 obtained by cutting along D_1 and D_2 is reversible, and each N_i ($i = 1, 2$) tiles the plane.*

Proof. By Theorem 3, there exists a D_2 for any D_1. Let four vertices of T be v_k ($k = 1, 2, 3, 4$). Draw both D_1 and D_2 on two Ts. Cut T along D_1, and the net N_1 inscribing D_2 is obtained. On the other hand, cut T along D_2, and the net N_2 inscribing D_1 is obtained (Fig. 11). As in Theorem 2, dissect N_1 along D_2 (or dissect N_2 along D_1) into four pieces P_1, P_2, P_3 and P_4, and join then in sequence by three hinges on the perimeter of N_1 like a chain. Fix one of the end pieces of the chain and rotate the remaining pieces, then they form the net N_2 which is obtained by cutting T along D_2. Since each of N_1 and N_2 is a net of an isotetrahedron, then both N_1 and N_2 are tessellative figures (Figs. 12 and 13).   □

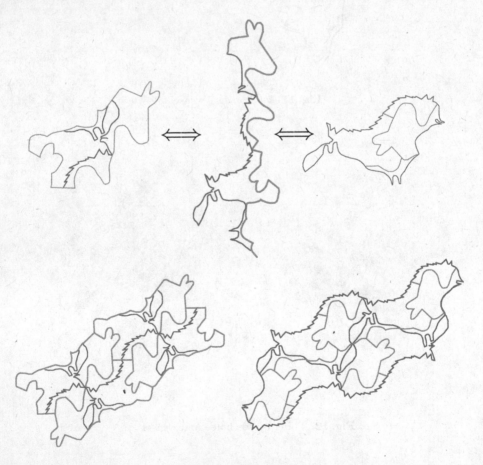

Fig. 13. donkey ⇔ fox

References

1. Abbott, T., Abel, Z., Charlton, D., Demaine, E.D., Demaine, M.L., Kominers, S.: Hinged dissections exist. Discrete Comput. Geom. **47**(1), 150–186 (2010)
2. Akiyama, J.: Tile-maker and semi-tile-maker. Am. Math. Mon. **114**, 602–609 (2007)
3. Akiyama, J., Matsunaga, K.: Treks into Intuitive Geometry. Springer, New York (2015)
4. Akiyama, J., Nakamura, G.: Congruent Dudeney dissections of triangles and convex quadrangles - all hinge points interior to the sides of the polygons. In: Pach, J., Aronov, B., Basu, S., Sharir, M. (eds.) Discrete and Computational Geometry, The Goodman-Pollack Festschrift. Algorithms and Combinatorics, vol. 25, pp. 43–63. Springer, New York (2003)
5. Akiyama, J., Rappaport, D., Seong, H.: A decision algorithm for reversible pairs of polygons. Discrete Appl. Math. **178**, 19–26 (2014)
6. Akiyama, J., Seong, H.: A criterion for a pair of convex polygons to be reversible. Graphs Comb. **31**(2), 347–360 (2015)
7. Akiyama, J., Tsukamoto, T.: Envelope magic (to appear)
8. Aronov, B., O'Rourke, J.: Nonoverlap of the star unfolding. Discrete Comput. Geom. **8**(3), 219–250 (1992)
9. Demaine, E.D., Demaine, M.L., Eppstein, D., Frederickson, G.N., Friedman, E.: Hinged dissection of polynominoes and polyforms. Comput. Geom. Theory Appl. **31**(3), 237–262 (2005)
10. Demaine, E.D., O'Rourke, J.: Geometric Folding Algorithms: Linkages, Origami, Polyhedra. Cambridge University Press, Cambridge (2007)
11. Frederickson, G.N.: Hinged Dissections: Swinging and Twisting. Cambridge University Press, Cambridge (2002)
12. Itoh, J., Nara, C.: Transformability and reversibility of unfoldings of doubly-covered polyhedra. In: Akiyama, J., Ito, H., Sakai, T. (eds.) JCDCGG 2013. LNCS, vol. 8845, pp. 77–86. Springer, Heidelberg (2014)

Geometric p-Center Problems with Centers Constrained to Two Lines

Binay Bhattacharya[1], Ante Ćustić[2], Sandip Das[3], Yuya Higashikawa[4], Tsunehiko Kameda[1(✉)], and Naoki Katoh[5]

[1] School of Computing Science, Simon Fraser University, Burnaby, Canada
tiko@sfu.ca
[2] Department of Mathematics, Simon Fraser University, Burnaby, Canada
[3] Advanced Computing and Microelectronics Unit,
Indian Statistical Institute, Kolkata, India
[4] Department of Information and System Engineering,
Chuo University, Tokyo, Japan
[5] School of Science and Technology, Kwansei University,
Sanda, Hyogo, Japan

Abstract. We first consider the weighted p-center problem, in which the centers are constrained to lie on two axis-parallel lines. Given a set of n points in the plane, which are sorted according to their x-coordinates, we show how to test in $O(n \log n)$ time if p piercing points placed on two lines, parallel to the x-axis, can pierce all the disks of different radii centered at the n given points. This leads to an $O(n \log^2 n)$ time algorithm for the weighted p-center problem. We then consider the unweighted case, where the centers are constrained to be on two perpendicular lines. Our algorithm runs in $O(n \log^2 n)$ time in this case as well.

1 Introduction

The *p-center problem* is one of the most intensively studied problems in computational geometry. The *geometric p*-center problem is defined as follows [17]. Given are a set $P = \{p_1, p_2, \ldots, p_n\}$ of n points in the plane, where point p_i ($i = 1, 2, \ldots, n$) has weight w_i, and a positive integer p. The objective is to find a set of p centers such that the maximum over all points in P of the weighted distance from a point to its nearest center is minimized. It was shown by Megiddo [16] that this problem is NP-hard in its general form. To find more tractable cases solvable in polynomial time, researchers have considered many variations of this problem in terms of the metric used, the number of centers, the weights (uniform vs. non-uniform), and constraints on the allowed positions (discrete points, lines, polygon, etc.) of the centers [10,14,15].

When the centers are constrained to a single line, Karmakar et al. showed that the *unweighted p*-center problem, where the vertices have the same weight, can be solved in $O(n \log n)$ time [11], and Wang and Zhang showed that the *weighted p*-center problem, can also be solved in $O(n \log n)$ time [18]. It was recently shown by Chen and Wang [4] that the weighted version of this problem

© Springer International Publishing AG 2016
J. Akiyama et al. (Eds.): JCDCGG 2015, LNCS 9943, pp. 24–36, 2016.
DOI: 10.1007/978-3-319-48532-4_3

has a lower bound of $\Omega(n \log n)$ on its time complexity. The $O(n \log n)$ time algorithms mentioned above are based on Megiddo's parametric search [13] with Cole's speed-up [5]. Since they use the AKS network [1], the coefficient hidden in the big-O notation is huge. If a more practical sorting network, such as *bitonic sorting network* [12], is used the time requirement increases to $O(n \log^2 n)$. The 1-dimensional p-center problem is discussed by Bhattacharya et al. [3], Chen and Wang [4], and Fournier and Vigneron [6]. If the l_1 or l_∞ distance metric is used instead of l_2, in other words, if the enclosing shape is an axis-parallel square, then we can apply a method due to Frederickson and Johnson [7, 8] to solve this problem in $O(n)$ time [6], provided the points are presorted.

In this paper we consider the p-center problem constrained to two parallel or perpendicular lines. We first show that for a given weighted distance ρ, we can test in $O(n \log n)$ time whether there exist p centers on two parallel lines such that each point is within weighted distance ρ from a center. This implies that p-center problem can be solved in $O(n \log^2 n)$ time, using Megiddo's parametric search [13] with Cole's speed-up [5]. In the unweighted case and under the l_1 or l_∞ distance metric, it is known that this problem can be solved in $O(n \log n)$ time without using a huge sorting network [2].

In the line-constrained problem, it is a standard practice to use the intersection of an object $o(p_i)$ around each point p_i and a line. This object $o(p_i)$ is a square under the l_∞ distance metric and a disk under the l_2 metric, which is the case in our model. When disks are used instead of squares, some results from [2] carry over, but a few complications arise. First, the end points of the intersections do not have the same order as the points' x-coordinates. Second, if a disk intersects both of the two lines, the intersection points do not have the same x-coordinate in general. These complications make the problem more challenging. If the centers are constrained to a single line (e.g., the x-axis), ρ-feasibility can be tested in linear time [11], provided that the given points are presorted by their x-coordinates. We will often make use of this linear time 1-line algorithm in this paper.

The rest of the paper is organized as follows. Section 2 discusses the case where the centers are constrained to two parallel lines, and presents an $O(n \log^2 n)$ time algorithm. In Sect. 3 we discuss the unweighted case, where the centers are constrained to two perpendicular lines. We then show that the problem can be solved in $O(n \log^2 n)$ time. Finally, Sect. 4 concludes the paper with a summary and open problems.

2 Centers Constrained to Two Parallel Lines

2.1 Preliminaries

Let $P = \{p_1, p_2, \ldots, p_n\}$ be a set of points in the plane, where point p_i has a positive weight w_i. We denote the x-coordinate (resp. y-coordinate) of point p_i by $p_i.x$ (resp. $p_i.y$), and assume that $p_i.x \leq p_{i+1}.x$ holds for $i = 1, 2, \ldots, n - 1$, i.e., the points are sorted according to their x-coordinates. If the x-coordinates of two points are the same, they can be ordered arbitrarily. Let $D_i(\rho)$ denote the disk with radius ρ/w_i centered at point p_i. Let L_1 and L_2 be the two given axis-parallel horizontal lines (L_1 above L_2). A problem instance is said to be ρ-*feasible* if p centers can be placed on the two lines in such a way that there is a center within distance ρ/w_i from every point p_i. Otherwise, it is ρ-*infeasible*. For $i = 1, 2, \ldots, n$ let $J_1^i(\rho) = D_i(\rho) \cap L_1$ (resp. $J_2^i(\rho) = D_i(\rho) \cap L_2$). We assume that at least one of $J_1^i(\rho)$ and $J_2^i(\rho)$ is non-empty, since otherwise the problem instance is not ρ-feasible. If $J_1^i(\rho) \neq \emptyset$ and $J_2^i(\rho) \neq \emptyset$, then they are called *buddy intervals* or just *buddies*, and the corresponding point p_i is called a *buddy point*. So, given a set P of weighted points and a radius ρ, we are interested in finding a minimum cardinality set S of points on L_1 and L_2 such that $S \cap J_1^i(\rho) \neq \emptyset$ or $S \cap J_2^i(\rho) \neq \emptyset$, for every $i = 1, 2, \ldots, n$. We call such S, a set of piercing points. From now on, up to Lemma 2, we assume that radius ρ is fixed.

Given i, $1 \leq i \leq n$, we call a set S_i of piercing points for all the disks around points $P_i = \{p_1, p_2, \ldots, p_i\}$ a *partial solution* on P_i. If $P_i = P$, a partial solution is a *complete solution*. Given a partial solution S_i with $|S_i| = z$ piercing points, let I_1 be an interval on L_1 that represents the section along which the rightmost point of $S_i \cap L_1$ can be moved so that all the disks for the points in P_i are still pierced. Analogously, let I_2 represents such an interval on L_2. Then the triple $c = (I_1, I_2; z)$ is said to be a *configuration* for P_i, where the non-negative integer z is called the *count* of the configuration [2]. Furthermore, associated with configuration c is a *piercing sequence* $\wp(c)$ of $\max\{z - 2, 0\}$ fixed piercing points on L_1 and L_2, i.e., S_i without the rightmost points on L_1 and L_2. In other words, configuration $(I_1, I_2; z)$ represents a class of partial solutions that consist of z piercing points that are identical, except for the rightmost points on L_1 and L_2, which can be anywhere on I_1 and I_2, respectively.

For the purpose of our algorithm, we would like to be able to dismiss partial solutions/configurations that cannot be the unique ones that lead to the complete solutions of minimum cardinality. To that end, we introduce the concept of domination. Given a partial solution S, a *complete extension* of S is a superset of S which is a complete solution. Let $c' = (I_1', I_2'; z')$ and $c'' = (I_1'', I_2''; z'')$ be two different configurations for P_i. We say that c' *dominates* c'' if, regardless of what the remaining points in P are, there cannot exist a complete extension of some partial solution represented by c'' which has a strictly smaller cardinality than all the complete extensions of all partial solutions represented by c'.

2.2   Algorithm

We scan the points in $P = \{p_1, p_2, \ldots, p_n\}$ from left to right, in such a way that in step i we generate all non-dominated configurations for P_i. We maintain the set F of such configurations, called the *frontier configurations*. After we scan all the n points, we know that the given instance is ρ-feasible if and only if there is a configuration in F with the count at most p. Given an interval I on either L_1 or L_2, let $l(I)$ (resp. $r(I)$) denote the left (resp. right) end point of I.

Algorithm 1

1. *Initialize $F = \{([-\ell, -\ell], [-\ell, -\ell]; 0)\}$, where ℓ is a very large number.*
2. *For $i = 1, 2, \ldots, n$, execute Step 3.*
3. *Set $F' = \emptyset$. For each configuration $c = (I_1, I_2; z) \in F$, do Steps (a)-(c) that apply, and put the generated configurations in F'.*
   - (a) *$[J_1^i(\rho) = \emptyset \wedge J_2^i(\rho) = \emptyset]$ The problem instance is not ρ-feasible (no point can pierce $D_i(\rho)$). Stop*
   - (b) *$[J_1^i(\rho) \neq \emptyset]$ If $I_1 \cap J_1^i(\rho) \neq \emptyset$ then convert c into $(I_1 \cap J_1^i(\rho), I_2; z)$ else convert it into $(J_1^i(\rho), I_2; z + 1)$ and add $r(I_1)$ into $\wp(c)$.*
   - (c) *$[J_2^i(\rho) \neq \emptyset]$ If $I_2 \cap J_2^i(\rho) \neq \emptyset$ then convert c into $(I_1, I_2 \cap J_2^i(\rho); z)$ else convert it into $(I_1, J_2^i(\rho); z + 1)$ and add $r(I_2)$ into $\wp(c)$.*

   Remove the dominated configurations from F', and replace F by F'.
4. *The problem instance is ρ-feasible if and only if there is a configuration in F whose count is no more than p.*   □

Theorem 1. *Algorithm 1 is correct.*

Proof (Sketch). In Step 3 of the algorithm, for all i we aim to calculate a set F of partial solutions on points P_i, such that, no mater what the rest of P is, there is one partial solution in F that can be extended to a complete solution with the minimum cardinality. If that is the case, the algorithm is correct.

Recall that we group partial solutions into configurations and treat them as a unit. One such configuration $c = (I_1, I_2; z)$ represents all partial solutions consisting of the points in $\wp(c)$ and one piercing point each on I_1 and I_2. Consider configuration $c = (I_1, I_2; z)$, and a new point p_i with the corresponding interval $J_1^i(\rho)$. If $J_1^i(\rho)$ is already pierced by a point from $\wp(c)$, then we have $l(J_1^i(\rho)) \leq l(I_1)$ by the definition of $\wp(c)$. In this case, from the fact that $p_i.x \geq p_j.x$ for $i > j$, we also have $r(I_1) \leq r(J_1^i(\rho))$. It thus follows that $I_1 \subseteq J_1^i(\rho)$, which implies that $J_1^i(\rho)$ is also pierced by every point in I_1. Therefore, if $I_1 \subseteq J_1^i(\rho)$ then we can ignore p_i, and otherwise, we compute $I_1 \cap J_1^i(\rho)$. In other words, we know whether $J_1^i(\rho)$ can be pierced by some partial solution corresponding to c without observing $\wp(c)$. Therefore, we do not lose any crucial information by restricting to configurations, i.e., only the number of piercing points (count) and positions of the rightmost piercing points on L_1 and L_2 play a role in the quality of a partial solution with respect to the remaining points in P.

In Step 3(b) of the algorithm, if $I_1 \cap J_1^i(\rho) \neq \emptyset$, only $(I_1 \cap J_1^i(\rho), I_2; z)$, and not $(J_1^i(\rho) \setminus I_1, I_2; z + 1)$, is generated. That is because the latter is dominated by the former. When creating $(J_1^i(\rho), I_2; z + 1)$, in addition we need to add an arbitrary point from I_1 to $\wp(c)$. Step 3(c) works analogously.

From the definition of domination, it follows that removing dominated configurations at the end of Step 3 will not affect the correctness.   □

By implementing Algorithm 1 directly, inefficiencies would be caused if searching F and removing dominated configurations in in Step 3 are blindly executed. In the following subsection we describe how Algorithm 1 can be implemented efficiently.

2.3   Implementation

Let us first discuss how to implement one round of Step 3 of Algorithm 1 efficiently. For this purpose we identify the interval I_1 (resp. I_2) of configuration $(I_1, I_2; z)$ with its right endpoint $r(I_1)$ (resp. $r(I_2)$). We can do so since I_1 intersects $J_1^i(\rho)$ if and only if $l(J_1^i(\rho)) \leq r(I_1)$, in which case $r(I_1 \cap J_1^i(\rho)) = \min\{r(I_1), r(J_1^i(\rho))\}$. This follows from $l(I_1) < r(J_1^i(\rho))$, which is a consequence of $p_i.x \geq p_j.x$ for $i > j$ and the fact that $l(I_1)$ is the left endpoint of some interval that has been processed so far. So in the rest of this section, a configuration will be represented by $(x_1, x_2; z)$, where $x_1 = r(I_1)$ and $x_2 = r(I_2)$.

To represent configuration $c = (x_1, x_2; z)$ visually, we draw a line between x_1 and x_2 and label it by its count z. We say that configurations $(x_1', x_2'; z')$ and $(x_1'', x_2''; z'')$ *cross each other* if either $x_1' < x_1''$ and $x_2' > x_2''$ or $x_1' > x_1''$ and $x_2' < x_2''$ hold. Configurations $(x_1', x_2'; z')$ and $(x_1'', x_2''; z'')$ are said to be *disjoint* if either $x_1' < x_1''$ and $x_2' < x_2''$ or $x_1' > x_1''$ and $x_2' > x_2''$ hold. See Fig. 1.

Lemma 1. *Let F be a set of configurations that represent partial solutions for point set P_i such that no configuration dominates another in F. Then*

(a) The counts of the configurations in F differ by at most one.
(b) Every pair of configurations in F cross each other if they have the same count.
(c) Every pair of configurations $(x_1', x_2'; z)$, $(x_1'', x_2''; z - 1) \in F$ are disjoint, and $x_1' > x_1''$ and $x_2' > x_2''$ hold. (See Fig. 2.)

Proof. (a) Let $c' = (x_1', x_2'; z')$ and $c'' = (x_1'', x_2''; z'')$ be two configurations for the points in P_i, such that $z' \geq z'' + 2$. Let S be a minimum cardinality set of piercing points that extends $\wp(c')$ to a complete solution. Then $S \cup \{(x_1'', L_1), (x_2'', L_2)\}$ extends $\wp(c'')$ to a complete solution also, where L_1 (resp. L_2) indicates the line

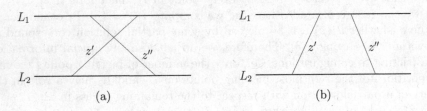

Fig. 1. Representations of configurations: (a) The two configurations cross each other (b) They are disjoint.

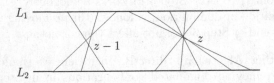

Fig. 2. Set of configurations without any dominating configuration

the piercing point is on. Since the cardinality of the latter complete solution does not exceed the cardinality of the former one, we have that c'' dominates c'.

(b) It suffices to prove that if $c' = (x_1', x_2'; z)$ and $c'' = (x_1'', x_2''; z)$ do not dominate each other, then they must cross each other. To prove the contrapositive, assume that they do not cross each other. Then without loss of generality we can assume that $x_1' \leq x_1''$ and $x_2' \leq x_2''$. But then from arguments in the proof of Theorem 1, it easily follows that c' cannot lead to a complete solution of a smaller cardinality, i.e., c'' dominates c'.

(c) Assume to the contrary that there are two disjoint configurations $c' = (x_1', x_2'; z)$ and $c'' = (x_1'', x_2''; z - 1)$ in F such that, without loss of generality, $x_2' \leq x_2''$ holds. Then again, if S is a minimum cardinality extension of $\wp(c')$ to a complete solution, then $S \cup \{(x_1'', L_1)\}$ is an extension of $\wp(c'')$ to a complete solution, hence c'' dominates c', a contradiction to the fact that no configuration in F dominates another.   □

Lemma 1 gives us a description of frontier configurations F. Now we describe how Step 3 of Algorithm 1 can be implemented efficiently, i.e., how F can be updated when intervals $J_1^i(\rho)$ and/or $J_2^i(\rho)$ corresponding to a new point p_i are introduced.

For simplicity, let us use symbols $l_k^i = l(J_k^i(\rho))$ and $r_k^i = r(J_k^i(\rho))$ for $k = 1, 2$. First, let the new point p_i be a non-buddy point such that $J_1^i(\rho) \neq \emptyset$ and $J_2^i(\rho) = \emptyset$. In order to check which configurations from F will be preserved, modified, or discarded, after $J_1^i(\rho) = [l_1^i, r_1^i]$ is introduced, we partition F into three groups. Let $F_l = \{(x_1, x_2; z) \in F \mid x_1 < l_1^i\}$, $F_m = \{(x_1, x_2; z) \in F \mid l_1^i \leq x_1 \leq r_1^i\}$ and $F_r = \{(x_1, x_2; z) \in F \mid r_1^i < x_1\}$. Step 3(b) of Algorithm 1 replaces each configuration $(x_1, x_2; z) \in F_l$ with $(r_1^i, x_2; z+1)$. Let $\bar{c} = (\bar{x}_1, \bar{x}_2, z)$ be an element of F_l with the smallest first component. By a similar argument to that in the proof of Lemma 1, one can easily see that $(r_1^i, \bar{x}_2; z + 1)$, created from $\bar{c}$, will dominate all other configurations created from F_l. Analogously, configuration $(r_1^i, \hat{x}_2; z)$, created in Step 3(b) from a configuration $(\hat{x}_1, \hat{x}_2; z)$ from F_r with the smallest first component, will dominate all other configurations created from the elements of F_r. Lastly, Step 3(b) leaves all elements of F_m unchanged. To summarize, in the case of non-buddy point p_i with $J_1^i(\rho) \neq \emptyset$, we can update frontier configurations in F by scanning it from left to right with respect to the first component, and removing all components until we hit the interval $J_1^i(\rho)$. Furthermore, in place of the removed configurations, two new configurations are considered to join F. This is illustrated in Fig. 3(a), where the solid lines represent the frontier configurations that are in F before p_i is processed. The configurations that survive are indicated by thick lines. Two dashed lines are new configurations considered, where the right one dominates in this example, and hence is admitted to F. We present the above implementation more formally as a procedure.

Procedure 1. NonBuddy($J_1^i(\rho)$)

1. *Scan the configurations in F from left to right and right to left on L_1, until l_1^i and r_1^i are reached, respectively. Delete all scanned configurations $(x_1, x_2; z)$ (except those with the first component (i.e., x_1) equal to l_1^i or r_1^i).*

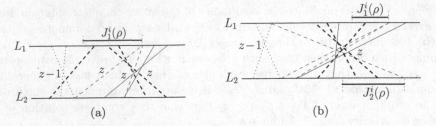

Fig. 3. Updating the frontier configurations: (a) for a non-buddy interval on L_1; (b) for buddy intervals.

2. Let $(\bar{x}_1, \bar{x}_2; \bar{z})$ and $(\hat{x}_1, \hat{x}_2; \hat{z})$ be the deleted configurations with smallest first components in the left to right and right to left scanning process, respectively. (Note that they do not exist if $F_l = \emptyset$ and $F_r = \emptyset$.)
3. If there is no configuration in F with the first component r_1^i, then insert in F the dominating configuration in the pair $\{(r_1^i, \bar{x}_2; \bar{z} + 1), (r_1^i, \hat{x}_2; \hat{z})\}$.   □

The case where the new point p_i is a non-buddy point such that $J_1^i(\rho) = \emptyset$ and $J_2^i(\rho) \neq \emptyset$ can be implemented by a similar procedure.

We are thus left with the case where p_i is a buddy point, i.e., $J_1^i(\rho) \neq \emptyset$ and $J_2^i(\rho) \neq \emptyset$. This case can be handled by applying the approach from Procedure NonBuddy on both L_1 and L_2 "simultaneously." We scan F in two different directions, from left to right and right to left on L_1 until we reach l_1^i and r_1^i, respectively, and then from left to right and right to left on L_2 until we reach l_2^i and r_2^i, respectively. We now delete all the configurations that were scanned twice. Furthermore, for L_1 and L_2 each we consider at most two new configurations as in Step 3 of Procedure 1. Among these at most four configurations, we add to F those which are not dominated. See Fig. 3(b), where non-bold lines must be deleted, and dashed lines are considered to be added to F. Thus, each point p_i causes is at most two net increase in the number of configurations.

Lemma 2. *Algorithm 1, if implemented with the procedures defined above, can test ρ-feasibility in $O(n \log n)$ time.*

Proof. Our procedures create at most two new configurations for each new point p_i introduced, hence the total number of configurations created is $O(n)$.

We can keep two lists of pointers to configurations of F, one which is sorted with respect to the first component, and second one with respect to the second component. We implement these lists using a data structure that allows search, insert and delete in $O(\log n)$ time, for example using a 2-3 tree. Then it is easy to see that the scanning part of NonBuddy procedures (Step 1) can be done in $O(m \log |F|)$ time, where m is the number of removed (scanned) configurations. Steps 2 and 3 can be done $O(\log |F|)$ time. Now let us consider the more complicated non-buddy case. As can be seen in Fig. 3(b), there could be multiple subsequences of configurations that need to be removed, some of which are surrounded by configurations that need to be preserved. To reach areas that need to be removed we need $O(\log |F|)$ time, and to delete configurations we need $O(m \log |F|)$ time, where m is the number of configurations that need to be removed. Hence, in both buddy and non-buddy cases, the processing of point p_i can be done in $O(m_i \log n)$ time, where m_i is the number of configurations deleted in the processing of p_i. Once deleted, a configuration will not be considered again, hence $\sum_{i=1}^{n} m_i = O(n)$. Therefore the complexity of Algorithm 1 is $O(n \log n)$.   □

2.4 Finding Optimal ρ^* for p-Center

We adopt Megiddo's parametric search to find the optimal weighted radius ρ^*. Here we give an intuitive idea, referring the reader to [13] for details. Suppose we start with a radius ρ such that a problem instance is not ρ-feasible, and q ($> p$) piercing points are required to pierce all the intervals on L_1 and L_2. If we increase ρ, we will reach a point where $q-1$ piercing points are sufficient to pierce all the intervals on L_1 and L_2. At this new value of ρ, at least one more pair of intervals must intersect. This indicates that whether intervals intersect or not plays an important role in decreasing/increasing the number of piercing points. Whether a pair of intervals intersects or not on a line (L_1 or L_2) clearly depends on the order of their end points on the line. If we can somehow determine their order for ρ^*, then we can find a p-center. Megiddo [13] proposed an ingenious may of implementing prune and search, using a sorting network to find the sorted order of the endpoints under ρ^* without knowing ρ^*.

Using the AKS sorting network [1], he showed that $O(\log^2 n)$ invocations of an ρ-feasibility test can determine the optimal radius ρ^*. Later Cole [5] showed that this number can be reduced to $O(\log n)$ invocations. Since each invocation takes $O(n \log n)$ time by Lemma 2, we have proved

Theorem 2. *When the centers are constrained to two parallel lines, we can solve the p-center problem in $O(n \log^2 n)$ time.*   □

3 Centers Placed on Both x- and y-axes

In this section we discuss only the *unweighted* case, where each point has unit weight. We assume that line L_1 (resp. L_2) is the x- (resp. y)-axis, denoted by X (resp. Y). For a given radius ρ, the points in P must lie within the horizontal and/or vertical bands of width 2ρ defined by four lines, $x = \pm\rho$ and $y = \pm\rho$. We first sort the points according to the x- and y-coordinates, separately. Let us call $A(\rho)$ the clover shaped area which is the union of the four disks of radius ρ whose centers are on the two axes at distance ρ from the origin. See Fig. 4(a). Clearly, any point outside $A(\rho)$ can be covered by a circle centered on either the x-axis or y-axis, not both. To cover all those points with the minimum number of enclosing circles, we can use the 1-dimensional algorithm four times "outside-in" on the x- and y-axes. Note that these circles may also cover some points in $A(\rho)$. Thus we need to consider only the remaining set $P(\rho)$ of uncovered points that lie within $A(\rho)$.

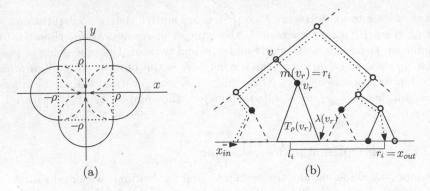

Fig. 4. (a) Area $A(\rho)$; (b) Updown search of tree T_ρ.

It is clear that all the points in $P(\rho)$ can be covered by four centers, as can be seen from Fig. 4(a). We thus want to test if they can be covered by less than four circles of radius ρ. By a greedy method, we can easily test if all the points in $P(\rho)$ can be covered by one, two or three circles on the same axis in $O(n)$ time. Therefore, without loss of generality, we assume that one center, named C_y, whose center, c_y, is on the y-axis, and the others (no more than two) are placed on the x-axis. We start with $c_y = (0, \rho)$, and push c_y downwards. The points in $P(\rho)$ that lie outside (resp. inside) C_y are said to be *active* (resp. *inactive*).

3.1 ρ-Feasibility Test

In this subsection we assume that radius ρ is given. For each point $p_i \in P$, let $D_i(\rho)$ be the disk of radius ρ centered at p_i, and define the interval $J_x^i(\rho) = [l_i, r_i] = D_i(\rho) \cap X$.[1] We call $J_x^i(\rho)$ *active* (resp. *inactive*) if p_i is active (resp. inactive). We can sort the $2n$ endpoints, $\{l_i, r_i \mid p_i \in P\}$, in $O(n \log n)$ time.

Let $|P(\rho)| = m$ ($\leq n$). We now construct, in $O(m)$ time, a balanced tree named T_ρ, whose leaves consist of the sorted elements $\{l_i \mid p_i \in P(\rho)\}$ arranged from left to right. For interval $J_x^i(\rho) = [l_i, r_i]$, we interpret r_i as the *value* of l_i. For any node u of T_ρ, let $m(u)$ denote the minimum value among the leaves of subtree $T_\rho(u)$ rooted at u. Tree T_ρ implements a heap as well based on the values of its leaves, and we can compute $m(u)$ for all nodes u in linear time.

For example, in Fig. 4(b), given a point x_{in} on the x-axis, suppose we are interested in the rightmost point x_{out} such that the values (defined above) of the leaves that lie between x_{in} and x_{out} are at or to the right of x_{out}. By this definition of x_{out}, we have $r_j \geq x_{out}$ for each leaf $l_j \in [x_{in}, x_{out}]$, and there exists a leaf $l_i \in [x_{in}, x_{out}]$ such that $r_i = x_{out}$. It is clear that x_{out} pierces all the intervals whose left endpoint lies in $[x_{in}, x_{out}]$, and it is the rightmost such piercing point.

---

[1] This interval can be empty. We refer to this interval as a *disk interval* to distinguish it from another kind of interval on the x-axis.

Let us design a procedure that finds x_{out} for a given x_{in}, assuming that we already have search tree T_ρ. For a node u of T_ρ, let $\lambda(u)$ denote the rightmost leaf that belongs to subtree $T_\rho(u)$. We assume that $\lambda(u)$ is also stored at u. Starting from the leaf nearest to x_{in} on its right, we trace the path towards the root of T_ρ. Let v be a node on this path and let v_r be its right child. See Fig. 4(b). We maintain a position variable $\underline{x}$, initialized appropriately, and update it by $\underline{x} = \min\{\underline{x}, m(v_r)\}$. If $\underline{x} > \lambda(v_r)$, it implies that $x_{out} > \lambda(v_r)$, so we move to the parent of v and perform a similar test for its right child. Eventually, we may have $\underline{x} \leq \lambda(v_r)$ for some v_r. If $\underline{x} = \lambda(v_r)$, then $x_{out} = \underline{x}$, and we are done. If $\underline{x} < \lambda(v_r)$, on the other hand, then we need to check the left subtree of v_r, and then the right subtree of v_r. See the dotted path in Fig. 4(b). A filled circle in Fig. 4(b) indicates the node u at which the test $\underline{x} > \lambda(u)$ succeeded. We call this search process *updown search*, which consists of the *Up phase* followed by the *Down phase*. The following procedure describes updown search more formally.

Procedure 2. UpDown(x_{in}, T_ρ, c_y)

1. Find the first leaf l_i of T_ρ that lies to the right of x_{in}. Set $v = l_i$ and $\underline{x} = r_i$.
2. [Up phase] While v is the right child of $p(v)$ (the parent of v) do $v = p(v)$. Let v_r be right child of v.
   (a) Update $\underline{x} = \min\{\underline{x}, m(v_r)\}$.
   (b) If $\underline{x} > \lambda(v_r)$ then set $v = p(v)$, and repeat Step 2. Else let $v = v_r$.
3. [Down phase] Let v_l be the left child of vertex of v.
   (a) Update $\underline{x} = \min\{\underline{x}, m(v_l)\}$.
   (b) If $\underline{x} \leq \lambda(v_l)$ then set $v = v_l$ else set $v = v_r$.
   (c) If v is a leaf l_j, set $x_{out} = \min\{\underline{x}, r_j\}$ and stop. Else repeat Step 3.   □

Once the first piercing point is identified, and if it doesn't pierce all the active intervals on the x-axis, then we invoke UpDown(x_{in}, T_ρ, c_y) again with x_{in} set to x_{out} returned from the first invocation, to find the second piercing point, and so forth. For a given position of c_y, the following algorithm finds the minimum number of piercing points for the active intervals on the x-axis, where z is the number of centers needed, including c_y. We assume $P(\rho) \neq \emptyset$, so that at least one center is needed to cover the points in $P(\rho)$.

Algorithm 2. Find-Centers(ρ, T_ρ, c_y)

1. Set $z = 2$, and x_{in} to a point that lies to the left of the leftmost leaf of T_ρ.
2. Call UpDown(x_{in}, T_ρ, c_y).
3. If the returned x_{out} lies at or to the right of the rightmost leaf of T_ρ, then stop and return $z - 1$ if $x_{out} = M \triangleq \max\{r_i \mid p_i \in P(\rho)\} + 1$, or z otherwise.
4. Increment z by 1. If $z = 5$ then stop. (Use the four circles in Fig. 4(a).) Else set $x_{in} = x_{out}$ and go to Step 3.   □

As C_y is pushed downward and c_y decreases, the set of active points changes. When an interval $J_x^i(\rho)$ becomes inactive, we change the value of l_i to M, and update the $m(\cdot)$ values associated with the nodes on the path $\pi(l_i)$ from l_i to the root of T_ρ, which takes $O(\log m)$ time, using standard heap operations.

When $J_x^i(\rho)$ becomes active again, we restore the value of l_i to r_i and update the $m(\cdot)$ values along $\pi(l_i)$. For each point $p_i \in P$, define the interval $J_y^i(\rho) = [d_i, u_i] = D_i(\rho) \cap Y$, where $d_i \leq u_i$.

Algorithm 3. Feasibility(ρ)

1. *Sort the end points of $\{J_x^i(\rho) \mid p_i \in P\}$ and $\{J_y^i(\rho) \mid p_i \in P\}$ separately.*
2. *Compute $z_{\overline{A}}$, the minimum number of centers needed to cover the points not in $A(\rho)$, and determine $P(\rho)$. If $z_{\overline{A}} > p$, output "ρ-infeasible" and stop.*
3. *Construct search tree T_ρ, assuming that all points are active.*
4. *Initialize c_y at $(0, \rho) \in Y$, determine all the active points, and update T_ρ. Call Find-Centers(ρ, T_ρ, c_y) and set z_A to the returned value z.*
5. *If pushing c_y to the next position in the sorted list of $\{d_i, u_i \mid p_i \in P(\rho)\}$ makes a point p_i with $J_x^i(\rho) = \emptyset$ active, then go to Step 6. Otherwise, while c_y is not at the bottom of the list, push c_y to the next position, and call Find-Centers(ρ, T_ρ, c_y). Update $z_A = \min\{z_A, z\}$, where z is the returned value.*
6. *If $z_{\overline{A}} + z_A \leq p$ then output "ρ-feasible." Else output "ρ-infeasible."*   □

In Step 5 Find-Centers(ρ, T_ρ, c_y) is called for $O(m)$ possible positions for c_y, and finds the minimum number of centers required among them.

Lemma 3. *For a given ρ, we can test ρ-feasibility for the points in $P(\rho)$ in $O(m \log m)$ time, where $|P(\rho)| = m$.*   □

Since Steps 1, 4, and 5 of Feasibility(ρ) each take $O(n \log n)$ time, Lemma 3 implies the following theorem.

Theorem 3. *When the centers are constrained to be on x- and y-axes, we can test ρ-feasibility in $O(n \log n)$ time.*   □

3.2   Optimization

We use the same idea as in Sect. 2.4 to find the minimum ρ, named ρ^*. Clearly, if we sort all the end points of the disk intervals with respect to ρ^*, then the left endpoints (the leaves of T_ρ) are automatically sorted. As in Theorem 2, we use Megiddo's method [13] to sort all these end points, using the AKS sorting network [1] with Cole's improvement [5]. Thus we need to perform ρ-feasibility tests only $O(\log n)$ times, and Theorem 3 implies

Theorem 4. *When the centers are constrained to be on two perpendicular lines, we can solve the p-center problem in $O(n \log^2 n)$ time.*   □

4   Conclusion

This paper considered two models of constrained p-center problem. In the first model, the given n points are weighted and the centers are constrained to lie on two parallel lines, and in the second model, the points are unweighted and the centers are constrained to lie on two perpendicular lines. We have presented $O(n \log^2 n)$ time algorithms for the above two p-center problems. An open problem is the weighted case for two perpendicular lines, which appears much more difficult. The case where the centers are constrained to more than two lines is also open. It is known that AKS sorting network is too huge to be practical, but the recent result by Goodrich [9] gives us hope that a practical $O(n \times \log n)$ sorting network may be designed in the not-too-distant future.

Acknowledgement. We would like to thank Hirotaka Ono of Kyushu University and Yota Otachi of JAIST for stimulating discussions on the topic of Sect. 3. This work was supported in part by Discovery Grant #13883 from the Natural Science and Engineering Research Council (NSERC) of Canada and in part by MITACS, both awarded to Bhattacharya.

References

1. Ajtai, M., Komlós, J., Szemerédi, E.: An $O(n \log n)$ sorting network. In: Proceedings of 15th ACM Symposium on Theory of Computing (STOC), pp. 1–9 (1983)
2. Bereg, S., Bhattacharya, B., Das, S., Kameda, T., Mahapatra, P.R.S., Song, Z.: Optimizing squares covering a set of points. Theoret. Comput. Sci. (2015). http://dx.doi.org/10.1016/j.tcs.2015.11.029
3. Bhattacharya, B., Shi, Q.: Optimal algorithms for the weighted p-center problems on the real line for small p. In: Dehne, F., Sack, J.-R., Zeh, N. (eds.) WADS 2007. LNCS, vol. 4619, pp. 529–540. Springer, Heidelberg (2007). doi:10.1007/978-3-540-73951-7_46
4. Chen, D.Z., Li, J., Wang, H.: Efficient algorithms for the one-dimensional k-center problem. Theoret. Comput. Sci. **592**, 135–142 (2015)
5. Cole, R.: Slowing down sorting networks to obtain faster sorting algorithms. J. ACM **34**, 200–208 (1987)
6. Fournier, H., Vigneron, A.: A deterministic algorithm for fitting a step function to a weighted point-set. Inf. Process. Lett. **113**, 51–54 (2013)
7. Frederickson, G., Johnson, D.: The complexity of selection and ranking in $X + Y$ and matrices with sorted columns. J. Comput. Syst. Sci. **24**, 197–208 (1982)
8. Frederickson, G., Johnson, D.: Finding kth paths and p-centers by generating and searching good data structures. J. Algorithms **4**, 61–80 (1983)
9. Goodrich, M.T.: Zig-zag sort: a simple deterministic data-oblivious sorting algorithm running in $O(n \log n)$ time [cs.DS] 11 March 2014. arXiv:1403,2777v1
10. Hurtado, F., Sacristn, V., Toussaint, G.: Constrained facility location, pp. 15–17. Studies of Location Analysis, Special Issue on Computational Geometry (2000)
11. Karmakar, A., Das, S., Nandy, S.C., Bhattacharya, B.: Some variations on constrained minimum enclosing circle problem. J. Comb. Opt. **25**(2), 176–190 (2013)
12. Knuth, D.: The Art of Computer Programming: Sorting and Searching, vol. 3, 3rd edn. Addison-Wesley, Boston (1997)
13. Megiddo, N.: Applying parallel computation algorithms in the design of serial algorithms. J. ACM **30**, 852–865 (1983)
14. Megiddo, N.: Linear-time algorithms for linear-programming in R^3 and related problems. SIAM J. Comput. **12**, 759–776 (1983)

15. Megiddo, N.: The weighted euclidian 1-center problem. Math. Oper. Res. **8**(4), 498–504 (1983)
16. Megiddo, N., Supowit, K.: On the complexity of some common geometric location problems. SIAM J. Comput. **14**, 182–196 (1984)
17. Preparata, F., Shamos, M.: Computational Geometry: An Introduction. Springer, Berlin (1990)
18. Wang, H., Zhang, J.: Line-constrained k-median, k-means, and k-center problems in the plane. In: Ahn, H.-K., Shin, C.-S. (eds.) ISAAC 2014. LNCS, vol. 8889, pp. 3–14. Springer, Heidelberg (2014). doi:10.1007/978-3-319-13075-0_1

Dissection with the Fewest Pieces is Hard, Even to Approximate

Jeffrey Bosboom[1]([⊠]), Erik D. Demaine[1], Martin L. Demaine[1], Jayson Lynch[1], Pasin Manurangsi[2], Mikhail Rudoy[1], and Anak Yodpinyanee[1]

[1] Computer Science and AI Laboratory, Massachusetts Institute of Technology, 32 Vassar St., Cambridge, MA 02139, USA
{jbosboom,edemaine,mdemaine,jaysonl,mrudoy,anak}@mit.edu
[2] University of California, Berkeley, CA 94720, USA
pasin@berkeley.edu

Abstract. We prove that it is NP-hard to dissect one simple orthogonal polygon into another using a given number of pieces, as is approximating the fewest pieces to within a factor of $1 + 1/1080 - \varepsilon$.

1 Introduction

We have known for centuries how to dissect any polygon P into any other polygon Q of equal area, that is, how to cut P into finitely many pieces and re-arrange the pieces to form Q [2,7,11,13,14]. But we know relatively little about how many pieces are necessary. For example, it is unknown whether a square can be dissected into an equilateral triangle using fewer than four pieces [6,8, pp. 8–10]. Only recently was it established that a pseudopolynomial number of pieces suffices [1].

In this paper, we prove that it is NP-hard even to approximate the minimum number of pieces required for a dissection, to within some constant ratio. While perhaps unsurprising, this result is the first analysis of the complexity of dissection. We prove NP-hardness even when the polygons are restricted to be simple (hole-free) and orthogonal. The reduction holds for all cuts that leave the resulting pieces connected, even when rotation and reflection are permitted or forbidden.

Our proof significantly strengthens the observation (originally made by the Demaines during JCDCG'98) that the second half of dissection—re-arranging given pieces into a target shape—is NP-hard: the special case of exact packing rectangles into rectangles can directly simulate 3-PARTITION [5]. Effectively, the challenge in our proof is to construct a polygon for which any k-piece dissection must cut the polygon at locations we desire, so that we are left with a rectangle packing problem.

Due to the lack of space, we omit the proofs of some lemmas from this current version of our paper. For missing proofs, see the full version of this paper [3].

P. Manurangsi—Part of this work was completed while the author was at Massachusetts Institute of Technology and Dropbox, Inc.

A. Yodpinyanee—Research supported by NSF grant CCF-1420692.

J. Akiyama et al. (Eds.): JCDCGG 2015, LNCS 9943, pp. 37–48, 2016.
DOI: 10.1007/978-3-319-48532-4_4

2   The Problems

2.1   Dissection

We begin by formally defining the problems involved in our proofs, starting with k-PIECE DISSECTION, which is the central focus of our paper.

Definition 1. k-PIECE DISSECTION *is the following decision problem:*
   INPUT: *two polygons P and Q of equal area, and a positive integer k.*
   OUTPUT: *whether P can be cut into k pieces such that these k pieces can be packed into Q (via translation, optional rotation, and optional reflection).*

To prevent ill-behaved cuts, we require every piece to be a *Jordan region (with holes)*: the set of points interior to a Jordan curve e and exterior to $k \geq 0$ Jordan curves $h_1, h_2, \ldots, h_k$, such that $e, h_1, h_2, \ldots, h_k$ do not meet. There are two properties of Jordan regions that we use in our proofs. First, Jordan regions are Lebesgue measurable; we will refer to the Lebesgue measure of each piece as its area. Second, a Jordan region is path-connected. For brevity, we refer to path-connected as connected throughout the paper.

Next we define the optimization version of the problem, MIN PIECE DISSECTION, in which the objective is to minimize the number of pieces.

Definition 2. MIN PIECE DISSECTION *is the following optimization problem:*
   INPUT: *two polygons P and Q of equal area.*
   OUTPUT: *the smallest positive integer k such that P can be cut into k pieces such that these k pieces can be packed into Q.*

2.2   5-Partition

Our NP-hardness reduction for k-PIECE DISSECTION is from 5-PARTITION, a close relative of 3-PARTITION.

Definition 3. 5-PARTITION *is the following decision problem:*
   INPUT: *a multiset $A = \{a_1, \ldots, a_n\}$ of $n = 5m$ integers.*
   OUTPUT: *whether A can be partitioned into $A_1, \ldots, A_m$ such that, for each $i = 1, \ldots, m$, $\sum_{a \in A_i} a = p$ where $p = \left(\sum_{a \in A} a \right) / m$.*

Throughout the paper, we assume that the partition sum p is an integer; otherwise, the instance is obviously a NO instance.

Garey and Johnson [9] originally proved NP-completeness of 3-PARTITION, a problem similar to 5-PARTITION except that 5 is replaced by 3. In their book [10], they show that 4-PARTITION is NP-hard; this result was, in fact, an intermediate step toward showing that 3-PARTITION is NP-hard. It is easy to reduce 4-PARTITION to 5-PARTITION and thus show it also NP-hard.[1]

Our reduction would work from 3-Partition just as well as 5-Partition. The advantage of the latter is that we can analyze the following optimization version.

---

[1] Given a 4-PARTITION instance $A = \{a_1, \ldots, a_n\}$, we can create a 5-PARTITION instance by setting $A' = \{na_1, \ldots, na_n, 1, \ldots, 1\}$ where the number of 1s is $n/4$.

Definition 4. MAX 5-PARTITION *is the following optimization problem:*
   INPUT: *a multiset* $A = \{a_1, \ldots, a_n\}$ *of* $n = 5m$ *integers.*
   OUTPUT: *the maximum integer* m' *such that there exist disjoint subsets* $A_1, \ldots, A_{m'}$ *of* A *such that, for each* $i = 1, \ldots, m'$, $\sum_{a \in A_i} a = p$ *where* $p = \frac{5}{n} \left(\sum_{a \in A} a \right)$.

2.3   Gap Problems

We show that our reductions have a property stronger than approximation preservation called *gap preservation*. Let us define the gap problem for an optimization problem, a notion widely used in hardness of approximation.

Definition 5. *For an optimization problem* P *and parameters* $\beta > \gamma$ *(which may be functions of* n*), the* GAP$_P[\beta, \gamma]$ *problem is to distinguish whether the optimum of a given instance of* P *is at least* β *or at most* γ. *The input instance is guaranteed to not have an optimum between* β *and* γ.

If GAP$_P[\beta, \gamma]$ is NP-hard, then it immediately follows that approximating P to within a factor of β/γ of the optimum is also NP-hard. This result makes gap problems useful for proving hardness of approximation.

3   Main Results

Now that we have defined the problems, we state our main results.

Theorem 1. k-PIECE DISSECTION *is NP-hard.*

We do not know whether k-PIECE DISSECTION is in NP (and thus is NP-complete). We discuss the difficulty of showing containment in NP in Sect. 7.
   We also prove that the optimization version, MIN PIECE DISSECTION, is hard to approximate to within some constant ratio:

Theorem 2. *There is a constant* $\varepsilon_{\mathrm{MPD}} > 0$ *such that it is NP-hard to approximate* MIN PIECE DISSECTION *to within a factor of* $1 + \varepsilon_{\mathrm{MPD}}$ *of optimal.*[2]

Both results are based on essentially the same reduction, from 5-PARTITION for Theorem 1 or from MAX 5-PARTITION for Theorem 2. We present the common reduction in Sect. 4. We then prove Theorems 1 and 2 in Sects. 5 and 6 respectively.
   Restricting the kinds of polygons given as input, the kinds of cuts allowed, and the ways the pieces can be packed gives rise to many variant problems. Section 7 explains for which variants our results continue to hold.

---

[2] The best $\varepsilon_{\mathrm{MPD}}$ we can achieve is $1/1080 - \varepsilon$ for any $\varepsilon \in (0, 1/1080)$.

4   The Reduction

This section describes a polynomial-time reduction from 5-PARTITION to k-PIECE DISSECTION and states a lemma crucial to both of our main proofs later in the paper. The proof of the lemma is deferred to the full version.

Reduction from 5-PARTITION *to* k-PIECE DISSECTION. Let $A = \{a_1, \ldots, a_n\}$ be the given 5-PARTITION instance and let $p = \frac{5}{n}\Sigma_{a \in A}a$ denote the target sum. Let $d_s = 12(\max_{a \in A} a + p)$ and $d_t = (n-1)d_s + \Sigma_{a \in A}a + 2\max_{a \in A} a$. We create a source polygon P consisting of *element rectangles* of width a_i and height 1 for each $a_i \in A$ spaced d_s apart, connected below by a rectangular *bar* of width $\Sigma_{a \in A}a + (\frac{n}{5}-1)d_t$ and height $\delta = \frac{1}{10\Sigma_{a \in A}a + 2(\frac{n}{5}-1)d_t}$. The first element rectangle's left edge is flush with the left edge of the bar; the bar extends beyond the last element rectangle. Our target polygon Q consists of $\frac{n}{5}$ *partition rectangles* of width p and height 1 spaced d_t apart, connected by a bar of the same dimensions as the source polygon's bar. The first partition rectangle's left edge and last partition rectangle's right edge are flush with the ends of the bar. The illustration of both polygons are given in Fig. 1. Both polygons' bars have the same area and the total area of the element rectangles equals the total area of the partition rectangles, so the polygons have the same area. Finally, let the number of pieces k be n.

Reduction from MAX 5-PARTITION *to* MIN PIECE DISSECTION. The optimization problem uses the same reduction as the decision problem, except that we do not specify k for the optimization problem.

The idea behind our reduction is to force any valid dissection to cut each element rectangle off the bar in its own piece.[3] When δ is small enough, the resulting packing problem is a direct simulation of 5-PARTITION.

Intuitively, each dissected piece should contain only one element rectangle. Our reduction sets d_s large enough that any piece containing parts of two element

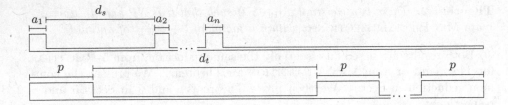

Fig. 1. The source polygon P (above) and the target polygon Q (below) are shown (not to scale). Length d_t is longer than the distance between the leftmost edge of the leftmost element rectangle and the rightmost edge of the rightmost element rectangle.

---

[3] Because $k = n$, a_1 will remain attached to the bar, forcing it to be the first element rectangle placed in the first partition rectangle. Because the order of and within partitions does not matter, this constraint does not affect the 5-PARTITION simulation.

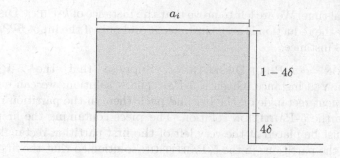

Fig. 2. The ith trimmed element rectangle.

rectangles does not fit in a partition rectangle. At the same time, we pick d_t large enough that no piece can be placed in more than one partition rectangle. Thus one could plausibly prove that each element rectangle must be in its own piece.

Unfortunately, we were unable to prove that each element rectangle must be in its own piece. For each element rectangle, we define the *trimmed element rectangle* corresponding to each element rectangle as the rectangle resulting from ignoring the lower 4δ of the element rectangle's height; see Fig. 2. In other words, for each a_i, the corresponding trimmed element rectangle is the rectangle that shares the upper left corner with the element rectangle and is of width a_i and height $1 - 4\delta$.

While we could not prove that each element rectangle is in its own piece, we can prove the corresponding statement about trimmed element rectangles:

Lemma 1. *If P can be cut into pieces that can be packed into Q, then each of these pieces intersect with at most one trimmed element rectangle.*

The proofs of both of our main theorems use this lemma. The intuition behind the proof of this lemma is similar to the intuitive argument for why each element rectangle should be in its own piece. As the details of the proof are not central to this paper, we defer the proof of this lemma to the full version [3].

5   Proof of NP-hardness of k-PIECE DISSECTION

Before we prove Theorem 1, we state the result from [10] for 5-PARTITION:

Theorem 3 ([10]). 5-PARTITION *is NP-hard.*[4]

We now prove Theorem 1.

Proof (of Theorem 1). We prove that the reduction described in the previous section is indeed a valid reduction from 5-PARTITION. The reduction clearly runs

---

[4] As stated earlier, the result from [10] is for 4-PARTITION, but 4-PARTITION is easily reduced to 5-PARTITION; see Sect. 2.

in polynomial time. We are left to prove that the instance of k-PIECE DISSECTION produced by the reduction is a YES instance if and only if the input 5-PARTITION is also a YES instance.

(5-PARTITION $\implies$ k-PIECE DISSECTION). Suppose that the 5-PARTITION instance is a YES instance. Given a 5-PARTITION solution, we can cut all but the first element rectangle off the bar and pack them in the partition rectangles according to the 5-PARTITION solution. The piece containing the first element rectangle must be placed at the very left of the first partition rectangle, but we can reorder the partitions in the 5-PARTITION solution so that the first element is in the first partition. As a result, the k-PIECE DISSECTION instance is also a YES instance.

(k-PIECE DISSECTION $\implies$ 5-PARTITION). Suppose that the k-PIECE DISSECTION instance is a YES instance, i.e., P can be cut into k pieces that can then be placed into Q. By Lemma 1, no two trimmed element rectangles are in the same piece. Because there are $n = k$ such rectangles, each piece contains exactly one whole trimmed element rectangle. Because these pieces can be packed into Q, we must also be able to pack all the trimmed element rectangles into Q (with some space in Q left over).

Let B_i be the set of all trimmed element rectangles (in the packing configuration) that intersect the ith partition rectangle. From our choice of d_t, each trimmed element rectangle can intersect with at most one partition rectangle. Moreover, no trimmed element rectangles fit entirely in the bar area, so each of them must intersect with at least one partition rectangle. This means that $B_1, \ldots, B_{n/5}$ is a partition of the set of all trimmed element rectangles. Let A_i be the set of all integers in A corresponding to the trimmed element rectangles in B_i. Observe that $A_1, \ldots, A_{n/5}$ is a partition of A.

We claim that $A_1, \ldots, A_{n/5}$ is indeed a solution for 5-PARTITION. Assume for the sake of contradiction that $A_1, \ldots, A_{n/5}$ is not a solution, that is, $\sum_{a \in A_i} a \neq p$ for some i. Because $\sum_{a \in A} a = p(n/5)$, there exists j such that $\sum_{a \in A_j} a > p$. Because all $a \in A$ are integers and p is an integer, $\sum_{a \in A_j} a \geq p + 1$.

Consider the jth partition rectangle. Define the *extended partition rectangle* as the area that includes a partition rectangle, the bar area directly below it, and the bar $\delta/2$ to the left and to the right of the partition rectangle. Figure 3 shows an extended partition rectangle enclosed in thick edges. (Ignore the shaded rectangle for the moment.)

Consider any trimmed element rectangle in the packing configuration that intersects with this partition rectangle. We claim that each such trimmed element rectangle must be wholly contained in the extended partition rectangle.

Consider the area of the trimmed element rectangle outside the partition rectangle and the bar below it. If this is not empty, this must be a right triangle with hypotenuse on the extension down to the bar of a vertical side of the partition rectangle (see Fig. 3). The hypotenuse of this triangle is of length at most δ, so the height of the triangle (perpendicular to the hypotenuse) is at most $\delta/2$. Thus, the triangle must be in the extended partition rectangle. Thus the

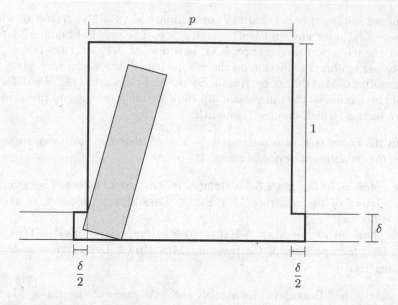

Fig. 3. The area enclosed by thick edges is the extended partition rectangle corresponding to this partition rectangle. In this configuration, the trimmed element rectangle, shown as the shaded area, is partially outside of the partition rectangle and the bar below it. This external area is a right triangle with hypotenuse on the extension of a vertical edge of the partition rectangle (shown as the dotted line segment), which is of length δ.

whole trimmed element rectangle must be in the extended partition rectangle, as claimed.

The area of the extended partition rectangle is $p + p\delta + \delta^2 < p + 1/2$. However, the total area of the trimmed element rectangles contained in this area is $\sum_{a \in A_j} a(1 - 4\delta) = \sum_{a \in A_j} a - 4\delta \sum_{a \in A_j} a \geq (p+1) - 4\delta \sum_{a \in A_j} a > p + 1/2$, which is a contradiction.

Thus we conclude that $A_1, \ldots, A_{n/5}$ is a solution to 5-PARTITION, which implies that the 5-PARTITION instance is a YES instance as desired.   □

6   Proof of Inapproximability of MIN PIECE DISSECTION

In this section, we show the inapproximability of MIN PIECE DISSECTION via a reduction from the intermediate problem MAX 5-PARTITION, whose inapproximability result is described in the following lemma.

Lemma 2. *There is a constant $\alpha_{M5P} > 1$ such that* $\text{GAP}_{\text{MAX-5-PARTITION}}[n(1 - \varepsilon)/5, n(1/\alpha_{M5P} + \varepsilon)/5]$ *is NP-hard for any sufficiently small constant $\varepsilon > 0$.[5]*

---

[5] The best α_{M5P} we can achieve here is 216/215.

Lemma 2 implies that it is hard to approximate MAX 5-PARTITION to within an $\alpha_{\text{M5P}} - \varepsilon$ ratio for any sufficiently small $\varepsilon > 0$. The proof of Lemma 2 largely relies on the reduction used to prove NP-hardness of 4-PARTITION in [10], but we apply our modified reduction on the inapproximability result of 4-UNIFORM 4-DIMENSIONAL MATCHING by Hazan, Safra, and Schwartz [12]. We defer the proof of this lemma to the full version [3]. Here we focus on the gap preservation of the reduction, which implies Theorem 2.

Lemma 3. *There is a constant $\alpha_{\text{MPD}} > 1$ such that the following properties hold for the reduction described in Sect. 4:*

- *if the optimum of the MAX 5-PARTITION instance is at least $n(1 - \varepsilon)/5$, then the optimum of the resulting MIN PIECE DISSECTION instance is at most $n(1 + \varepsilon/5)$; and*
- *if the optimum of the MAX 5-PARTITION instance is at most $n(1/\alpha_{\text{M5P}} + \varepsilon)/5$, then the optimum of the resulting MIN PIECE DISSECTION is at least $n(\alpha_{\text{MPD}} + \varepsilon/5)$.*

Because it is NP-hard to distinguish the two cases of the input MAX 5-PARTITION instance, it is also NP-hard to approximate MIN PIECE DISSECTION to within an $\alpha_{\text{MPD}} - \varepsilon$ ratio for any sufficiently small constant $\varepsilon > 0$. Thus, Lemma 3 immediately implies Theorem 2. It remains to prove Lemma 3:

Proof (of Lemma 3). We will show that both properties are true when we choose α_{MPD} to be $1 + (1 - 1/\alpha_{\text{M5P}})/5$.

(MAX 5-PARTITION $\Longrightarrow$ MIN PIECE DISSECTION). Suppose that the input MAX 5-PARTITION instance has optimum at least $n(1 - \varepsilon)/5$. Let $A_1, \ldots, A_{m'}$ be the optimal partition where $m' \geq n(1 - \varepsilon)/5$. We cut P into pieces as follows (see Fig. 4):

1. First, we cut every element rectangle except the first one from the bar.
2. Next, let the indices of the elements in $A - (A_1 \cup A_2 \cup \cdots \cup A_{m'})$ be $i_1, \ldots, i_l$ where $1 \leq i_1 < i_2 < \cdots < i_l \leq n$.
3. For each $i = 1, \ldots, n/5 - m'$, let j be the smallest index such that $a_{i_1} + \cdots + a_{i_j} \geq ip$. Cut the piece corresponding to a_{i_j} vertically at position $ip - (a_{i_1} + \cdots + a_{i_{j-1}})$ from the left. (If the intended cut position is already the right edge of the piece, then do nothing.)

To pack these pieces into Q, we arrange all pieces whose corresponding elements are in partitions in the optimal MAX 5-PARTITION solution, then pack the remaining pieces into the remaining partition rectangles using the additional cuts made in step 3. We leave the piece containing the first element rectangle (and the bar) at its position in P, but this does not constrain our solution because the other pieces and the partitions can be freely reordered.

The number of cuts in step 1 is $n - 1$ and in step 3 is at most $n/5 - m' \leq \varepsilon n/5$. Thus the total number of cuts is at most $n - 1 + \varepsilon n/5$, so the number of pieces is at most $1 + (n - 1 + \varepsilon n/5) = n(1 + \varepsilon/5)$ as desired.

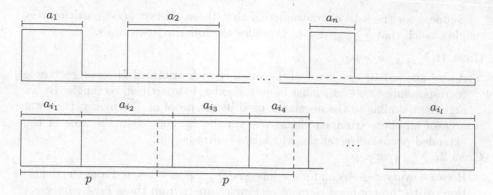

Fig. 4. An illustration of how the source polygon P is cut. The cuts from step 1 are shown as dashed lines on the top figure; every element rectangle except the first one is cut from the bar. On the bottom, the cuts from step 3 are demonstrated. We can think of the cutting process as first arranging $a_{i_1}, \ldots, a_{i_l}$ consecutively and then cutting at $p, 2p, \ldots$.

(MIN PIECE DISSECTION $\implies$ MAX 5-PARTITION). We prove this property in its contrapositive form. Suppose that the resulting MIN PIECE DISSECTION has an optimum of $k < n(\alpha_{\mathrm{MPD}} + \varepsilon/5)$. Let us call these k pieces $R_1, \ldots, R_k$.

For each $i = 1, \ldots, k$, let R_i' denote the intersection between R_i with the union of all trimmed element rectangles. By Lemma 1, each trimmed element rectangle can intersect with only one piece. This means that each R_i' is a part of a trimmed element rectangle. (Note that R_i' can be empty; in this case, we say that it belongs to the first trimmed element rectangle.)

Consider $R_1', \ldots, R_k'$. Because each of them is a part of a trimmed rectangle and there are n trimmed rectangles, at most $k - n$ trimmed rectangles contain more than one of the R_i'. In other words, there are at least $n - (k - n) = 2n - k$ indices i such that R_i' is a whole trimmed element rectangle. Without loss of generality, suppose that $R_1', \ldots, R_{2n-k}'$ are entire trimmed element rectangles.

We call a partition rectangle a *good partition rectangle* if it does not intersect with any of $R_{2n-k+1}', \ldots, R_n'$ in the packing configuration. From our choice of d_t, each R_i' which is part of a trimmed element rectangle can intersect with at most one partition rectangle. As a result, there are at least $n/5 - (k - n)$ good partition rectangles.

For each good partition rectangle O, let A_O be the subset of all elements of A corresponding to R_i's that intersect O. (Because O is a good partition rectangle, each R_i' that intersects O is always a whole trimmed element rectangle.)

We claim that the collection of T_O's for all good partition rectangles O is a solution to the MAX 5-PARTITION instance. We will show that this is indeed a valid solution. First, observe again that, because each R_i' intersects with at most one partition rectangle, all A_O's are mutually disjoint. Thus, we now only need to prove that the sum of elements of A_O is exactly the target sum p.

Suppose for the sake of contradiction that there exists a good partition rectangle O such that $\sum_{a \in A_O} a \neq p$. Consider the following two cases.

Case 1: $\sum_{a \in A_O} a > p$.

As we showed in the proof of Theorem 1, each trimmed element rectangle corresponding to $a \in A_O$ must be in the extended partition rectangle. By an argument similar to the argument used in the proof of Theorem 1, the total area of all these trimmed element rectangles is more than the area of the extended partition rectangle, which is a contradiction.

Case 2: $\sum_{a \in A_O} a < p$.

Because every $a \in A_O$ and p are integers, $\sum_{a \in A_O} a + 1 \leq p$. From the definition of A_O, no trimmed element rectangles apart from those in A_O intersect O. Hence the total area that trimmed element rectangles contribute to O is at most

$$\left(\sum_{a \in A_O} a \right) (1 - 4\delta) < \sum_{a \in A_O} a \leq p - 1.$$

This means that an area of at least 1 unit square in O is not covered by any of the trimmed element rectangles. However, the area of the source polygon outside of all the trimmed element rectangles is

$$\delta \left(\left(\frac{n}{5} - 1 \right) d_t + \sum_{a \in A} a \right) + 4\delta \left(\sum_{a \in A} a \right) < 1,$$

which is a contradiction.

Hence, the solution defined above is a valid solution. Because the number of good partition rectangles is at least $n/5 - (k - n) > n/5 - n(\alpha_{\text{MPD}} + \varepsilon/5 - 1) = n(1/\alpha_{\text{M5P}} - \varepsilon)/5$, the solution contains more than $n(1/\alpha_{\text{M5P}} - \varepsilon)/5$ subsets, which completes the proof of the second property. $\qquad\square$

7 Variations and Open Questions

Table 1 lists variations of k-PIECE DISSECTION and whether our proofs of NP-hardness and inapproximability continue to hold. Because it is obvious from the proofs, we do not give detailed explanations as to why the proofs still work (or do not work) in these settings. Specifically:

1. Our proofs remain valid when the input polygons are restricted to be simple (hole-free) and orthogonal with all edges having integer length.[6]
2. Our results still hold under any cuts that leave each piece connected and Lebesgue measurable.
3. Our proofs work whether or not rotations and/or reflections are allowed when packing the pieces into Q.

While we have proved that the k-PIECE DISSECTION is NP-hard and that its optimization counterpart is NP-hard to approximate, we are far from settling

---

[6] Our reduction uses rational lengths, but the polygons can be scaled up to use integer lengths while still being of polynomial size.

Table 1. Variations on the dissection problem.

Variation on	Variation description	Do our results hold?
Input Polygons	Polygons must be orthogonal	YES
	Polygons must be simple (hole-free)	YES
	Edges must be of integer length	YES
	Polygons must be convex	NO
Cuts Allowed	Cuts must be straight lines	YES
	Cuts must be orthogonal	YES
	Pieces must be simple (hole-free)	YES
	Pieces may be disconnected	NO
Packing Rules	Rotations are forbidden	YES
	Reflections are forbidden	YES

the complexity of these problems and their variations. We pose a few interesting remaining open questions:

– Is k-PIECE DISSECTION in NP, or even decidable? We do not know the answer to this question even when only orthogonal cuts are allowed and rotations and reflections are forbidden. In particular, there exist two-piece orthogonal (staircase) dissections between pairs of rectangles which seem to require a cut comprised of arbitrarily many line segments [7, p. 60].
  If we require each piece to be a polygon with a polynomial number of sides, then problem becomes decidable. In fact, we can place this special case in the complexity class $\exists \mathbb{R}$, that is, deciding true sentences of the form $\exists x_1 : \cdots : \exists x_m : \varphi(x_1, \ldots, x_m)$ where φ is a quantifier-free formula consisting of conjunctions of equalities and inequalities of real polynomials. To prove membership in $\exists \mathbb{R}$, use $x_1, \ldots, x_m$ to represent the coordinates of the pieces' vertices in P and Q. Then, use φ to verify that the pieces are well-defined partitions of P and Q and that each piece in P is a transformation of a piece in Q; these conditions can be written as polynomial (in)equalities of degree at most two. $\exists \mathbb{R}$ is known to be in PSPACE [4].
– Is k-PIECE DISSECTION still hard when one or both of the input polygons are required to be convex?
– Can we prove stronger hardness of approximation, or find an approximation algorithm, for MIN PIECE DISSECTION? The current best known algorithm for finding a dissection is a worst-case bound of a pseudopolynomial number of pieces [1].
– Is k-PIECE DISSECTION solvable in polynomial time for constant k? Membership in FPT would be ideal, but even XP would be interesting.

Acknowledgment. We thank Greg Frederickson for helpful discussions.

References

1. Aloupis, G., Demaine, E.D., Demaine, M.L., Dujmović, V., Iacono, J.: Meshes preserving minimum feature size. In: Márquez, A., Ramos, P., Urrutia, J. (eds.) EGC 2011. LNCS, vol. 7579, pp. 258–273. Springer, Heidelberg (2012). doi:10.1007/978-3-642-34191-5_25
2. Bolyai, F.: Tentamen juventutem studiosam in elementa matheseos purae, elementaris ac sublimioris, methodo intuitiva, evidentiaque huic propria, introducendi. Typis Collegii Refomatorum per Josephum et Simeonem Kali, Maros Vásárhely (1832–1833)
3. Bosboom, J., Demaine, E.D., Demaine, M.L., Lynch, J., Manurangsi, P., Rudoy, M., Yodpinyanee, A.: Dissection with the fewest pieces is hard, even to approximate. CoRR abs/1512.06706 (2015). http://arxiv.org/abs/1512.06706
4. Canny, J.: Some algebraic and geometric computations in PSPACE. In: Proceedings of the Twentieth Annual ACM Symposium on Theory of Computing, STOC 1988, pp. 460–467. ACM, New York (1988). http://doi.acm.org/10.1145/62212.62257
5. Demaine, E.D., Demaine, M.L.: Jigsaw puzzles, edge matching, and polyomino packing: connections and complexity. Graphs Comb. **23**(Suppl.), 195–208 (2007). Special issue on Computational Geometry and Graph Theory: The Akiyama-Chvatal Festschrift. Preliminary version presented at KyotoCGGT 2007
6. Dudeney, H.E.: Puzzles and prizes. Weekly Dispatch (1902), the puzzle appeared in the April 6 issue of this column. A discussion followed on April 20, and the solution appeared on May 4
7. Frederickson, G.N.: Dissections: Plane and Fancy. Cambridge University Press, Cambridge (1997)
8. Frederickson, G.N.: Hinged Dissections: Swinging & Twisting. Cambridge University Press, Cambridge (2002)
9. Garey, M.R., Johnson, D.S.: Complexity results for multiprocessor scheduling under resource constraints. SIAM J. Comput. **4**(4), 397–411 (1975)
10. Garey, M.R., Johnson, D.S.: Computers and Intractability: A Guide to the Theory of NP-Completeness. W. H. Freeman & Co., New York (1979)
11. Gerwien, P.: Zerschneidung jeder beliebigen Anzahl von gleichen geradlinigen Figuren in dieselben Stücke. Journal für die reine und angewandte Mathematik (Crelle's Journal) **10**, 228–234 (1833). Taf. III
12. Hazan, E., Safra, S., Schwartz, O.: On the hardness of approximating k-dimensional matching. Electronic Colloquium on Computational Complexity (ECCC) **10**(020) (2003). http://eccc.hpi-web.de/eccc-reports/2003/TR03-020/index.html
13. Lowry, M.: Solution to question 269, [proposed] by Mr. W. Wallace. In: Leybourn, T. (ed.) Mathematical Repository Part 1, pp. 44–46. W. Glendinning, London (1814)
14. Wallace, W. (ed.): Elements of Geometry, 8th edn. Bell & Bradfute, Edinburgh (1831)

Mario Kart Is Hard

Jeffrey Bosboom[1], Erik D. Demaine[1], Adam Hesterberg[1], Jayson Lynch[1(✉)],
and Erik Waingarten[2]

[1] MIT Computer Science and Artificial Intelligence Laboratory,
32 Vassar St., Cambridge, MA 02139, USA
{jbosboom,edemaine,achester,jaysonl}@mit.edu
[2] Department of Computer Science, Columbia University,
1214 Amsterdam Avenue, New York, NY 10027, USA
eaw@cs.columbia.edu

Abstract. Nintendo's Mario Kart is perhaps the most popular racing
video game franchise. Players race alone or against opponents to finish in
the fastest time possible. Players can also use items to attack and defend
from other racers. We prove two hardness results for generalized Mario
Kart: deciding whether a driver can finish a course alone in some given
time is NP-hard, and deciding whether a player can beat an opponent
in a race is PSPACE-hard.

1 Introduction

Mario Kart is a popular racing video game series published by Nintendo, starting
with Super Mario Kart on SNES in 1992 and since adapted to eleven platforms,
most recently Mario Kart 8 on Wii U in 2014; see Table 1. The series has sold over
100 million game copies, and contains the best-selling racing game ever, Mario
Kart Wii [Gui14]. The games feature characters from the classic Nintendo series
Super Mario Bros. and Donkey Kong.

In this paper, we analyze the computational complexity of most Mario
Kart games, showing that optimal gameplay is computationally intractable. Our
results follow a series of recent work on the computational complexity of video
games, including the broad work of Forisek [For10] and Viglietta [Vig14] as well
as the specific analyses of classic Nintendo games [ADGV15].

In Mario Kart, each player picks a character and a race track. There are three
modes of play: players race against each other (racing), a player races alone to
finish in the fastest time possible (time trial), and players battle in an arena
(battle). We focus here on the first two modes. Each race track features its own
set of obstacles and geometry.

A particularly distinctive feature of Mario Kart is that players may acquire
items (also known as power-ups). Items temporarily give players special abilities.
Each Mario Kart game has its own set of items, but two items are common to
all Mario Kart games: Koopa shells and bananas. Koopa shells come in multiple

E. Waingarten—Work performed while at MIT.

J. Akiyama et al. (Eds.): JCDCGG 2015, LNCS 9943, pp. 49–59, 2016.
DOI: 10.1007/978-3-319-48532-4_5

Table 1. History and total sales [Sal] of Mario Kart. Our results apply to all games with 3D tracks.

	Game title	Game system	Release date	Sales	3D?
1	Super Mario Kart	Super NES	August 27, 1992	8.76M	no
2	Mario Kart 64	Nintendo 64	December 14, 1996	9.87M	yes
3	Mario Kart: Super Circuit	Game Boy Advance	July 21, 2001	5.47M	no
4	Mario Kart: Double Dash!!	Nintendo GameCube	November 7, 2003	6.95M	yes
5	Mario Kart DS	Nintendo DS	November 14, 2005	23.56M	yes
6	Mario Kart Wii	Wii	April 10, 2008	35.53M	yes
7	Mario Kart 7	Nintendo 3DS	December 1, 2011	12.19M	yes
8	Mario Kart 8	Wii U	May 29, 2014	5.87M	yes
9	Mario Kart: Arcade GP	arcade	October 2005	?	yes
10	Mario Kart: Arcade GP 2	arcade	March 14, 2007	?	yes
11	Mario Kart: Arcade GP DX	arcade	July 25, 2013	?	yes

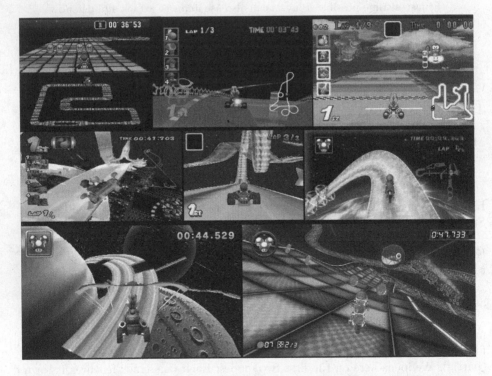

Fig. 1. Screenshots of Rainbow Road tracks from Mario Kart 1–8 (Table 1).

colors; our reduction only uses the green shells, which we refer to simply as shells. Shells are shot at other players and, upon contact, temporarily stun them, reducing their speed and control. Bananas can be dropped by players along the track, and any player who runs over a banana becomes temporarily stunned. Crucially, shells can destroy bananas.

In this paper, we consider generalized versions of time trial and racing. We allow race tracks to be any size and have carefully placed items on the track. We more precisely define our model of the game in Sect. 2. In Sect. 3, we show that time trial is NP-hard, that is, it is NP-hard to decide whether a lone player can finish a race track in time at most t. In Sect. 4, we show PSPACE-hardness of racing: it is PSPACE-hard to decide whether a player can win the race against even a single opposing player. Finally, Sect. 5 considers upper bounds.

The items used in our reductions are present in all Mario Kart games. Our reductions use the "Rainbow Road" style of racetrack. These tracks are present in every game, but our reductions require them to be three-dimensional, which they are in Mario Kart 64 and in every game since Mario Kart: Double Dash!!. The proofs thus apply to nine of the Mario Kart games (Games 2 and 4–11 in Table 1). Super Mario Kart and Mario Kart Super Circuit lack tracks with multiple altitudes, presumably from the lack of power in the Super NES and Game Boy Advance systems, and so our proofs do not apply to them.

2   Model

In our mathematical model of Mario Kart, each player's state consists of a position, orientation, and speed. The track is a two-dimensional surface in Euclidean 3-space. The player generally controls their acceleration, with limits on speed and position imposed by the track. Leaving the bounds of the racetrack does not result in death, with players being respawned on the track after a significant speed and time penalty.

Computationally, we assume that we can compute the optimal traversal of a track described by a constant number of real parameters, and that this optimal traversal time typically changes continuously with the real parameters. This allows us to, for example, tweak multiple pieces of the track to have nearly identical optimal traversal times. In fact, we require that these assumptions hold only up to an error factor of $1 + O(1/n^c)$, that is, up to $O(\log n)$ bits. We leave to future work the careful analysis of the physics and geometry of actual Mario Kart implementations, and the evaluation of the validity of our assumptions.[1]

Players obtain items from item boxes which are at fixed locations on the track, and regenerate after a fixed amount of time. We use two kinds of items common to all Mario Kart games to date, each of which can be used only once:

---

[1] We conjecture that implementations model the position and velocity vector of a player by floating-point numbers, discretize time into fixed-duration intervals, and ·model the track by a collection of succinctly describable segments and turns. For a sufficiently fine discretization of time, this model should approach our continuous model. To compute the optimal traversal time of a constant-complexity track, we can finitely sample the position/velocity space and search the resulting state graph. We conjecture that a polynomial-resolution sampling suffices to approximate the optimal traversal time to the needed $1 + O(1/n^c)$ accuracy for our reductions.

1. **Bananas.** Bananas are slippery. When a player drives over a banana (or is hit by one), the driver slips and spins temporarily out of control, resulting in a temporary slowdown. Bananas can be dropped immediately behind the player, or thrown up and ahead with a fixed trajectory. Once a banana lands on the track, there are two ways to remove it: either a player drives over it, or the banana is hit by a shell (described below).
2. **Green Shells.** A green shell is one of the many attacks in Mario Kart. The player can shoot a green shell like a projectile. If a green shell hits a driver, the driver is temporarily knocked out. A green shell can also remove a banana if the banana is hit first. (Green shells should not be confused with red shells, which can lock onto a target driver.) Green shells follow a particular direction, are subject to gravity, and bounce off of walls. After some time, green shells become inactive and disappear.

A driver can possess only one item at a time. For example, if a driver picks up a green shell, s/he cannot pick up another item until s/he uses the green shell. However, in most Mario Kart games (with the notable exception of Mario Kart 8), it is possible to "use" a green shell or banana without throwing it: a driver can hold a green shell or banana behind the car before throwing it, allowing them to pick up one additional item. The items still must be used in order.

In our reductions, we will assume that some bananas have already been placed on the track, but this does not occur in any real Mario Kart tracks. In fact, we assume that the game has already been played for some time, e.g., previous laps of the track, and the computational question is whether Player 1 can win within one final lap from the given track configuration. We can easily add "initialization" paths and banana item boxes to the track, ensuring that the initial configuration of placed bananas would actually be reachable from an initially empty track. By making these initialization paths very long, they will not affect optimal play of the final lap under consideration.

In this way, we can also assume that two players start at very different positions on the track. The finish line is shared between the two players, but is fairly wide. Thus we can cross the finish line with two equally elevated and separated paths for the two players, guaranteeing no interaction near the finish, to effectively allow distinct goal locations for the two players.

3   Time Trial is NP-Hard

First we study the following solo ("time trial") variant of Mario Kart:

Theorem 1. *It is NP-hard to determine whether a driver can finish a given course in at most t time, in the absence of opponents.*

3.1 Proof Structure

The reduction is from 3SAT. Given a Boolean formula ϕ with variables $x_1, x_2, \ldots, x_n$, we build a level with the "Rainbow Road" style. The driver first drives through each variable gadget in sequence. In each variable gadget, the player can decide whether to set each variable to true or false. After setting all the variables, the driver must traverse each clause gadget. The driver will be able to complete the level without delay if and only if the variable assignments chosen in the gadgets form a satisfying assignment for ϕ.

Figure 2 gives a schematic overview of the reduction. Each node labeled x_i corresponds to a variable gadget, and each node labeled c_i corresponds to a clause gadget. The solid lines correspond to the path in the level. The dashed lines indicate that a variable or its negation is contained in a given clause. In our case, the dashed lines also correspond to clause gadgets being reachable by green shells when thrown from the variable gadgets. We prevent players from following the dashed paths.

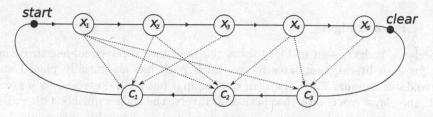

Fig. 2. General reduction structure. Dashed lines correspond to reachability of green shells.

3.2 Variable Gadget

For each variable x_i, we have one variable gadget as shown in Fig. 3. The variable gadget first splits the road into two. The driver must choose which of the two directions to follow, corresponding to the truth setting of x_i. We refer to the two split roads as *literal roads* x_i and $\overline{x_i}$. Both literal roads have the same optimal travel time.

Each literal road has a sequence of visits to clause gadgets corresponding to clauses containing the literal. Literal road x_i goes above the clauses containing the literal x_i, and similarly for $\overline{x_i}$. Each road has a green shell item which can be fired into the clause gadget. When a literal road is above each clause, the driver can pick up a green shell and shoot it down to the clause, where it will remove a banana.

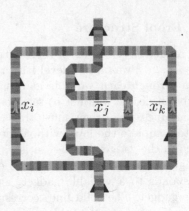

Fig. 3. A variable gadget where Player 1 assigns x_i. Player 1 goes left to set x_i to true, and goes right to set $\overline{x_i}$ to true.

Fig. 4. Clause gadgets split into three literals. They are considered false if a banana remains on the path.

3.3 Clause Gadget

The clause gadget, seen in Fig. 4 splits the road into three equal-length paths, one for each literal, that later merge. Each path has an initially placed and unavoidable banana. Thus, if any of the bananas has been destroyed by a green shell, the player can choose that path and traverse the gadget quickly. Otherwise, the player must hit a banana and incur a speed and time penalty—assuming that the player is not carrying any green shells.

3.4 Clearing Held Items

To guarantee that the player traverses the sequence of clause gadgets without any green shells, we add a *clearing gadget* between the sequence of variable gadgets and the sequence of clause gadgets. The clearing gadget, shown in Fig. 5, forces the driver to afterward hold no items (behind the car or otherwise).

There are actually two different gadgets, depending on whether the Mario Kart game permits carrying a second item behind the car. For games where this is impossible (currently just Mario Kart 8), the gadget consists of a single green shell item box followed by an already placed banana. Otherwise, we have two green shell item boxes followed by two already placed bananas. The distance between the item boxes and bananas is longer than the lifetime of a shell. Thus, to avoid slowdown from the bananas, the player must use all storable green shells (either just picked up or stored from before) and be left holding nothing.

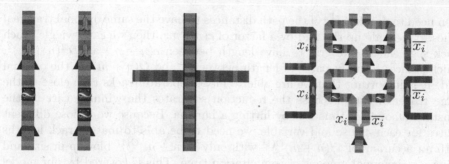

Fig. 5. Two types of Clear Gadgets.

Fig. 6. A Crossover Gadget. The vertical path is placed higher in the level with a wall along the track.

Fig. 7. Variable gadget being able to unlock clause. Once Player 1 assigns x_i, it can shoot shells to unlock clauses where x_i appears.

3.5 Crossover Gadget

Crossover gadgets are relatively simple given the three-dimensional nature of Rainbow Road levels, so one road can pass over another road; see Fig. 6. To ensure that the player does not jump from the upper road to the lower road, and that the player does not throw a shell from the upper road to the lower road, we surround the sides of the upper road with vertical walls, for sufficient length before and after the intersection.

3.6 Putting Gadgets Together

Figure 7 shows how a literal road of a variable gadget interacts with each clause gadget containing the literal. By bringing the variable road somewhat close and above the clause road, the player can shoot the green shell from the variable and destroy the banana in the clause, without slowing down. This action "unlocks" the clause gadget for later traversal, corresponding to satisfying the clause.

However, we cannot place the roads too close to each other, or else the player could jump from the variable road to the lower clause road. Fortunately, there is a suitable distance traversable by shells but not by players, because shells move faster than players. (Alternatively, even if players could move as fast as shells, this property could be arranged by having the shell bounce off of a floating vertical wall, which the player could not do.)

Finally we describe how to lay out the gadgets. Because there is a constant maximum speed that can be attained on a flat track, there a constant size of gadget with straight tracks as inputs/outputs that guarantees two properties: (1) the player cannot traverse from a gadget to a gadget not logically connected to it, and (2) the player normalizes to a standard maximum straight-away speed before entering the next gadget. We use this constant gadget size as our unit size. The literals, crossovers, and their connecting lines can be laid out orthogonally on an $O(n+m) \times O(n+m)$ unit square grid in polynomial time [BK94]. We may then need to tweak some of the path distances to have the same optimal traversal times. If we scale up the grid by a factor of $c(n + m)$, then we can "wiggle" each track segment on the grid to have length between $c(n + m)$ and $c^2(n + m)^2$, which suffices to unify paths of length between 1 and $O(n + m)$ on the original grid. It is important that we are able to make separate tracks take close to the same traversal time because the reduction separates the winning kart by the constant amount of time lost by hitting a banana. Because we choose different routes for each clause and variable, we need to be able to match track lengths with an accuracy of $1/(n + m)^{O(1)}$ with only a $(n + m)^{O(1)}$ blowup in size and using a polynomial amount of computation time. This is covered by our model assumptions in Sect. 2. Thus we can lay out the gadgets in a polynomial-time reduction.

4   Racing is PSPACE-Hard

We now study the following two-player variant of Mario Kart, where players race against each other:

Theorem 2. *It is* PSPACE-*hard to decide whether Player 1 has a forced win in a two-player Mario Kart race from given starting positions for the players.*

4.1   Proof Structure

The reduction is from Q3SAT: decide a quantified Boolean formula $\phi = \exists x_1 : \forall y_1 : \exists x_2 : \forall y_2 : \cdots \exists x_{n/2} : \forall y_{n/2} : \phi'(x_1, \ldots, x_{n/2}, y_1, \ldots, y_{n/2})$ where ϕ' is in 3CNF, has a satisfying assignment. We construct the track similar to the NP-hardness proof, but with Player 1 setting the existentially quantified variables and Player 2 setting the universally quantified variables; refer to Fig. 8. As in the proof for NP-hardness, Player 1 will shoot shells from an elevated road to clear bananas from clause gadgets. Player 2, who sets the universal quantified variables is on a separate elevated road throwing bananas into clause gadgets. While each player sets a variable, the other player is forced along a higher road of the same traversal time, within visual range so that both players know the variable setting; see Fig. 10. This way, we get the alternating behavior and perfect information while setting variables. The overall path Player 1 takes is slightly shorter than Player 2. So if Player 1 can get through the clauses without hitting any bananas, s/he will win. If Player 1 runs over any bananas and slips, Player 2 will win.

Player 2 can "cheat" in a variety of ways, but all of them consume time. For these cases, Player 1 has an alternative winning path that bypasses all clauses, but takes longer than if Player 2 plays "straight". This threat prevents Player 2 from cheating (in optimal play).

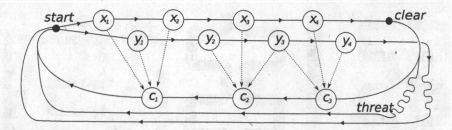

Fig. 8. General reduction structure for 2 players. Dashed lines correspond to reachability of green shells and bananas.

4.2 Clause Gadget

As shown in Fig. 11, the clause gadget is a road that splits into one road per literal, as in the NP-hardness proof. The literals of existentially quantified variables are initially blocked by a banana, as in the NP-hardness proof, while literals of universally quantified variables are initially empty.

4.3 Variable Gadget

Player 1's (existential) variable gadgets are the same as in the NP-hardness proof (Fig. 3): each gadget forks to make the player choose between setting x_i or $\overline{x_i}$ to true, with each fork passing by all the clauses containing that literal, so the player can shoot a shell down to remove the banana from that existential variable's literal instance.

Player 2's (universal) variable gadgets have the same structure, but as shown in Fig. 9, the player instead sets y_i or $\overline{y_i}$ to *false* by shooting bananas (picked up from item boxes in the variable) down into literal instances in the clause gadgets, filling what was initially empty.

4.4 Putting Gadgets Together

Existential variable gadgets and clause gadgets interact as in the NP-hardness proof. Universal variable gadgets interact with clause gadgets at a closer distance, given the lobbed trajectory of bananas. To prevent Player 2 from jumping down to the clause gadget in this situation, we can use a vertical wall or rail that is tall enough to block the player but not tall enough to block a thrown banana.

We use the same crossover gadgets as the NP-hardness proof (Fig. 6), and the same clearing gadget (Fig. 5) before Player 1 enters the sequence of clause gadgets. Everywhere else, whenever a player would be helped by an item, that item is presented by an item box, so it never helps to hold onto an item for later. (Note that it does not help to block a literal with two bananas instead of just one. A single banana penalty is enough for Player 2 to win.)

After all variables have been set, Player 1 drives through the clause gadgets while Player 2 drives along a winding road slightly longer to traverse than the

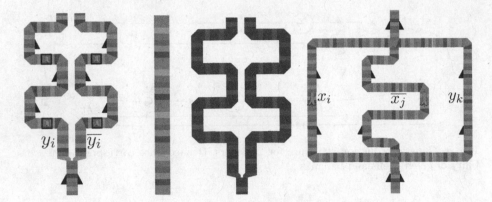

Fig. 9. Variable gadget for Player 2. Player 2 assigns y_i and grabs bananas to throw to the clause gadgets.

Fig. 10. Observation of other player. The variable gadget (grayed out) appears below in 3-dimensional space.

Fig. 11. Clause gadget split into literals. A clause splits into the three literals which comprise the clause. Note that since y_k is a variable set by Player 2, there is no banana on the path until Player 2 throws a banana down.

road through the clause gadgets. If all clauses are satisfied (have at least one literal branch without a banana), Player 1 wins; otherwise, Player 1 must drive through at least one banana and slow down. In this case, Player 2 wins, by setting the "slightly longer" amount to strictly less than the banana penalty. (For a more comfortable construction, we can repeat every clause k times, allowing the difference to be strictly less than k times the banana penalty.)

Player 2 can attempt to "cheat" in a couple of ways: traversing both sides of a universal variable gadget, or waiting to choose the value of a universal variable gadget until after Player 1 chooses the next variable (breaking the quantifier structure). In this case, Player 2 will fall behind relative to the intended traversal. This would be worthwhile if Player 2 could slow down Player 1 substantially as a result, but the availability of the slightly longer threat path means that Player 1 can avoid all clauses and thus all slowdowns in this case. Player 1 also cannot afford to cheat in these ways, because s/he starts with only a small advantage, and is unable to slow down Player 2.

Gadget layout can be done analogous to Sect. 3.

5   Conclusion

In practice, players in Mario Kart generally make forward progress on the track, other than short aberrations caused by attacks, and have knowledge (via the minimap) of the state of all players. These assumptions imply a polynomial bound on the length of solutions, which in turn implies that our results are tight: time trial is NP-complete and racing is PSPACE-complete. Without the game-length assumption, however, we only know containment in PSPACE and EXPTIME, respectively, and it is plausible that we could establish corresponding hardness. With hidden information (unknown state of the track or items held by opponents), Mario Kart racing is potentially as hard as 2EXPTIME.

References

[ADGV15] Aloupis, G., Demaine, E.D., Guo, A., Viglietta, G.: Classic Nintendo games are (computationally) hard. Theoret. Comput. Sci. **586**, 135–160 (2015)

[BK94] Biedl, T., Kant, G.: A better heuristic for orthogonal graph drawings. In: Leeuwen, J. (ed.) ESA 1994. LNCS, vol. 855, pp. 24–35. Springer, Heidelberg (1994). doi:10.1007/BFb0049394

[For10] Forišek, M.: Computational complexity of two-dimensional platform games. In: Boldi, P., Gargano, L. (eds.) FUN 2010. LNCS, vol. 6099, pp. 214–227. Springer, Heidelberg (2010). doi:10.1007/978-3-642-13122-6_22

[Gui14] Guinness World Records: Best-selling racing videogame (2014). http://www.guinnessworldrecords.com/world-records/best-selling-racing-video-game/

[Sal] Sales figures based on http://www.polygon.com/2014/5/15/5718168/mario-kart-series-sales, http://www.nintendo.co.jp/ir/en/sales/software/3ds.html, and http://www.nintendo.co.jp/ir/en/sales/software/wiiu.html, November 2015

[Vig14] Viglietta, G.: Gaming is a hard job, but someone has to do it!. Theory Comput. Syst. **54**(4), 595–621 (2014)

Single-Player and Two-Player Buttons & Scissors Games

(Extended Abstract)

Kyle Burke[1], Erik D. Demaine[2], Harrison Gregg[3], Robert A. Hearn[11],
Adam Hesterberg[2], Michael Hoffmann[4], Hiro Ito[5], Irina Kostitsyna[6],
Jody Leonard[3], Maarten Löffler[7], Aaron Santiago[3], Christiane Schmidt[8],
Ryuhei Uehara[9], Yushi Uno[10], and Aaron Williams[3(✉)]

[1] Plymouth State University, Plymouth, USA
kgburke@plymouth.edu
[2] Massachusetts Institute of Technology, Cambridge, USA
{edemaine,achester}@mit.edu
[3] Bard College at Simon's Rock, Great Barrington, USA
{hgregg11,jleonard11,asantiago11,awilliams}@simons-rock.edu
[4] ETH Zürich, Zürich, Switzerland
hoffmann@inf.ethz.ch
[5] The University of Electro-Communications, Chofu, Japan
itohiro@uec.ac.jp
[6] Université libre de Bruxelles, Brussels, Belgium
irina.kostitsyna@ulb.ac.be
[7] Universiteit Utrecht, Utrecht, Netherlands
m.loffler@uu.nl
[8] Linköping University, Norrköping, Sweden
christiane.schmidt@liu.se
[9] Japan Advanced Institute of Science and Technology, Nomi, Japan
uehara@jaist.ac.jp
[10] Osaka Prefecture University, Sakai, Japan
uno@mi.s.osakafu-u.ac.jp
[11] Portola Valley, CA, USA
bob@hearn.to

Abstract. We study the computational complexity of the Buttons &
Scissors game and obtain sharp thresholds with respect to several parameters. Specifically we show that the game is NP-complete for $C = 2$ colors but polytime solvable for $C = 1$. Similarly the game is NP-complete
if every color is used by at most $F = 4$ buttons but polytime solvable for
$F \leq 3$. We also consider restrictions on the board size, cut directions,
and cut sizes. Finally, we introduce several natural two-player versions
of the game and show that they are PSPACE-complete.

I. Kostitsyna—Supported in part by NWO project no. 639.023.208.
C. Schmidt—Supported in part by grant 2014-03476 from Sweden's innovation
agency VINNOVA.

J. Akiyama et al. (Eds.): JCDCGG 2015, LNCS 9943, pp. 60–72, 2016.
DOI: 10.1007/978-3-319-48532-4_6

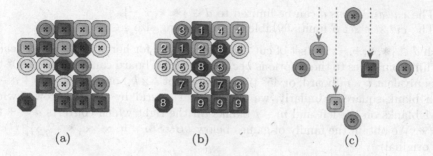

Fig. 1. (a) Level 7 in the Buttons & Scissors app is an $m \times n = 5 \times 5$ grid with $C = 5$ colors, each used at most $F = 7$ times; (b) a solution using nine cuts with sizes in $S = \{2, 3\}$ and directions $d = *$ (no vertical cut is used); (c) a gadget used in Theorem 5.

1 Introduction

Buttons & Scissors is a single-player puzzle by KyWorks. The goal of each level is to remove every button by a sequence of horizontal, vertical, and diagonal cuts, as illustrated by Fig. 1. It is NP-complete to decide if a given level is solvable [2]. We study several restricted versions of the game and show that some remain hard, whereas others can be solved in polynomial time. We also consider natural extensions to two player games which turn out to be PSPACE-complete.

Section 2 begins with preliminaries, then we discuss one-player puzzles in Sect. 3 and two-player games in Sect. 4. Open problems appear in Sect. 5. Due to space restrictions, some proofs are sketched or omitted. A full version of this article can be found on arXiv.

2 Preliminaries

A Buttons & Scissors *board* B is an $m \times n$ grid, where each grid position is either empty or occupied by a button with one of C different colors. A *cut* is given by two distinct buttons b_1, b_2 of the same color c that share either the x-coordinate, the y-coordinate, or are located on the same diagonal ($45°$ and $-45°$). The *size* s of a cut is the number of buttons on the line segment $\overline{b_1 b_2}$ and so $s \geq 2$. A cut is *feasible* for B if $\overline{b_1 b_2}$ only contains buttons of a single color.

When a feasible cut is applied to a board B, the resulting board B' is obtained by substituting the buttons of color c on $\overline{b_1 b_2}$ with empty grid entries. A *solution* to board B is a sequence of boards and feasible cuts $B_1, x_1, B_2, x_2, \ldots, B_t, x_t, B_{t+1}$, where B_{t+1} is empty, and each cut x_i is feasible for B_i and creates B_{i+1}.

Each instance can be parameterized as follows (see Fig. 1 for an example):

1. The *board size* $m \times n$.
2. The *number of colors* C.
3. The *maximum frequency* F of an individual color.

4. The *cut directions* d can be limited to $d \in \{*, \varkappa, +, -\}$.
5. The *cut size set* S limits feasible cuts to having size $s \in S$.

Each $d \in \{*, \varkappa, +, -\}$ is a set of cut directions (i.e. $+$ for horizontal and vertical). We limit ourselves to these options because an $m \times n$ board can be rotated $90°$ to an equivalent $n \times m$ board, or $45°$ to an equivalent $k \times k$ board for $k = m + n - 1$ with blank squares. Similarly, we can shear the grid by padding row i with $i - 1$ blanks on the left and $m - i$ blanks on the right which converts $d = +$ to $d = \varkappa$. We obtain the family of games below ($B\&S[n \times n, \infty, \infty, *, \{2,3\}](B)$ is the original):

Decision Problem: $B\&S[m \times n, C, F, d, S](B)$.
Input: An $m \times n$ board B with buttons of C colors, each used at most F times.
Output: True $\iff$ B is solvable with cuts of size $s \in S$ and directions d.

Now we provide three observations for later use. First note that a single cut of size s can be accomplished by cuts of size $s_1, s_2, \ldots, s_k$ so long as $s = s_1 + s_2 + \cdots + s_k$ and $s_i \geq 2$ for all i. Second note that removing all buttons of a single color from a solvable instance cannot result in an unsolvable instance.

Remark 1. A board can be solved with cut sizes $S = \{2, 3, \ldots\}$ if and only if it can be solved with cut sizes $S' = \{2, 3\}$. Also, $\{3, 4, \ldots\}$ and $\{3, 4, 5\}$ are equivalent.

Remark 2. If board B' is obtained from board B by removing every button of a single color, then $B\&S[m \times n, C, F, d, S](B) \implies B\&S[m \times n, C, F, d, S](B')$.

3   Single-Player Puzzle

3.1   Board Size

We solve one row problems below, and give a conjecture for two rows in Sect. 5.

Theorem 1. $B\&S[1 \times n, \infty, \infty, -, \{2,3\}](B)$ *is polytime solvable.*

Proof. Consider the following context-free grammar,

$$S \to \varepsilon \mid \square \mid SS \mid xSx \mid xSxSx$$

where $\square$ is an empty square and $x \in \{1, 2, \ldots, C\}$. By Remark 1, the solvable $1 \times n$ boards are in one-to-one correspondence with the strings in this language. $\square$

3.2   Number of Colors

Hardness for 2 Colors. We begin with a straightforward reduction from 3SAT. The result will be strengthened later by Theorem 7 using a more difficult proof.

Theorem 2. $B\&S[n \times n, 2, \infty, +, \{2,3\}](B)$ *is NP-complete.*

Proof Sketch: A *variable gadget* has its own row with exactly three buttons. The middle button is alone in its column, and must be matched with at least one of the other two in the variable row. If the left button is not used in this match, we consider the variable set to *true*. If the right button is not used, we consider

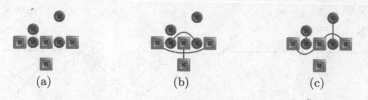

Fig. 2. Split gadget (a) and the two possible ways to clear it (b) and (c).

the variable set to *false*. A button not used in a variable is an *available output*, and can then serve as an *available input* to be used in other gadgets.

Every *clause gadget* has its own column, with exactly four buttons. The topmost button (*clause button*) is alone in its row; the others are inputs. If at least one of these is an available input, then we can match the clause button with all available inputs. We construct one clause gadget per formula clause, connecting its inputs to the appropriate variable outputs. Then, we can clear all the clauses just when we have made variable selections that satisfy the formula.

The variables are connected to the clauses via a multi-purpose *split gadget* (Fig. 2(a)). Unlike the variable and the clause, this gadget uses buttons of two colors. The bottom button is an input; the top two are outputs. If the input button is available, we can match the middle row of the gadget as shown in Fig. 2(b), leaving the output buttons available. But if the input is not available, then the only way the middle row can be cleared is to first clear the red buttons in vertical pairs, as shown in Fig. 2(c); then the output buttons are not available.

We provide a further description of the split gadget and complete the proof in the full version of this article. □

Polynomial-Time Algorithm for 1-Color and Any Cut Directions.

Given an instance B with $C = 1$ color and cut directions $d \in \{*, *, +, -\}$, we construct a hypergraph G that has one node per button in B. A set of nodes is connected with a hyperedge if the corresponding buttons lie on the same horizontal, vertical, or diagonal line whose direction is in d, i.e., they can potentially be removed by the same cut. By Remark 1 it is sufficient to consider a hypergraph with only 2- and 3-edges. A solution to B corresponds to a perfect matching in G. For clarity, we shall call a 3-edge in G a *triangle*, and a 2-edge simply an *edge*.

Cornuéjols et al. [1] showed how to compute a perfect K_2 and K_3 matching in a graph in polynomial time. However, their result is not directly applicable to our graph G yet, as we need to find a matching that consists only of edges and proper triangles, and avoids K_3's formed by cycles of three edges.

To apply [1] we construct graph G' by adding vertices to eliminate all cycles of three edges as follows (see top of Fig. 3). Start with $G' = G$. Consider an $e = (v, w) \in G'$ in a 3-cycle (a cycle of three edges). There are two cases: e is not adjacent to any triangle in G', or e is adjacent to some triangles in G'. In the first case we add vertices u_1 and u_2 that split e into three edges (v, u_1), (u_1, u_2),

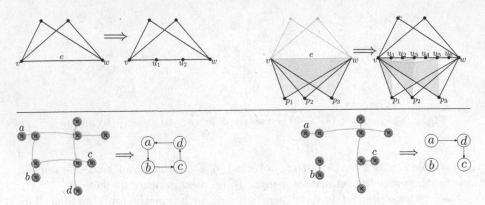

Fig. 3. Top-left: splitting 3-cycles when there are no adjacent triangles to edge e; top-right: splitting 3-cycles when e has adjacent triangles (shaded). Bottom-left: constructing G_c from four cuts blocking each other in a cycle; bottom-right: constructing G_c from the same cuts after reassigning the blocking buttons

and (u_2, w). In the second case, when e is adjacent to k triangles, we add $2k$ vertices $u_1, u_2 \ldots, u_{2k}$ along e, and replace every $\triangle p_i v w$ with $\triangle p_i v u_{2i-1}$.

Lemma 1. *There exists a perfect edge- and triangle-matching in G' iff there exists perfect edge- and triangle-matching in G.*

Proof. Given a perfect matching M in G, we construct a perfect matching M' in G'. Consider $e = (v, w)$ in G. If e is not adjacent to any triangles in G, then

- if $e \in M$ then add edges (v, u_1) and (u_2, w) of G' to M' (both v and w are covered by e, and all v, w, u_1, and u_2 are covered by M');
- if $e \notin M$ then add edge (u_1, u_2) of G' to M' (v and w are not covered by e, and u_1 and u_2 are covered by M').

In both cases above the extra nodes in G' are covered by edges in M', and if v and w in G are covered by e in M then v and w are covered by (v, u_1) and (u_2, w) in G'. If e is adjacent to some triangles in G,

- if $e \in M$ then in G' add edges (v, u_1), (u_{2k}, w), and (u_{2j}, u_{2j+1}) to M', for $1 \leq j < k$;
- if $\triangle p_i v w \in M$ for some i then add $\triangle p_i v u_{2i-1}$, edges (u_{2j-1}, u_{2j}) for $1 \leq j < i$, (u_{2j}, u_{2j+1}) for $i \leq j < k$, and (u_{2k}, w) of G' to M';
- if neither e nor any triangle adjacent to e is in M then add edges (u_{2j-1}, u_{2j}) of G' to M', for $1 \leq j \leq k$.

In all the above cases the extra nodes in G' are covered by edges in M', and if v and w in G are covered by e or a triangle in M then v and w are also covered by (v, u_1) and (u_2, w) or by a corresponding triangle in G'.

Refer to the full version of this article for the details on how to create a perfect matching in G from one in G'. □

Thus, a perfect edge- and triangle-matching in G that does not use a 3-cycle (if it exists) can be found by first converting G to G' and applying the result in [1] to G'. A solution of B consisting of 2- and 3-cuts can be reduced to a perfect edge- and triangle-matching in G; however, the opposite is not a trivial task. A perfect matching in G can correspond to a set of cuts C_M in B that are blocking each other (see bottom of Fig. 3). To extract a proper order of the cuts we build another graph G_c that has a node per cut in C_M and a directed edge between two nodes if the cut corresponding to the second node is blocking the cut corresponding to the first node. If G_c does not have cycles, then there is a partial order on the cuts. The cuts that correspond to the nodes with no outgoing edges can be applied first, and the corresponding nodes can be removed from G_c. However, if G_c contains cycles, there is no order in which the cuts can be applied to clear up board B. In this case we will need to modify some of the cuts in order to remove cycles from G_c. We provide the details in the full version of this article.

By Lemma 1 and by the construction above we obtain the following theorem.

Theorem 3. $B\&S[n \times n, 1, \infty, d, \{2,3\}](B)$ *is polytime solvable for all* $d \in \{*, \divideontimes, +, -\}$.

3.3   Frequency of Colors

Theorem 4. $B\&S[n \times n, \infty, 3, *, \{2,3\}](B)$ *is polytime solvable.*

Proof. A single cut in any solution removes a color. By Remark 2, these cuts do not make a solvable board unsolvable. Thus, a greedy algorithm suffices.   □

Hardness was established for maximum frequency $F = 7$ in [2]. We strengthen this to $F = 4$ via the modified clause gadget in Fig. 1 (c). In this gadget the leftmost circular button can be removed if and only if at least one of the three non-circular buttons is removed by a vertical cut. Thus, it can replace the clause gadget in Sect. 4.1 of [2]. Theorem 5 is proven in the full version of this article.

Theorem 5. $B\&S[n \times n, \infty, 4, *, \{2,3\}](B)$ *is NP-complete.*

3.4   Cut Sizes

Section 3.2 provided a polytime algorithm for 1-color. However, if we reduce the cut size set from $\{2,3,4\}$ to $\{3,4\}$ then it is NP-complete. We also strengthen Theorem 2 by showing that 2-color puzzles are hard with cut size set $\{2\}$.

Hardness for Cut Sizes {3,4} and 1-Color

Theorem 6. $B\&S[n \times n, 1, \infty, *, \{3,4\}](B)$ *is NP-complete.*

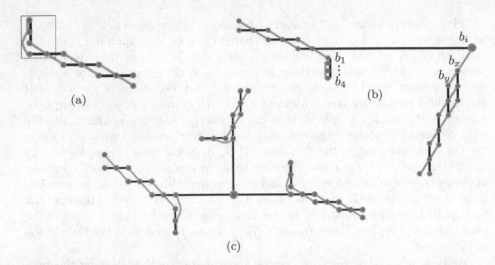

Fig. 4. (a) The only two cut possibilities in the variable gadget (shown in black and gray), corresponding to truth assignments of "true" and "false", respectively. (b) The bend gadget for the 1-color case. (c) The clause gadget for the 1-color case.

Proof. We show $B\&S[n \times n, 1, \infty, *, \{3, 4\}](B)$ to be NP-hard by a reduction from PLANAR 3-SAT, which was shown to be NP-complete by Lichtenstein [3].

An instance F of the PLANAR 3-SAT problem is a Boolean formula in 3-CNF consisting of a set $\mathcal{C}$ of m clauses over n variables $\mathcal{V}$ The variable-clause incidence graph $G = (\mathcal{C} \cup \mathcal{V}, E)$ is planar, and all variables are connected in a cycle. The PLANAR 3-SAT problem is to decide whether there exists a truth assignment to the variables such that at least one literal per clause is true.

We turn the planar embedding of G into a Buttons & Scissors board, i.e., we present variables, clauses and edges by single-color buttons that need to be cut. We provide detailed descriptions of each gadget in the full version of this article.

The **variable gadget**, shown in Fig. 4(a), enables us to associate horizontal and diagonal cut patterns with "true" and "false" values, respectively.

The **bend gadget**, shown in Fig. 4(b), enables us to bend a wire to match the bends in G's embedding while enforcing that the same values are propagated through the bent wire.

The **split gadget**, shown in Fig. 5(b), enables us to increase the number of wires leaving a variable and propagating its truth assignment.

The **not gadget**, shown in Fig. 5(a), enables us to reverse the truth assignment in a variable wire.

The **clause gadget** is shown in Fig. 4(c). This gadget simulates a conjunction of literals.

Thus, the resulting Buttons & Scissors board has a solution if and only if at least one of the literals per clause is set to true, that is, if and only if the original PLANAR 3-SAT formula F is satisfiable. It is easy to see that this reduction is possible in polynomial time. In addition, given a Buttons & Scissors board

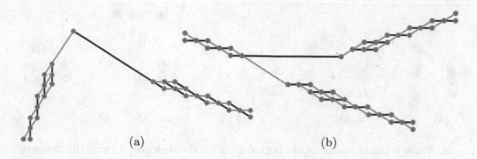

Fig. 5. (a) The not gadget, negating the input truth assignment, for the 1-color case.
(b) The split gadget for the 1-color case.

and a sequence of cuts, it is easy to check whether those constitute a solution,
i.e., whether all cuts are feasible and result in a board with only empty grid
entries. Hence, $B\&S[n \times n, 1, \infty, *, \{3, 4\}](B)$ is in the class NP. Consequently,
$B\&S[n \times n, 1, \infty, *, \{3, 4\}](B)$ is NP-complete. $\qquad\qquad\qquad\qquad\qquad$ □

Hardness for Cut Size $\{2\}$ and 2-Colors. An intermediate problem is below.

Decision Problem: Graph Decycling on (G, S).
Input: Directed graph $G = (V, E)$ and a set of disjoint pairs of vertices $S \subseteq V \times V$.
Output: True, if we can make G acyclic by removing either s or s' from G for
every pair $(s, s') \in S$. Otherwise, False.

Lemma 2. *Graph Decycling reduces to Buttons & Scissors with 2 colors.*

Proof. Consider an instance (G, S) to graph decycling. First, we observe that
we can assume that every vertex in G has degree 2 or 3, and more specifically,
in-degree 1 or 2, and out-degree 1 or 2. Indeed, we can safely remove any vertices
with in- or out-degree 0 without changing the outcome of the problem. Also, we
can replace a node with out-degree k by a binary tree of nodes with out-degree
2. The same applies to nodes with in-degree k.

Furthermore, we can assume that every vertex that appears in S has degree 2.
Indeed, we can replace any degree 3 vertex by two vertices of degree 2 and 3,
and use the degree 2 vertex in S without changing the outcome. Similarly, we
can assume that no two vertices of degree 3 are adjacent. Finally, we can assume
that G is bipartite, and furthermore, that all vertices that occur in S are in the
same half of V, since we can replace any edge by a path of two edges.

Now, we discuss how to model such a graph in a Buttons & Scissors instance.
Each node will correspond to a pair of buttons, either a red or a green pair according to a bipartition of V. These pairs of buttons will be mapped to locations
in the plane on a common (horizontal for red, vertical for green) line, and such
that any two buttons of the same color that are not a pair are not on a common
(horizontal, vertical, or diagonal) line (unless otherwise specified). If two nodes

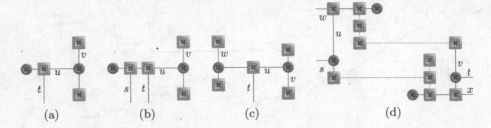

Fig. 6. Three types of nodes: (a) in-degree 1 (tu) and out-degree 1 (uv); (b) in-degree 2 (su and tu) and out-degree 1 (uv); (c) in-degree 1 (tu) and out-degree 2 (uv and uw). In (d) the nodes u and v are linked in S and we can choose to remove u or v.

of opposite colors u and v are connected by an edge in G, we say that u *blocks* v. In this case, one of the buttons of u will be on the same line as the buttons of v, and more specifically, it will be between the two buttons of v. That is, v can only be cut if u is cut first. Buttons of opposite colors that are not connected by an edge will not be on any common lines either.

As discussed above, we can assume we have only three possible types of nodes. Figure 6(a) illustrates the simplest case, of a node u with one incoming edge tu and one outgoing edge uv. Clearly, t blocks u and u blocks v. To model a node with in-degree 2, we need to put two buttons of different same-colored nodes on the same line (see Fig. 6(b)). As long as the other endpoints of these two edges are not on a common line this is no problem: we never want to create a cut that removes one button of s and one of t, since that would create an unsolvable instance. Finally, to model a node with out-degree 2, we simply place a vertical edge on both ends of u (see Fig. 6(c)). Note that is it important here that we do not connect two nodes with out-degree 2 to the same two nodes with in-degree 2, since then we would have both pairs of endpoints on a common line; however, we assumed that nodes of degree 3 are never adjacent so this does not occur.

It remains to create a mechanism to remove vertices from G as dictated by S. This is illustrated in Fig. 6(d) with details in the full version of this article. □

A proof of Lemma 3 is in the full version of this article. Lemmas 2 and 3 give Theorem 7.

Lemma 3. *SAT reduces to Graph Decycling.*

Theorem 7. $B\&S[n \times n, 2, \infty, +, \{2\}](B)$ *is NP-complete.*

4   Two-Player Games

We consider three two-player Buttons & Scissors variants. First we consider color restricted games where (a) each player can only cut specific colors, and (b) players are not restricted to specific colors. For (a) player blue may only cut Blue buttons, while the red player may only cut Red buttons. For (b) we distinguish by winning criterion: for (IMPARTIAL) the last player who makes a feasible cut wins; for (SCORING) players keep track of the total number of buttons they've cut. When no cuts can be made, the player with the most buttons cut wins.

In the following sections, we show that all variants are PSPACE-complete.

4.1 Cut-By-Color Games

In this section the first player can only cut blue buttons, the second player can only cut red buttons, and the last player to make a cut wins.

Theorem 8. *The partisan LAST two-player Buttons & Scissors game, where one player cuts blue buttons, the other red buttons, is PSPACE-complete.*

Proof. The proof is by reduction from $G_{\%free}(\text{CNF})$ [5]: given a boolean formula $\Phi(x_1, \ldots, x_n)$ in CNF and a partition of the variables into two disjoint subsets of equal size V_b and V_r, two players take turns in setting values of the variables, the first (Blue) player sets the values of variables in V_b, and the second (Red) player sets the values of variables in V_r. Blue wins if, after all variables have been assigned some values, formula Φ is satisfied, and loses otherwise.

For a given instance of formula Φ we construct a Buttons & Scissors board B, such that Blue can win the game on B if and only if he can satisfy formula Φ. We will prove this statement in different formulation: Red wins the game on B if and only if formula Φ cannot be satisfied. For a complete example see the full version of this article.

The **red variable gadget** is shown in Fig. 7(a). Red "sets the value" of the corresponding variable by choosing the first cut to be a (false) or b (true), and thus unlocking one of the two cuts, c or d, respectively, for Blue to follow up (and to propagate the value of the variable).

The **blue variable gadget** is shown in Fig. 7(b). Blue "sets the value" of the corresponding variable by choosing the first cut to be a (false) or b (true),

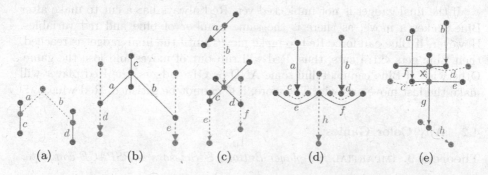

(a)   (b)   (c)   (d)   (e)

Fig. 7. (a) The red (dashed) variable gadget, (b) the blue (solid) variable gadget, (c) the split gadget, (d) the OR gadget, and (e) the AND gadget. Lines (or arcs used for clarity) indicate which buttons are aligned. (Color figure online)

and thus unlocking one of the two cuts, d or e, respectively, for the red player to follow up. Blue has one extra cut c that is used to pass the turn to Red. Alternatively, Blue can choose to start with the 3-button cut c and disallow Red from making any cuts in the gadget. In that case the corresponding variable cannot be used to satisfy Φ.

Figure 7(d) depicts the **OR** gadget: if Blue cuts a or b (or both), Red can leave the gadget with cut h. Cuts a and b unblock cuts c and d, respectively, which in turn unblock e and f, respectively.

Figure 7(e) depicts the **AND** gadget for two inputs. The proper way of passing the gadget: Blue makes both cuts a and b, and Red makes cuts c and d when they get unblocked, thus enabling Blue to make cut g and exit the gadget. However, Red could also take an "illegal" cut x, thus, unblocking two extra cuts, e and f, for the blue player, and, hence, putting Red at a disadvantage. Thus, if at any point in the game Red chooses (or is forced to) make cut x in any of the AND gadgets, the game result is predetermined, and Red cannot win on B.

Figure 7(c) shows the **split** gadget; it enables us to increase the number of cuts leaving a variable and propagating its truth assignment. Blue's cut a unblocks Red's cut b, which unblocks both c and d. If Blue cuts c and d this enables Red to cut e and f, respectively. The gadget also exists with Blue and Red reversed.

A variant of the split gadget evaluates the formula Φ: cuts e and f are deleted. If the variable values are propagated to this gadget and Red is forced to make the cut b, Blue then gets extra cuts which Red will not be able to follow up.

The game progresses as follows: Blue selects an assignment to a blue variable. This unlocks a path of red-blue cuts that goes through some AND and OR gadgets and leads to the final gadget. As the order of the cuts in such a path is deterministic, and does not affect the choice of values of other variables, w.l.o.g., we assume that Red and Blue make all the cuts in this path (until it gets "stuck") before setting the next variable. The path gets stuck when it reaches some AND gadget for which the other input has not been cleared. The last cut in such a path was made by Red, thus afterwards it will be Blue's turn, and he may choose to make the leftover cut c from the variable gadget to pass the turn to Red.

If the final gadget is not unblocked yet, Red always has a cut to make after Blue makes a move, as there is the same number of blue and red variables. However, if Blue can force Red to make moves until the final gadget is reached, then Blue gets extra cuts; thus, Red will run out of moves and lose the game. Otherwise, if Blue cannot fulfill some AND or OR gadgets, the Red player will make the last move and win. Therefore, if Φ cannot be satisfied, Red wins.   □

4.2   Any Color Games

Theorem 9. IMPARTIAL *two-player Buttons & Scissors is PSPACE-complete.*

Theorem 10. SCORING *two-player Buttons & Scissors is PSPACE-complete.*

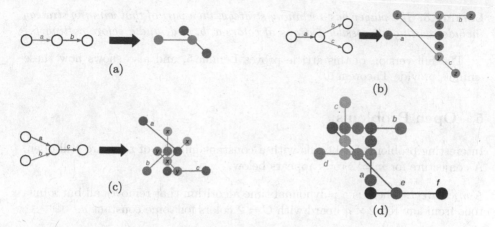

Fig. 8. Reduction gadgets for vertex with (a) one incoming arc and one outgoing arc, (b) one incoming arc and two outgoing arcs, and (c) two incoming arcs and one outgoing arcs. (d) The starting gadget for SCORING.

We show that IMPARTIAL is PSPACE-complete, then use one more gadget to show SCORING is PSPACE-complete. We reduce from GEOGRAPHY[1], (PSPACE-complete [4]). We use Lemma 4 to start with low-degree GEOGRAPHY instances.

Lemma 4. GEOGRAPHY *is PSPACE-complete even when vertices have max degree 3 and the max in-degree and out-degree of each vertex is 2.*

The full version of this article proves Lemma 4 and Theorem 9 with these gadgets:

– **In-degree 1, out-degree 1**: The gadget for this is a pair of buttons such that removing the first pair frees up the second, as in Fig. 8(a).
– **In-degree 1, out-degree 2**: See Fig. 8(b) and the full version of this article.
– **In-degree 2, out-degree 1**: See Fig. 8(c) and the full version of this article.
– **In-degree 0**: The gadgets for this look just like the gadgets for the analogous in-degree 1 gadgets, but without the button pair for the incoming edge.
– **Out-degree 0**: Each edge is a button pair that won't free up other buttons.

To show SCORING is hard, we create a reduction where after each turn, that player will have cut the most buttons; the last player to move wins. This alternating-advantage situation is caused by an initial gadget. The optimal play sequence begins by cutting two buttons, then three, then three, then three a final time. After these four moves, the first player will have five points and the second player six. Each subsequent cut removes two buttons so each turn ends with the current player ahead.

Figure 8(d) shows the starting gadget that sets up this initial back-and-forth. The color-f buttons will be the last two cut; the right-hand f button must be blocking the next gadget. Lemma 5 postulates that f will be last.

---

[1] Specifically, DIRECTED VERTEX GEOGRAPHY, usually called GEOGRAPHY.

Lemma 5. *If a player has a winning strategy, then part of that winning strategy includes cutting all possible buttons of colors a, b, c, d, and e before cutting f.*

The full version of this article proves Lemma 5, and also shows how these lemmas provide Theorem 10.

5   Open Problems

Interesting problems for boards with a constant number of rows are still open. A conjecture for $m = 2$ rows appears below.

Conjecture 1. There is a polynomial time algorithm that removes all but s buttons from any full $2 \times n$ board with $C = 2$ colors for some constant s.

References

1. Cornuéjols, G., Hartvigsen, D., Pulleyblank, W.: Packing subgraphs in a graph. Oper. Res. Lett. **1**(4), 139–143 (1982)
2. Gregg, H., Leonard, J., Santiago, A., Williams, A.: Buttons & scissors is NP-complete. In Proceedings of the 27th Canadian Conference on Computational Geometry (2015)
3. Lichtenstein, D.: Planar formulae and their uses. SIAM J. Comput. **11**(2), 329–343 (1982)
4. Lichtenstein, D., Sipser, M.: Go is polynomial-space hard. J. ACM **27**(2), 393–401 (1980)
5. Schaefer, T.J.: On the complexity of some two-person perfect-information games. J. Comput. Syst. Sci. **16**(2), 185–225 (1978)

Fitting Spherical Laguerre Voronoi Diagrams to Real-World Tessellations Using Planar Photographic Images

Supanut Chaidee[✉] and Kokichi Sugihara

Graduate School of Advanced Mathematical Sciences, Meiji University,
4-21-1 Nakano, Nakano-ku, Tokyo 164-8525, Japan
{schaidee,kokichis}@meiji.ac.jp

Abstract. There are many natural phenomena displayed as polygonal tessellations on curved surfaces, typically found in fruit skin patterns. The paper proposes a method to fit given tessellations with spherical Laguerre Voronoi diagrams. The main target of this paper is fruit skin patterns such as jackfruit and lychee covered by tessellation patterns in which each cell contains a unique spike dot that can be considered as a generator. The problem of estimating the weights is reduced to an optimization problem, and can be solved efficiently. The experiments were done with ideal data and real fruit skin data, which show the validity of the method. We also propose related problems for further studies.

Keywords: Spherical Laguerre Voronoi diagram · Voronoi approximation · Tessellation fitting · Fruit skin patterns

1 Introduction

Many natural phenomena display as polygonal patterns, for example, animal territories, flower inflorescences, and some fruit skin patterns. From a mathematical viewpoint, verifying whether a given polygonal pattern can be considered a Voronoi diagram, the tessellation partitioning a space into cells corresponding to the set of generators where each point in the space is assigned to the nearest generator, is useful for constructing mathematical models of polygonal pattern formation.

The concept of Voronoi diagrams, their generalization, and applications have been studied widely [6,21]. One of the research directions is focused on the Voronoi recognition and approximation problems. Suppose that we are given a polygonal tessellation. We firstly determine whether the given tessellation is a Voronoi diagram. If it is, we determine the set of Voronoi generators. This problem (Voronoi recognition) was studied in [1,2,5,11,18,22]. Otherwise, we approximate the Voronoi generators to find the Voronoi diagram that best fits the given tessellation [10,12,13,26]. This problem (Voronoi approximation) is more useful in practice in the real world because real-world tessellations are not exact Voronoi diagrams.

© Springer International Publishing AG 2016
J. Akiyama et al. (Eds.): JCDCGG 2015, LNCS 9943, pp. 73–84, 2016.
DOI: 10.1007/978-3-319-48532-4_7

As a generalization of Voronoi diagrams, we can consider weighted Voronoi diagrams for which each Voronoi generator has a weight. One interesting weighted Voronoi diagram is a Laguerre Voronoi diagram (alternatively called a power diagram), because it consists of straight line edges. Laguerre Voronoi diagrams were introduced in [3,14]. Some geometrical properties related to the polyhedron were defined in [4]. Most importantly for the present work, spherical Laguerre Voronoi diagrams, which were analogously defined using geodesic distance, and algorithms were introduced by Sugihara [24,25].

Unlike the Laguerre Voronoi construction, the Laguerre Voronoi recognition problem has not been much studied. Recently, Duan *et al.* [9] gave a framework to determine the Laguerre generators with weights from a planar Laguerre Voronoi tessellation. Laguerre Voronoi diagrams, including tessellation fitting using Laguerre Voronoi diagrams, have been applied to real-world problems such as those of the 3D material structure (e.g., [15–17,19]), tomographic image data (e.g., [23]), and molecular chemistry (e.g., [20]).

Recently, we [7] defined a class of objects called spike-containing objects which intuitively originate from fruits covered by spikes. We gave a framework for finding the spherical Voronoi diagrams which are best fit to the spike-containing tessellations extracted from the photos showing spikes. The problem is considered a Voronoi approximation problem. We [7] employed optimization techniques for adjusting the approximated sphere radius and spike height. The sphere center position is also adjusted in [8]. We applied this method to both jackfruit and lychee. Although it works well for jackfruit, it does not work well for lychee, suggesting that for lychee the weights of the tessellation cells are not equal, and hence, it may be not appropriate to fit a spherical Voronoi diagram in this case.

Using weighted Voronoi diagrams is one of the possible ways to resolve the problem. However, most weighted Voronoi diagrams contain complicated curved Voronoi edges, whereas real-world tessellations consist of almost straight line edges. Therefore, we focus on Laguerre Voronoi diagrams as fits to spike-containing tessellations.

In this study, we assume that we have a spike-containing tessellation and would like to find the spherical Laguerre Voronoi diagram fitting it. Unlike [9], our method finds the appropriated spherical Laguerre Voronoi diagram by approximating the weight of each spherical Voronoi generator.

This paper is organized as follows. In Sect. 2, we give the fundamental definitions and theorems which are related to our work. The problem is also formulated with some assumptions. In Sect. 3, we give the framework for finding the approximated spherical Laguerre Voronoi diagram of a given tessellation. We also introduce the optimization problem which is used for approximating weights of generators before presenting the fitting algorithm, which we analyze. In Sect. 4, we show experimental results for artificially generated data and real data, which confirm the validity of our framework. Finally, we summarize the results of the study and show related problems for future works.

2   Modeling Assumptions

In this section, we first provide the basic definitions and theorems related to spherical Laguerre Voronoi diagrams. Then we state the assumptions of the model. We also present a criterion for judging whether there is a difference between the given tessellations and the considered Voronoi diagrams.

2.1   Fundamental Definitions and Theorems

Briefly, it is well known that we can define Voronoi diagrams by spaces, generators, and distances. One example is a Voronoi diagram on a sphere S in $\mathbb{R}^3$. Basically, the ordinary distance on the unit sphere is the geodesic distance defined by

$$\tilde{d}(p, p_i) = \arccos(\mathbf{x}^{\mathrm{T}}\mathbf{x}_i) \leq \pi.$$

where $\mathbf{x}, \mathbf{x}_i$ are position vectors of two distinct points $p, p_i \in S$. Note that the geodesic distance $\tilde{d}(p, p_i)$ is the shorter distance between p and p_i on the great circle passing through p and p_i.

We can generalize Voronoi diagrams by putting a weight on each generator; these generalized diagrams are called weighted Voronoi diagrams. In this study, our focus is on Laguerre Voronoi diagrams. Starting from [3,14,24], we can intuitively consider an ordinary Laguerre Voronoi diagram on $\mathbb{R}^2$ to be a set of n circles $G = \{c_1, ..., c_n\}$ in $\mathbb{R}^2$, where $c_i = (p_i, r_i)$ are circles with centers and radii p_i and r_i, respectively. For a circle c_i and arbitrary point $p \in \mathbb{R}^2$, the Laguerre distance (or power distance) is defined by

$$d_{\mathrm{L}}(p, c_i) = d(p, p_i)^2 - r_i^2,$$

where $d(p, p_i)$ denotes the Euclidean distance between p and p_i.

The Laguerre Voronoi region associated with c_i is defined as

$$V_L(c_i) = \{p \in \mathbb{R}^2 : d_{\mathrm{L}}(p, c_i) \leq d_{\mathrm{L}}(p, c_j), j \in I_n \setminus \{i\}\},$$

where $I_n = \{1, ..., n\}$. In $\mathbb{R}^2$, the locus of Voronoi bisectors form a straight line.

We can also extend the concept of Laguerre Voronoi diagrams to the sphere. In [24,25], the Laguerre distance and its related objects were considered on a unit sphere U. In the present paper, we analogously use those definitions with a sphere $\mathcal{U}$ with radius R, where the sphere center is located at $O(0,0,0)$ in the Cartesian coordinate system.

Definition 1. *Let P_i be a point on the sphere $\mathcal{U}$. A spherical circle $\tilde{c}_i$ on sphere $\mathcal{U}$ is defined by*

$$\tilde{c}_i = \{P \in \mathcal{U} : \tilde{d}(P, P_i) = R_i\},$$

where $\tilde{d}(P, P_i)$ is the geodesic distance between P and P_i, and $0 \leq R_i/R < \pi/2$.

We can define the Laguerre Voronoi diagram on the sphere $\mathcal{U}$ of $\mathbb{R}^3$ corresponding to the set of spherical circles $\mathcal{G} = \{\tilde{c}_1, ..., \tilde{c}_n\}$ by using the Laguerre proximity

$$\tilde{d}_L(P, \tilde{c}_i) = \frac{\cos\left(\tilde{d}(P, P_i)/R\right)}{\cos\left(R_i/R\right)},$$

and the Laguerre bisector of two adjacent circles $\tilde{c}_i, \tilde{c}_j$ is defined by

$$B_L(\tilde{c}_i, \tilde{c}_j) = \{P \in \mathcal{U} : \tilde{d}_L(P, \tilde{c}_i) = \tilde{d}_L(P, \tilde{c}_j)\}.$$

In [25], the following theorem characterizes the spherical Laguerre Voronoi diagram bisectors.

Theorem 1 ([25])**.** *The Laguerre bisector $B_L(\tilde{c}_i, \tilde{c}_j)$ is a great circle, and it crosses the geodesic arc connecting the two centers P_i and P_j at a right angle.*

Algorithms for constructing a spherical Laguerre Voronoi diagram and a spherical Laguerre Delaunay diagram were reported in [24, 25].

2.2   Problem Formulation and Assumptions

In previous study [7], we defined a *spike-containing object* as an object consisting of a convex surface that can be approximated by a sphere that is covered by spikes of approximately uniform height and has a polygon-like tessellation on its surface with each cell containing exactly one spike.

We photograph a spike-containing object and assume that the photo is an orthogonal projection onto the XZ-plane. From this photo of the spike-containing object, we extract the *spike-containing tessellation T*, which is the convex tessellation composed of a planar 3-regular straight graph. Each tessellation cell contains a unique generator called a spike dot, the dot projected from the tip of the spike. Let $\mathcal{B} = \{s_1, ..., s_n\}$ be the set of spike dots of the spike-containing tessellation T.

In [7,8], we proposed methods for finding the ordinary spherical Voronoi diagram which is the best fit to the spike-containing tessellation T with respect to the spike dots set $\mathcal{B}$. The sphere radius R and spike height h are determined by the algorithms.

To judge whether a difference exists between two tessellations on the plane, we use the discrepancy as defined in [7,8], which is as follows.

Let T be a given tessellation and V be a Voronoi diagram. Denote by A_T, A_V the areas of the given tessellation and the Voronoi diagram, respectively. Define $D_T = A_T - A$ and $D_V = A_V - A$, where A is the sum of areas of intersection between cells of T and V corresponding to the same spike dot. The discrepancy $\Delta_{T,V}$, the ratio of the differences between two tessellations to the sum of the areas of the tessellations, is defined by

$$\Delta_{T,V} = \left(\frac{D_T + D_V}{2}\right) \bigg/ \left(\frac{A_T + A_V}{2}\right) = \frac{D_T + D_V}{A_T + A_V}.$$

3   Main Framework

In this study, we would like to find the spherical Laguerre Voronoi diagram fitting the spike-containing tessellation T with spike dots set $\mathcal{B}$. Assume that the radius R of a sphere $\mathcal{U}$ and spike height h are acquired from the frameworks in [7,8]. Our main concern is to approximate the spherical circle radius of each generator in the set $\mathcal{B}$ projected onto the sphere $\mathcal{U}$.

To obtain the fitted spherical Laguerre Voronoi diagram, we consider the discrepancy by comparing the spike-containing tessellation T with a projected spherical Laguerre Voronoi diagram, say V. We divide the processes into three main steps for obtaining V: projecting a tessellation T onto a sphere $\mathcal{U}$, approximating the spherical circle radii, and constructing the spherical Laguerre Voronoi diagram and projecting it onto the XZ-plane.

3.1   Projecting a Tessellation T onto a Sphere

Starting from a spike-containing tessellation T with spike dots set $\mathcal{B}$, we orthogonally project spike dots onto a sphere with radius $R + h$, and then radially project them to the sphere $\mathcal{U}$ with radius R.

Rigorously, let $s_i = (x_i, z_i)$ be a spike dot in the XZ-plane. The coordinates of s_i projected onto a sphere with radius $R + h$ are taken to be $s_i(R + h) = (x_i, \sqrt{(R + h)^2 - (x_i^2 + z_i^2)}, z_i)$. Then the central projection of $s_i(R + h)$ onto the sphere $\mathcal{U}$ is $s_i(R) = (R/(R + h)) s_i(R + h)$. The set of spike dots on sphere $\mathcal{U}$ is written as $\mathcal{B}(R) = \{P_1, ..., P_n\}$, where $P_i := s_i(R)$.

Differently from [7,8], we also project the spike-containing tessellation T onto sphere $\mathcal{U}$. To do so, we define the tessellation T as $T = \{T_1, ..., T_n\}$ where T_i is a set of polygonal vertices of the i-th cell corresponding to the spike dot s_i. Note that a cell T_i is adjacent to a cell T_j if and only if $|T_i \cap T_j| = 2$. This implies that there exists a tessellation edge e_{ij} partitioning the tessellation cell i, j.

For each vertex $v_k \in \cup_{m=1}^n T_m$, we project $v_k \in \cup_{m=1}^n T_m$ onto the sphere; we will denote this projection by $v_k(R)$. For each pair i, j, where $i \neq j$, if $T_i \cap T_j = \{v_{ijk_1}, v_{ijk_2}\}$ with $v_{ijk_1}, v_{ijk_2} \in \cup_{m=1}^n T_m$, then we construct a geodesic arc joining $v_{ijk_1}(R)$ and $v_{ijk_2}(R)$, say $\widehat{e}_{ij}$, which is the bisector of the i-th and j-th cells. The vertex set of the spherical polygon with respect to spike dot P_i is denoted by $T_i(R)$, and the vertex set of tessellation T projected onto sphere $\mathcal{U}$ is written as $T(R)$.

3.2   Radii Approximation

In this step, we approximate the weight R_i of each generator P_i for all $i \in I_n$. R_i is an unknown variable satisfying $0 \leq R_i/R < \pi/2$.

From a spike-containing tessellation projected onto a sphere $T(R)$, suppose that two adjacent cells $T_i(R)$ and $T_j(R)$ share a geodesic arc $\widehat{e}_{ij}$. To satisfy the requirements of Theorem 1, we draw the geodesic arc from P_i perpendicular to geodesic arc $\widehat{e}_{ij}$. Let the foot point be Q_{ij}, and similarly let the foot point corresponding to P_j be Q_{ji}. We then compute $\tilde{d}(P_i, Q_{ij})$ and $\tilde{d}(P_j, Q_{ji})$.

Now, we let $\mathcal{R} = \{R_1, ..., R_n\}$ be the set of spherical circle radii with respect to $\mathcal{B}(R)$, and spherical circle radii are defined as variables. To obtain an appropriate set $\mathcal{R}$, we desire to satisfy the condition $\tilde{d}_L(P, \tilde{c}_i) = \tilde{d}_L(P, \tilde{c}_j)$ in Theorem 1 for each adjacent pair i, j assumed to be the spherical Laguerre bisector.

Let $A_{ij} = \cos\left(\frac{\tilde{d}(P_i, Q_{ij})}{R}\right)$ and $A_{ji} = \cos\left(\frac{\tilde{d}(P_j, Q_{ji})}{R}\right)$. We define the residual

$$r(R_i, R_j) := \cos\left(\frac{R_j}{R}\right) A_{ij} - \cos\left(\frac{R_i}{R}\right) A_{ji}.$$

We employ least-squares optimization; specifically, we minimize the sum of squared residuals among all adjacent pairs i, j as the objective function

$$f(R_1, ..., R_n) := \sum_{i,j} \left[\cos\left(\frac{R_j}{R}\right) A_{ij} - \cos\left(\frac{R_i}{R}\right) A_{ji}\right]^2. \tag{1}$$

From Definition 1, we note that $0 \le R_i < (R\pi)/2$ for all $1 \le i \le n$. The set $\mathcal{R}$ which minimizes $f(R_i, ..., R_n)$ is defined as $\mathcal{R}_{\text{opt}}$.

3.3   Construction of the Projected Spherical Laguerre Voronoi Diagram V

After the step of radii approximation, we now have the set of optimal spherical circle radii $\mathcal{R}_{\text{opt}}$. Using this set along with the spike dot set $\mathcal{B}(R)$, we can construct the spherical circles which belong to the set $\tilde{G} = \{\tilde{c}_1, ..., \tilde{c}_n\}$. We are now ready to construct the spherical Laguerre Voronoi diagram or the spherical Laguerre Delaunay diagram.

We can directly apply Algorithm 1 of [25] to construct the spherical Laguerre Voronoi diagram.

In the case of the spherical Laguerre Delaunay diagram, we can instead apply Algorithm 2 of [25] by firstly constructing a set of points when a generator point $P_i = (x_i, y_i, z_i)$ on the sphere $\mathcal{U}$ is inversely mapped to the point $P_i^*(x_i/t, y_i/t, z_i/t)$, where $t = \cos(R_i/R)$. We then construct the convex hull from the set $\tilde{G}^* = \{P_1^*, ..., P_n^*\}$ and project it onto the sphere $\mathcal{U}$ radially to obtain the spherical Laguerre Delaunay diagram.

After we obtain a spherical Laguerre Voronoi diagram, say V_L, we project this spherical Laguerre Voronoi diagram orthogonally onto the XZ-plane. Let the projected planar tessellation be V. We compute the discrepancy between V and T.

We can now summarize the algorithm as follow.

Algorithm 1. Fitting to a Spherical Laguerre Voronoi Diagram
Input: Spike-containing tessellation T with the spike dot set $\mathcal{B}$.
Output: The spherical Laguerre Voronoi diagram that fits T, and the discrepancy $\Delta_{T,V}$
Procedure:

1. **for** $i = 1$ **to** $|\mathcal{B}|$;
      project s_i to $s_i(R + h)$;
      project $s_i(R + h)$ to $s_i(R) =: P_i$;
   end for;
2. **for** $k = 1$ **to** $|\cup_{m=1}^n T_m|$;
      project $v_k \in \cup_{m=1}^n T_m$ to $v_k(R)$;
   end for;
3. **for** each edge $\{v_{ijk_1}, v_{ijk_2}\} \subset \cup_{m=1}^n T_m$;
      construct the geodesic arc $\widehat{e}_{ij}$;
      construct the geodesic arc from P_i perpendicular to $\widehat{e}_{ij}$ at Q_{ij};
      construct the geodesic arc from P_j perpendicular to $\widehat{e}_{ij}$ at Q_{ji};
      compute $\tilde{d}(P_i, Q_{ij})$, $\tilde{d}(P_j, Q_{ji})$;
      set $r(R_i, R_j) := \cos\left(\frac{R_j}{R}\right) \cos\left(\frac{\tilde{d}(P_i, Q_{ij})}{R}\right) - \cos\left(\frac{R_i}{R}\right) \cos\left(\frac{\tilde{d}(P_j, Q_{ji})}{R}\right)$;
      define $f(R_1, ..., R_n) := \sum_{i,j} \left(r(R_i, R_j)\right)^2$;
   end for;
4. find the set $\mathcal{R}$ which minimizes $f(R_1, ..., R_n)$ with $0 \le R_i < (R\pi)/2$;
   $\mathcal{R}_{\text{opt}} \leftarrow \mathcal{R}$;
5. construct V_L by Algorithm 1 in [26];
6. project V_L to V;
7. intersect T and V cell by cell;
8. compute D_T, D_V, A_T, A_V;
9. compute $\Delta_{T,V}$.

end Procedure

We now analyze the complexity of Algorithm 1. Steps 1, 2, 3, and 6 to 8 require $O(n)$ computation, whereas Step 9 requires $O(1)$ computations. For the computation in Step 5, Algorithms 1 and 2 in [25] have complexity $O(n \log n)$, which is worst-case optimal. The complexity of Algorithm 1 depends on which solution method is used for the optimization of Step 4 employed in (1). However, the complexity does not affect the computation practically due to the smallness of the number of spike dots n of the tessellation T.

4   Experimental Results

The algorithm was applied both to ideal data, namely, an artificially generated spike-containing tessellation T, and to real data obtained from photos. We use Wolfram Wolfram Mathematica®10.0 for the algorithm implementation, and we employed the Nelder-Mead method for optimizing the objective function (1) which is provided by Wolfram Mathematica®10.0.

4.1   Experiments with Ideal Data

In the case of ideal data, we firstly constructed a spherical Laguerre Voronoi diagram from the generated data of the fixed values R, h and spike dot set $\mathcal{B}$

with a randomly generated radii set $\mathcal{R}$. We then projected the spherical Laguerre Voronoi diagram onto the XZ-plane. This tessellation is denoted by T.

In the experiment, we used a tessellation that consists of 34 spike dots, while 17 cells were used for comparing two tessellations. R and h were fixed as $R = 3.989$, $h = 0.0384$. For our convenience, the set $\mathcal{R}$ was generated from real numbers randomized in the interval $[0, 0.4]$.

From the data set, the resulting discrepancy was 0.0171, which is small but not exactly equal to zero, and the set $\mathcal{R}_{Opt}$ differed from the original $\mathcal{R}$. This situation can be understood by the following reason.

Let V be a spherical Laguerre Voronoi diagram. From [25], there exists a polyhedron P corresponding to the spherical Laguerre Voronoi diagram V. Using a transformation in the projective space of $\mathbb{R}^3$, the polyhedron is transformed in such a way that the central projection of P onto the sphere coincides with V. Although the plane alignment is fixed due to the position of spike dots, the planes can be shifted due to the adjustment of spherical circle radii. Thus, the set $\mathcal{R}_{Opt}$ is not necessary to converge to the correct answer. In addition, the case in which discrepancy is not exactly equal to zero can occur because the constraints of spherical circle radii reduce the freedom of polyhedron adjustment. In this case, the intepretation of imaginary spherical circle radii is necessary, and it will be considered in the approaching study.

4.2 Experiments with Real Data

In the case of real data, we performed our experiments by applying our framework to the tessellations in [7]. Note that for R and h of the sphere of each tessellation, the values estimated in [7] were used.

For each tessellation, we performed experiments for three pairs of R and h values obtained in [7] using their initial values. Next, we applied our method to those tessellations. The experimental results for lychee and jackfruit are shown in Tables 1 and 2, respectively. Specifically, they show the average discrepancies

Table 1. Average discrepancies from fitting tessellations to the ordinary spherical Voronoi diagrams (OSVD) in [7] and to spherical Laguerre Voronoi diagrams (SLVD) for Lychee.

Tessellation	Fitting with OSVD		Fitting with SLVD	
	Mean	S.D.	Mean	S.D.
1	0.187	0.0×10^{-17}	0.0790	7.08×10^{-6}
2	0.311	0.0×10^{-17}	0.0703	1.73×10^{-7}
3	0.146	0.0×10^{-17}	0.0586	2.52×10^{-7}
4	0.182	7.37×10^{-6}	0.110	7.31×10^{-5}
5	0.0859	1.70×10^{-17}	0.0721	4.58×10^{-7}
6	0.200	0.0×10^{-17}	0.0660	2.30×10^{-5}
7	0.199	0.0×10^{-17}	0.101	3.21×10^{-6}

Table 2. Average discrepancies from fitting tessellations to the ordinary spherical Voronoi diagrams (OSVD) in [7] and to spherical Laguerre Voronoi diagrams (SLVD) for jackfruit.

Tessellation	Fitting with OSVD		Fitting with SLVD	
	Mean	S.D.	Mean	S.D.
1	0.161	0.0×10^{-17}	0.0711	3.21×10^{-7}
2	0.100	1.70×10^{-17}	0.0539	5.77×10^{-8}
3	0.0956	1.70×10^{-17}	0.0810	4.51×10^{-7}
4	0.0617	0.0×10^{-17}	0.0434	0.0×10^{-17}
5	0.0969	0.0×10^{-17}	0.0547	2.08×10^{-7}

Fig. 1. Areas of difference from the tessellation extracted from a lychee photo: (left) Fitted spherical Voronoi diagram; (right) Fitted spherical Laguerre Voronoi diagram. The red and blue areas show the region of the difference of two tessellations. (Color figure online)

 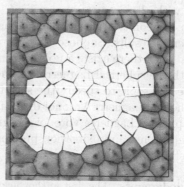

Fig. 2. Areas of difference from the tessellation extracted from a jackfruit photo: (left) Fitted spherical Voronoi diagram; (right) Fitted spherical Laguerre Voronoi diagram. The red and blue areas show the region of the difference of two tessellations. (Color figure online)

and their standard deviations for each tessellation. Examples of tessellations extracted from photos of lychee and jackfruit are shown in Figs. 1 and 2.

From the experimental results, we found that the discrepancy for a lychee tessellation significantly decreased relative to fitting with ordinary spherical Voronoi diagrams. Thus, it is reasonable to say that it is more appropriate to consider lychee skin patterns as spherical Laguerre Voronoi diagrams. Similarly, jackfruit skin patterns, while well fitted to ordinary spherical Voronoi diagrams, are also better fitted with spherical Laguerre Voronoi diagrams. These results support our conjecture that we should consider spike-containing fruit skin patterns as weighted Voronoi diagrams.

5   Concluding Remarks and Future Work

We proposed a framework to fit a spike-containing tessellation extracted from a spike-containing object to a spherical Laguerre Voronoi diagram. The results from the experiments with ideal data and real data verify the validity of our framework.

In the case of the experimental results from real data, The experimental results show that the discrepancies from fitting the given tessellations using spherical Laguerre Voronoi diagrams decrease when we compare to fitting using ordinary spherical Laguerre Voronoi diagram both jackfruit and lychee cases. This implies that it is more appropriate to fit the spike-containing object tessellations using the spherical Laguerre Voronoi diagram. The results confirm that the proposed method is worked well for approximating the real world tessellations using the spherical Laguerre Voronoi diagrams.

In the experiments, we optimized the spherical circle radii when the radius of sphere and spike height were fixed. For finding the better fitted spherical Laguerre Voronoi diagram, those parameters should be optimized, together with adjusting the spherical circle radii again. This case can be the other study when we cope with all parameters.

From a theoretical viewpoint, we made the following observations from the experimental results. Suppose that we are given a spherical Laguerre Voronoi diagram. Then there exists a polyhedron whose projection onto the sphere coincides with the spherical Laguerre Voronoi diagram [24,25]. This leads us to the next research topic to specify the set of all polyhedra which correspond to the same spherical Laguerre Voronoi diagram. Resolving this will be useful not only for the spherical Laguerre Voronoi recognition problem but also for the spherical Laguerre Voronoi approximation problem when the generators are not given in the tessellation.

Acknowledgments. The first author acknowledges the support of the MIMS Ph.D. Program of the Meiji Institute for Advanced Study of Mathematical Sciences, Meiji University, and the DPST of IPST, Ministry of Education, Thailand. This research is partly supported by Grant-in-Aid for Basic Research No. 24360039 of MEXT.

References

1. Aloupis, G., Pérez-Rosés, H., Pineda-Villavicencio, G., Taslakian, P., Trinchet-Almaguer, D.: Fitting voronoi diagrams to planar tesselations. In: Lecroq, T., Mouchard, L. (eds.) IWOCA 2013. LNCS, vol. 8288, pp. 349–361. Springer, Heidelberg (2013). doi:10.1007/978-3-642-45278-9_30
2. Ash, P.F., Bolker, E.D.: Recognizing dirichlet tessellations. Geom. Ded. **19**, 175–206 (1985)
3. Aurenhammer, F.: Power diagram: properties, algorithms, and applications. SIAM J. Comput. **16**, 78–96 (1987)
4. Aurenhammer, F.: A criterion for the affine equivalence of cell complexes in $\mathbb{R}^d$ and convex polyhedra in $\mathbb{R}^{d+1}$. Discrete Comput. Geom. **2**, 49–64 (1987)
5. Aurenhammer, F.: Recognising polytopical cell complexes and constructing projection polyhedra. J. Symbolic. Comput. **3**, 249–255 (1987)
6. Aurenhammer, F., Klein, R., Lee, D.T.: Voronoi Diagrams and Delaunay Triangulations. World Scientific Publishing Company, Singapore (2013)
7. Chaidee, S., Sugihara, K.: Approximation of fruit skin patterns using spherical voronoi diagram. Pattern Anal. Appl. (2016). DOI:10.1007/s10044-016-0534-2
8. Chaidee, S., Sugihara, K.: Numerical fitting of planar photographic images with spherical voronoi diagram. In: 10th Asian Forum on Graphic Science (2015). DOI:10.13140/RG.2.1.1398.1924
9. Duan, Q., Kroese, D.P., Brereton, T., Spettl, A., Schmidt, V.: Inverting laguerre tessellations. Comput. J. **57**, 1431–1440 (2014)
10. Evans, D.G., Jones, S.M.: Detecting voronoi (area-of-influence) polygons. Math. Geol. **19**, 523–537 (1987)
11. Hartvigsen, D.: Recognizing voronoi diagrams with linear programming. ORSA. J. Comput. **4**, 369–374 (1992)
12. Honda, H.: Description of cellular patterns by dirichlet domains: the two-dimensional case. J. Theor. Biol. **72**, 523–543 (1978)
13. Honda, H.: Geometrical models for cells in tissues. Int. Rev. Cytol. **81**, 191–246 (1983)
14. Imai, H., Iri, M., Murota, K.: Voronoi diagram in the laguerre geometry and its applications. SIAM J. Comput. **14**, 93–105 (1985)
15. Lautensack, C.: Random Laguerre Tessellations. Dissertation (2007)
16. Lautensack, C.: Fitting three-dimensional laguerre tessellations to foam structures. J. Appl. Stat. **35**, 985–995 (2008)
17. Liebscher, A.: Laguerre approximztion of random foams. Philos. Mag. **95**, 2777–2792 (2015)
18. Loeb, L.: Space Structures: Their Harmony and Counterpoint. Addison Wesley, Reading (1976)
19. Lyckegaard, A., Lauridsen, E.M., Ludwig, W., Fonda, R.W., Poulsen, H.F.: On the use of laguerre tessellations for representations of 3D grain structures. Adv. Eng. Mater. **13**, 165–170 (2011)
20. Mach, P., Koehl, P.: An analytical method for computing atomic contact areas in biomolecules. J. Comput. Chem. **34**, 105–120 (2013)
21. Okabe, A., Boots, B., Sugihara, K., Chiu, S.N.: Spatial Tessellations: Concepts and Applications of Voronoi Diagrams, 2nd edn. Wiley, Chichester (2000)
22. Schoenberg, F.P., Ferguson, T., Lu, C.: Inverting dirichlet tessellations. Comput. J. **46**, 76–83 (2003)

23. Spettl, A., Breregon, T., Duan, Q., Werz, T., Krill Ill, C.E., Kroese, D.P., Schmidt, V.: Fitting laguerre tessellation approximations to tomographic image data. Philos. Mag. **96**, 166–189 (2016). doi:10.1080/14786435.2015.1125540
24. Sugihara, K.: Three-dimensional convex hull as a fruitful source of diagrams. Theor. Comput. Sci. **235**, 325–337 (2000)
25. Sugihara, K.: Laguerre voronoi diagram on the sphere. J. Geom. Graph. **6**, 69–81 (2002)
26. Suzuki, A., Iri, M.: Approximation of a tessellation of the plane by a voronoi diagram. J. Oper. Res. Soc. Jpn. **29**, 69–97 (1986)

Continuous Flattening of Orthogonal Polyhedra

Erik D. Demaine[1], Martin L. Demaine[1], Jin-ichi Itoh[2], and Chie Nara[3(✉)]

[1] MIT Computer Science and Artificial Intelligence Laboratory, 32 Vassar St., Cambridge, MA 02139, USA
{edemaine,mdemaine}@mit.edu
[2] Faculty of Education, Kumamoto University, Kumamoto 860-8555, Japan
j-itoh@kumamoto-u.ac.jp
[3] Meiji Institute for Advanced Study of Mathematical Sciences, Meiji University, Nakano, Tokyo 164-8525, Japan
cnara@jeans.ocn.ne.jp

Abstract. Can we flatten the surface of any 3-dimensional polyhedron P without cutting or stretching? Such continuous flat folding motions are known when P is convex, but the question remains open for nonconvex polyhedra. In this paper, we give a continuous flat folding motion when the polyhedron P is an orthogonal polyhedron, i.e., when every face is orthogonal to a coordinate axis (x, y, or z). More generally, we demonstrate a continuous flat folding motion for any polyhedron whose faces are orthogonal to the z axis or the xy plane.

Keywords: Folding · Continuous flattening · Orthogonal polyhedra

1 Introduction

We routinely crush polyhedral boxes to lie flat, but is this possible mathematically? It is known that every polyhedron has a multilayered flat folded state, meaning that it can be *instantaneously* folded to lie in a (multilayer) plane [2,5]. But is there a continuous motion that does not stretch or rip the material?

In 2001, E. Demaine, M. Demaine, and A. Lubiw [1,4,5] asked whether there is a continuous motion of the surface of a polyhedron down to a multilayered flat folded state. For example, J.-i. Itoh and C. Nara [6] showed that the box in Fig. 1(a) continuously folds flat by pushing four side faces in, where the shapes of those four faces are changed continuously by infinitely many creases showed by dashed line segments in Fig. 1(b) and the box reaches the multilayered flat folded state in Fig. 1(c).

An important limitation to continuous flattening is the Bellows Theorem [3]: the volume of any polyhedron with rigid faces is invariant even if it can flex at

E.D. Demaine and M.L. Demaine—Supported in part by NSF ODISSEI grant EFRI-1240383 and NSF Expedition grant CCF-1138967.
J. Itoh—Supported by Grant-in-Aid for Scientific Research(B)(15KT0020) and Scientific Research(C)(26400072).
C. Nara—Supported by Grant-in-Aid for Scientific Research(C)(16K05258).

© Springer International Publishing AG 2016
J. Akiyama et al. (Eds.): JCDCGG 2015, LNCS 9943, pp. 85–93, 2016.
DOI: 10.1007/978-3-319-48532-4_8

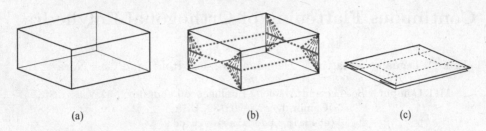

(a)   (b)   (c)

Fig. 1. (a) A box. (b) Mountain and valley creases are shown by bold segments and bold dotted segments respectively for the final flat folded state, together with dashed segments for moving creases. (c) The final flat folded state.

finitely many additional edges. Flattening a polyhedron necessarily changes the volume (from nonzero to zero), so some faces cannot be rigid, e.g., by changing their shapes continuously by infinitely moving/rolling creases.

Continuous flattenings are known for all convex polyhedra. J.-i. Itoh, C. Nara, and C. Vîlcu [7] gave a method using the cut locus and Alexandrov gluing theorem. The authors et al. [1] showed a surprisingly simple method using the straight skeleton gluing. But it remains an open problem to find continuous flattening motions for nonconvex polyhedra.

Our Results. Our main result is the continuous flattening of orthogonal polyhedra (not necessary convex or genus zero); see Fig. 2(a). A polyhedron is called *orthogonal* if the dihedral angle of each edge is ±90° (see [5]). By an appropriate choice of x, y, z axes for Euclidean space, we can equivalently define a polyhedron to be orthogonal if every face is orthogonal to the x, y, or z axis.

Theorem 1. *Every orthogonal polyhedron in $\mathbb{R}^3$ can be continuously folded flat so that all faces orthogonal to the z axis remain rigid and translated along the z axis throughout the motion.*

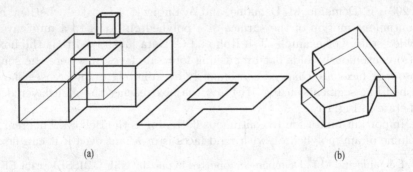

(a)   (b)

Fig. 2. (a) An example $\mathcal{P}$ of an orthogonal polyhedron with the figure of the base floor of $\mathcal{P}$. (b) An example of a "semi-orthogonal" polyhedron.

More generally, call a polyhedron *semi-orthogonal* if every face is orthogonal to the z axis or the xy plane; see Fig. 2(b). Any orthogonal polyhedron also satisfies this definition because being orthogonal to the x or y axis implies being orthogonal to the xy plane. We prove more generally that semi-orthogonal polyhedra have flat folding motions:

Theorem 2. *Every semi-orthogonal polyhedron in $\mathbb{R}^3$ can be continuously flat folded such that all faces orthogonal to the z axis remain rigid and translated along the z axis throughout the motion.*

2   Zig-Zag Belts and the Rhombus Property

Before we prove the two theorems, we need some tools for constructing continuous folding motions. We denote by uv the line segment joining points u and v.

Rhombus Property. A rhombus $R = abcd$ is a convex quadrilateral with sides of equal length. Rhombi have a very special property useful for flattening a polyhedron, as proposed in [6] and extended to the kite property in [8]. Denote the center of R by h and choose any point q on bh. Consider folding $\triangle acd$ into halves by a valley crease on hd, and folding $\triangle abc$ by a mountain crease on hq and valley creases on aq, cq and qb; see Fig. 3(a). The resulting figure is flexible; see Fig. 3(b, c). Furthermore, if we choose the distances of the two pairs $\{a, c\}$ and $\{b, d\}$ in the resulting 3D figure, such that these distances are not greater than the 2D lengths ac and cd respectively, then there exists a unique point q on hd and unique folding such that the resulting figure satisfies those distances. We call this property the *rhombus property*; see [6,8] for details.

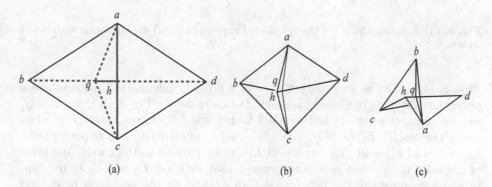

(a)   (b)   (c)

Fig. 3. The rhombus property: (a) a rhombus with mountain creases (grey bold line segments) and valley creases (grey bold dotted line segments); (b) the resulting figure; (c) a view of the resulting figure from a different direction.

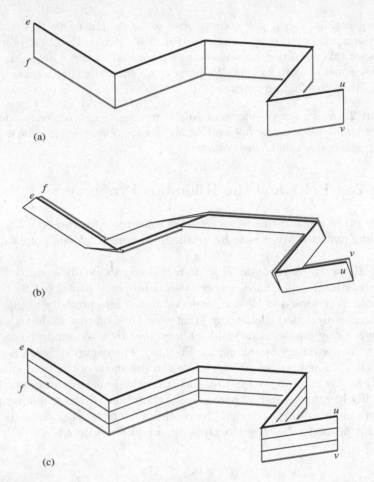

Fig. 4. (a) A zig-zag belt; (b) the flat folded zig-zag belt; (c) zig-zag belts with small widths.

Zig-zag Belts. Define a *zig-zag belt* B to be a finite orthogonal extrusion in z of a polygonal line lying in a plane parallel to the xy plane; see Fig. 4(a). Equivalently, we can think of a zig-zag belt as a 3D folded state B of a rectangle $T = efvu$ along creases $E_1, E_2, \ldots, E_{n-1}$ parallel (and congruent) to the opposite sides $E_0 = ef$ and $E_n = uv$. We require all E_i's to be parallel to the z axis, and none of the sides to intersect each other, except that we allow E_0 and E_n to overlap, in which case we call the belt *closed*. Call ef and uv the *end sides* of B, and call the line segments in B corresponding to eu and fv the *zig-zag sides* of B. (Either zig-zag side could be the polygonal line that was extruded to form B.) Define the *width* of the belt to be the common length of the E_i's.

We show how to continuously flatten zig-zag belts into a flat folded state where the zig-zag sides overlap each other; see Fig. 4(b).

Lemma 1. *Consider a zig-zag belt B, and if B is closed, assume that the number of faces is even. Then B can be continuously flattened so that the two zig-zag sides remain rigid and translate only in z.*

Proof. We will show Lemma 1 assuming that B's width is sufficiently small (less than a quantity to be defined in terms of the geometry of either zig-zag side). Then Lemma 1 follows, by slicing B along many uniformly spaced planes parallel to the xy plane, dividing B into congruent zig-zag belts that are z translations of each other and having arbitrarily small width; see Fig. 4(c). Continuously flattening each of these belts in sequence (or in parallel) proves Lemma 1.

Now assume that the belt's width is sufficiently small. We will show Lemma 1 for $n = 2$. This lets us analyze the local folding behavior of the two faces incident to each edge E_i. By synchronizing all of these parallel folding motions to match on the z offsets of the zig-zag sides, the motions will also be consistent on each face, and we obtain a continuous folding of the entire band B. By setting the width sufficiently small (smaller than half the minimum feature size of either zig-zag side), any self-intersections during the band folding must occur locally between one or two adjacent faces, and thus is prevented by the $n = 2$ case.

Now assume that $n = 2$, with $E_0 = ef$, $E_1 = ac$, and $E_2 = E_n = uv$; see Fig. 5(a). Let b and d be points on B so that the quadrilateral $abcd$ is a rhombus with angle $\angle abc = 180° - \angle eau$; see Fig. 5(b). Fold B by mountain creases on

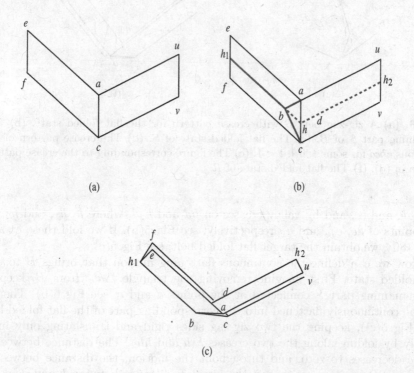

(a)   (b)

(c)

Fig. 5. (a) A zig-zag belt B; (b) B with a crease pattern; (c) the flat folded state of B.

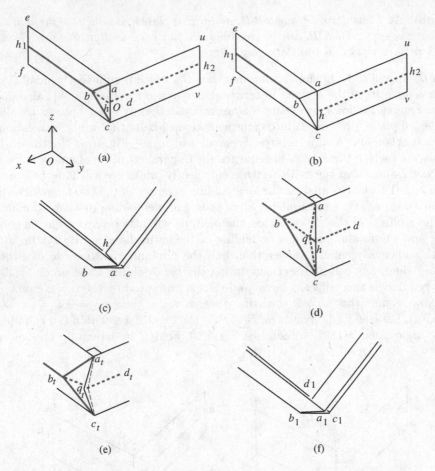

Fig. 6. (a) A zig-zag belt B with crease pattern for the flat folded state. (b) The remaining part S of B. (c) The flat folded state of S. (d) The crease pattern of the rhombus $abcd$ for some t, $0 < t < 1$. (e) The figure corresponding to the crease pattern shown in (d). (f) The flat folded state of R.

h_1b, ab, and bc, and by valley creases on hd and h_2d, where h, h_1, and h_2 are midpoints of ac, ef, and uv respectively; see Fig. 5(b). If we fold these creases by $\pm 180°$, we obtain the target flat folded belt; see Fig. 5(c).

Now we can define the continuous flattening motion that brings B to this flat folded state. First imagine removing the triangle $\triangle abc$ from B, keeping the remaining part S connected at the points a and c; see Fig. 6(b). Then S can be continuously flattened into the corresponding part of the flat folded belt (see Fig. 6(c)), keeping the two zig-zag sides rigid and translating only in z, simply by folding along the two creases h_1b and hh_2. The distance between a and c decreases to zero, and throughout the motion, the distance between b and d in 3D is not greater than the intrinsic distance between b and d in the

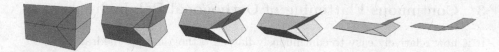

Fig. 7. Continuous flattening animation of the orthogonal corner from Lemma 1, produced with Mathematica.

original 2D figure. Finally, apply the rhombus property to fold the rhombus $abcd$ continuously, synchronizing the motion to match the motion of S.

To be more precise, we give a concrete continuous map for the case when $\angle eau = 90°$ and the edge length of ac is 2; see Figs. 6(d, e, f) and 7. Orient so that, before folding, $a = (0, 0, 1)$, $b = (0, -1, 0)$, $c = (0, 0, -1)$, and $d = (-1, 0, 0)$. Move a and c to the origin along the z axis with the same speed. The line segment hd remains in the xy plane and translates in the $-y$ direction, and the point b_t remains in the xy plane and translates in the $+x$ direction. Precisely, for t with $0 \le t \le 1$, we have

$$a_t = (0, 0, 1 - t), \qquad b_t = (\sqrt{2t - t^2}, -1, 0) = (\alpha, -1, 0),$$
$$c_t = (0, 0, -1 + t), \quad \text{and} \quad d_t = (-1, -\sqrt{2t - t^2}, 0) = (-1, -\alpha, 0).$$

where $\alpha = \sqrt{2t - t^2}$. Let p_t be the midpoint of b_t and d_t. Then $p_t = ((\alpha - 1)/2, (-1 - \alpha)/2, 0)$. The point q_t corresponding to q in Fig. 3(b, c) is the intersection of the plane bisecting $b_t d_t$ and the line segment $h_t d_t$. Because b_t and h_t are on the xy plane, for the sake of simplicity, we omit z coordinates. The line passing through p_t and bisecting $b_t d_t$ is

$$y - (-1 - \alpha)/2 = -\frac{\alpha + 1}{-1 + \alpha}(x - (\alpha - 1)/2).$$

The intersection between this line and the line $y = -\alpha$ is the point q_t. The x coordinate of q_t, denoted x_t, solves to

$$x_t = \frac{\alpha(\alpha - 1)}{\alpha + 1}.$$

As t increases from 0 to 1, α increases from 0 to 1, and hence the absolute value of x_t increases first and then decreases to 0. The maximum absolute value $|x|_{max}$ is attained at $\alpha = \sqrt{2} - 1$, where $|x|_{max} = 3 - 2\sqrt{2}$. Therefore, the area of moving creases is $3 - 2\sqrt{2}$, where the width of the belt is 2.

When $\angle eau \ne 90°$, by using oblique coordinates, we can calculate similarly, however it is a little tedious, so we omit the details.

This motion requires the two edges of a zig-zag side of B to be folded in opposite directions relative to the xy projections of the middle lines. Thus we require a global alternation of fold directions along the zig-zag. If B is closed, the number of faces is even by assumption, and hence we can continuously flatten B as required.   □

3 Continuous Flattening of Orthogonal Polyhedra

It is now relatively easy to continuously flatten orthogonal polyhedra:

Proof (of Theorem 1). Let $\mathcal{P}$ be an orthogonal polyhedron in $\mathbb{R}^3$. Conceptually remove all faces orthogonal to the z axis, and divide the resulting set of faces by planes orthogonal to the z axis that pass through each vertex of $\mathcal{P}$. The result is a collection of zig-zag belts B_i, for $1 \leq i \leq n$, where each belt is closed. Because $\mathcal{P}$ is an orthogonal polyhedron, each belt B_i has an even number of faces. By Lemma 1, each belt B_i can be continuously flattened so that its zig-zag sides remain rigid and translate only in the z direction. Composing these motions sequentially or in parallel, and re-attaching the faces orthogonal to the z axis to the zig-zag sides, we obtain a continuous flattening of $\mathcal{P}$ where the faces orthogonal to the z axis remain rigid and translate only in z.   □

4 Continuous Flattening of Semi-orthogonal Polyhedra

For semi-orthogonal polyhedra, we need to show how to continuously flatten a closed zig-zag belt with an odd number of faces.

Lemma 2. *A zig-zag belt B with two faces and sufficiently small width can be continuously flattened so that the two zig-zag sides remain rigid and translate only along z, and moreover, the zig-zag sides are folded to the same direction of the xy projection of the middle line segments.*

Proof. We use the same notations for points e, f, u, v, a, b, c, d, and h as the proof of Lemma 1, where $\angle abc = \angle adc = 180° - \angle eau$ and the quadrilateral $abcd$ is a rhombus. We move the points b and d toward the convex side of the angle formed by the zig-zag sides. Fold B with mountain creases on bh and valley creases on ab, bc, and hd. Then we obtain a flat folded state that satisfies all requirements; see Fig. 8(a, b).

Consider folding each face of B into halves with valley creases on the middle line segments. Then these faces will intersect each other. The intersection point of the two middle line segments is q_t; see Fig. 8(c, d). Thus the intersection gets resolved by the rhombus $abcd$. In this case, both the distance between a and b, and the distance between b and d, decrease to zero.   □

Proof (of Theorem 2). If the number of faces of B is odd, fold one corner by the method proposed in Lemma 2, and fold the other corners by the method proposed in Lemma 1. Then we obtain a continuous motion that satisfies all required conditions.   □

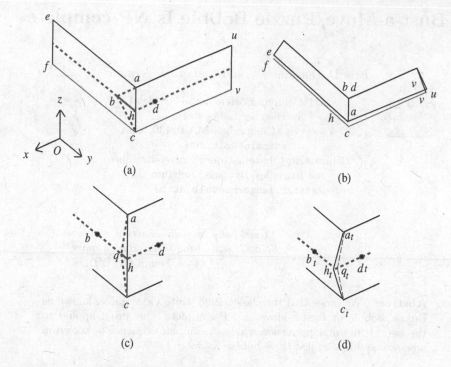

Fig. 8. (a) A zig-zag belt B with crease pattern for a flat folded state. (b) The resulting flat folded state of B. (c) The crease pattern of the rhombus $abcd$ for some t, $0 < t < 1$. (d) The resulting 3D figure.

References

1. Abel, Z., Demaine, E.D., Demaine, M.L., Itoh, J.-I., Lubiw, A., Nara, C., O'Rourke, J.: Continuously flattening polyhedra using straight skeletons. In: Proceedings of the 30th Annual Symposium on Computational Geometry (SoCG), pp. 396–405 (2014)
2. Bern, M., Hayes, B.: Origami embedding of piecewise-linear two-manifolds. Algorithmica **59**(1), 3–15 (2011)
3. Connelly, R., Sabitov, I., Walz, A.: The bellows conjecture. Beiträge Algebra Geom. **38**, 1–10 (1997)
4. Demaine, E.D., Demaine, M.L., Lubiw, A.: Flattening polyhedra (2001). Unpublished manuscript
5. Demaine, E.D., O'Rourke, J.: Geometric Folding Algorithms: Linkages, Origami, Polyhedra. Cambridge University Press, Cambridge (2007)
6. Itoh, J., Nara, C.: Continuous flattening of platonic polyhedra. In: Akiyama, J., Bo, J., Kano, M., Tan, X. (eds.) CGGA 2010. LNCS, vol. 7033, pp. 108–121. Springer, Heidelberg (2011). doi:10.1007/978-3-642-24983-9_11
7. Itoh, J., Nara, C., Vîlcu, C.: Continuous flattening of convex polyhedra. In: Márquez, A., Ramos, P., Urrutia, J. (eds.) EGC 2011. LNCS, vol. 7579, pp. 85–97. Springer, Heidelberg (2012). doi:10.1007/978-3-642-34191-5_8
8. Nara, C.: Continuous flattening of some pyramids. Elem. Math. **69**(2), 45–56 (2014)

Bust-a-Move/Puzzle Bobble Is NP-complete

Erik D. Demaine[1](✉) and Stefan Langerman[2]

[1] MIT Computer Science and Artificial
Intelligence Laboratory,
32 Vassar St., Cambridge, MA 02139, USA
edemaine@mit.edu
[2] Départment d'Informatique, Université Libre
de Bruxelles, Brussels, Belgium
stefan.langerman@ulb.ac.be

> *"A girl runs up with somethin' to prove.*
> *So don't just stand there. Bust a move!"*
> — *Young MC* [YDD89]

Abstract. We prove that the classic 1994 Taito video game, known as Puzzle Bobble or Bust-a-Move, is NP-complete. Our proof applies to the perfect-information version where the bubble sequence is known in advance, and it uses just three bubble colors.

1 Introduction

Erik grew up playing the action platform video game *Bubble Bobble* (バブルボブル), starring cute little brontosauruses Bub and Bob,[1] on the Nintendo Entertainment System. (The game was first released by Taito in 1986, in arcades [Thea].) Some years later (1994), Bub and Bob retook the video-game stage with the puzzle game *Puzzle Bobble* (パズルボブル), known as *Bust-a-Move* in the United States [Theb,Wik]. This game essentially got Stefan through his Ph.D.: whenever he needed a break, he would play as much as he could with one quarter. To celebrate the game's 21-year anniversary, we analyze its computational complexity, retroactively justifying the hours we spent playing. The gadgets and example reduction described here can be played in an accurate clone of the game we wrote for the web.[2]

In Puzzle Bobble, the game state is defined by a hexagonal grid, each cell possibly filled with a *bubble* of some color. In each turn, the player is given a bubble of some color, which can be fired in any (upward) direction from the pointer at the bottom center of the board. The fired bubble travels straight, reflecting off the left and right walls, until it hits another bubble or the top wall,

---

S. Langerman—Directeur de recherches du F.R.S.–FNRS.

[1] Spoiler: if you finish Bubble Bobble in super mode in co-op, then the true ending reveals that Bub and Bob are in fact human boys, transformed into brontosauruses by the evil whale Baron Von Blubba [Hun11].

[2] http://erikdemaine.org/bustamove/.

© The Author(s) 2016
J. Akiyama et al. (Eds.): JCDCGG 2015, LNCS 9943, pp. 94–104, 2016.
DOI: 10.1007/978-3-319-48532-4_9

in which case it terminates at the nearest grid-aligned position. If the bubble is now in a connected group of at least three bubbles of the same color, then that group disappears ("pops"), and any bubbles now disconnected from the top wall also pop.

Here we study the perfect-information (generalized) form of Puzzle Bobble. We are given an initial board of bubbles and the entire sequence of colored bubbles that will come. The goal is to clear the board using the given sequence of bubbles. (The actual game has an infinite, randomly generated sequence of bubbles, like Tetris [BDH+04].) The game also has a falling ceiling, where all bubbles descend every fixed number of shots; and if a bubble hits the floor, the game ends. We assume that the resolution of the input is sufficiently fine to hit any discrete cell that could be hit by an (infinitely precise) continuous shot. (This assumption seems to hold in the original game, so it is natural to generalize it.)

Theorem 1. *Puzzle Bobble is NP-complete.*

Membership in NP is easy: specify where to shoot each of the n given bubbles. The rest of this paper establishes NP-hardness.

Our reduction applies to all versions of Puzzle Bobble. Viglietta [Vig12] proved that Puzzle Bobble 3 is NP-complete, by exploiting "rainbow" (wild-card) bubbles. Our proof shows that this feature is unnecessary.

2   NP-hardness

The reduction is from Set Cover: given a collection $\mathcal{S} = \{S_1, S_2, \ldots, S_s\}$ of sets where each $S_i \subseteq U$, and given a positive integer k, are there k of the sets $S_{i_1}, S_{i_2}, \ldots, S_{i_k}$ whose union covers all elements of U?

Figure 1 shows the overall structure of the reduction. The bulk of the construction is in the central small square, which is aligned on the top side of an $m \times m$ square above the floor. By making the central square small enough, the angles of direct shots at the square are close to vertical (which we will need to solve most gadgets), and the rebound angles that hit the square are all approximately 45° (which we will need to solve the crossover gadget below), even after the ceiling falling caused by the shots in the reduction. The player could do multiple rebounds (or destroy bubbles to cause the ceiling to lower prematurely) to make shot angles more horizontal, but this will only make it harder to solve the gadgets.

2.1   Bubble Sequence

The sequence of bubbles given to the player is as follows. The very first color appears only k times, where k is the desired set-cover size. Each remaining color appears sufficiently many times ($\Theta(s|U|)$ times, which we will refer to as ∞). Unneeded bubbles can be discarded by forming isolated groups of size 3, 4, or 5 off to the side.

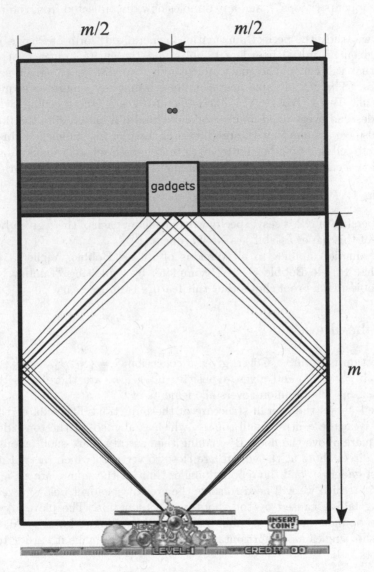

Fig. 1. Overall structure of the reduction. All other gadgets lie within a small square at the top of an $m \times m$ square, where m is the width of the game. Red horizontal lines separate the gadgets into layers, with blue fill in between. At the top is a huge rectangle of yellow bubbles with one red bubble and one blue bubble in the middle (Color figure online).

k blue ●, ∞ yellow ○, ∞ blue ●, ∞ red ●;
∞ blue ●, ∞ yellow ○, ∞ blue ●, ∞ red ●;
∞ blue ●, ∞ yellow ○, ∞ blue ●, ∞ red ●;

$\vdots$ $\vdots$ $\vdots$ $\vdots$

∞ blue ●, ∞ yellow ○, ∞ blue ●, ∞ red ●;
∞ red ●, ∞ red ●,   ∞ red ●, ...

The rough idea is the following. Red bubbles separate vertical layers that unravel sequentially, as enforced by blue buffers. Blue and yellow bubbles form triggers to communicate signals into the next layers, alternately. Blue triggers cup yellow triggers in the next level, and vice versa.

2.2 Gadgets

First we have one instance of the choice gadget, shown in Fig. 2, which allows triggering k sets (whichever the player chooses).

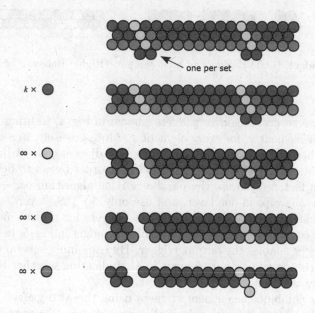

Fig. 2. Choice gadget, shown here with $s = 2$ sets. (Left) Behavior of a chosen set. (Right) Behavior of an unchosen set.

Then we use several split gadgets, shown in Fig. 3, to split each trigger for set S_i into $|S_i|$ triggers.

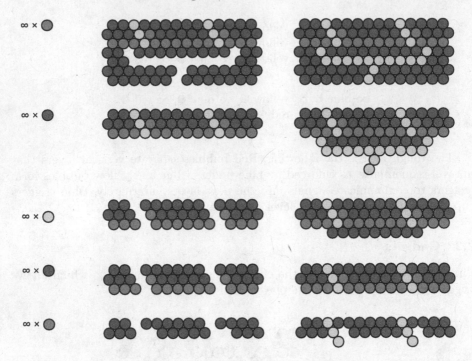

Fig. 3. Split gadget. (Left) Behavior of a chosen set. (Right) Behavior of an unchosen set.

Then we use several crossover gadgets, shown in Fig. 4, to bring together all the triggers for element x, for every element x. More precisely, Fig. 4 shows how to copy all other wire values while swapping an adjacent pair. Adjacent swaps suffice to bubble sort $S_1, S_2, \ldots, S_s$ from being in order by set to being in order by element. In fact, we can use the parallel sorting algorithm *odd–even sort* by executing several swaps in one layer, and use only $2\sum_i |S_i|$ layers.

Next, for each element, we merge all the triggers for that element (coming from sets that contain the element), using the OR gadget in Fig. 5. In this gadget, any input trigger enables the output trigger. By combining several OR gadgets, we end up with one trigger per element in U, indicating whether that element was covered by the k chosen sets.

Finally, we combine the element triggers using the AND gadget in Fig. 6. In this gadget, the output triggers only if all inputs trigger. By combining several AND gadgets, we end up with one trigger indicating that all elements are covered, i.e., we found a set cover of size k.

This trigger is connected to a huge ($n^{1-\varepsilon}$-area) rectangle of yellow bubbles at the top of the board, with one red bubble in the middle, as shown in Fig. 1. If the yellow triggers, the player wins the game immediately (as the red falls).

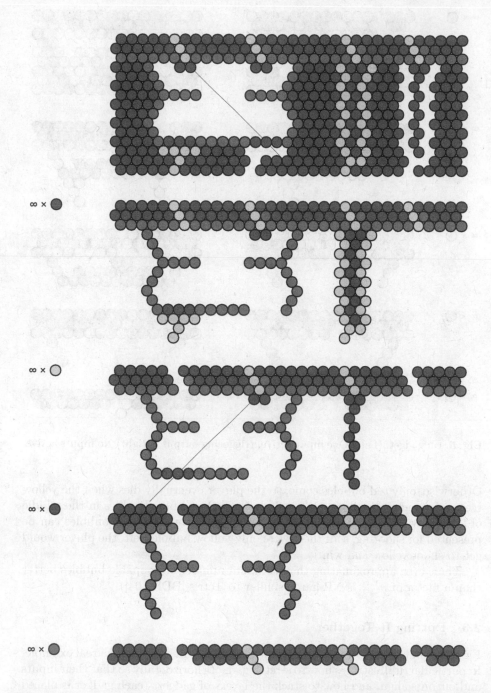

Fig. 4. Crossover gadget, shown here with left side inactive and right side active. On the right are other wires whose values are simply copied.

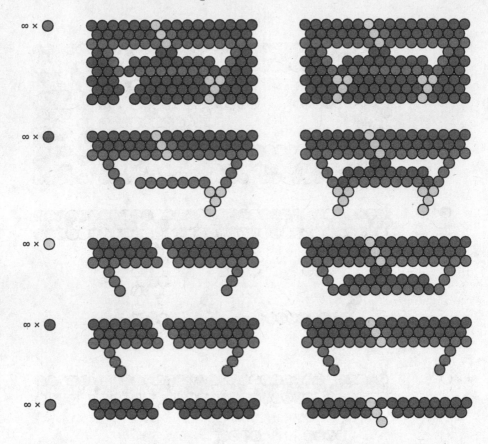

Fig. 5. OR gadget. (Left) One input active, triggering output. (Right) No inputs active.

Otherwise, only red bubbles come, so the player eventually dies when the yellow rectangle reaches the floor. (We include the red and blue bubbles in the middle of the yellow bubbles because, in the actual game, only present bubbles can be presented for shooting, so if there were only yellow bubbles left, the player would get to shoot yellow and win.)

Thus, even approximating the maximum number of poppable bubbles better than a factor of $n^{1-\varepsilon}$ is NP-hard (similar to Tetris [BDH+04]).

2.3 Putting It Together

Figures 7 and 8 show an example of how the gadgets fit together in a real example. In particular, it illustrates how to stretch gadgets horizontally so that their inputs and outputs align, and how to stack the layers of gadgets (each gadget is placed on the row immediately after the previous).

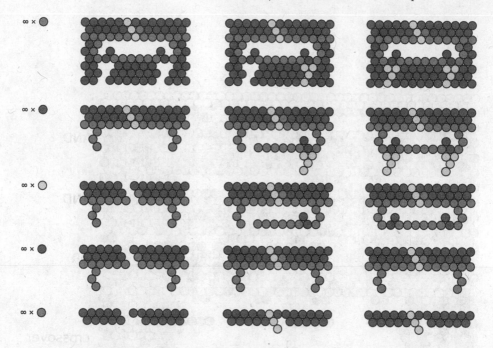

Fig. 6. AND gadget. (Left) Two inputs active, triggering output. (Middle) One input active. (Right) No inputs active.

The bijection between solutions of the Puzzle Bobble instance and the Set Cover instance come from which triggers get popped by the first k blue shots on the Choice gadget. (Fewer than k triggers could be popped, corresponding to smaller-than-k set covers.) The correctness follows from the claimed properties of the gadgets, which can be verified from the figures implementing a greedy algorithm of popping all possible bubbles of each provided color (which can only help for these instances).

A key lemma for correctness is that, during the ith blue–yellow–blue–red phase of the bubble sequence, only bubbles in the ith layer of the construction can be directly popped, with spillover into the next layer only from triggered yellow bubble wires. The ith red layer prevents any nonred bubbles from physically reaching the next layer in the ith phase, because the gaps between red bubbles are designed to be strictly less than one bubble width. (Precisely, the gap width is $\sqrt{3} - 1 \approx 0.73$.) At the end of the phase when firing red bubbles, the blue in the next layer uses the same < 1 gaps (when the yellow has been triggered) to prevent any red bubbles from reaching the next layer. So the lemma follows.

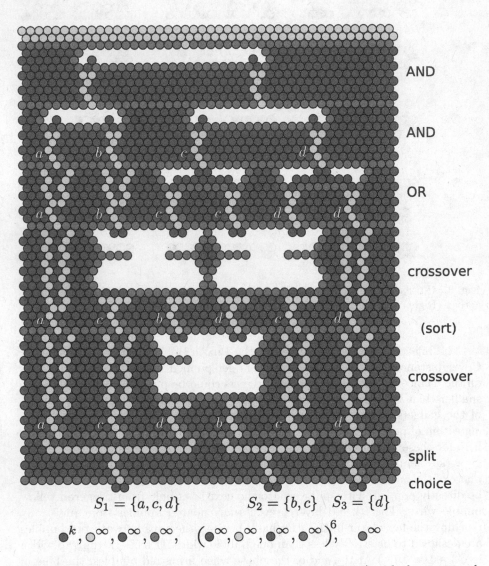

Fig. 7. Example of the main construction (the gray box in Fig. 1) with three sets and four elements. The bubble sequence at the bottom can solve the puzzle for $k = 2$ and $k = 3$, but not for $k = 1$.

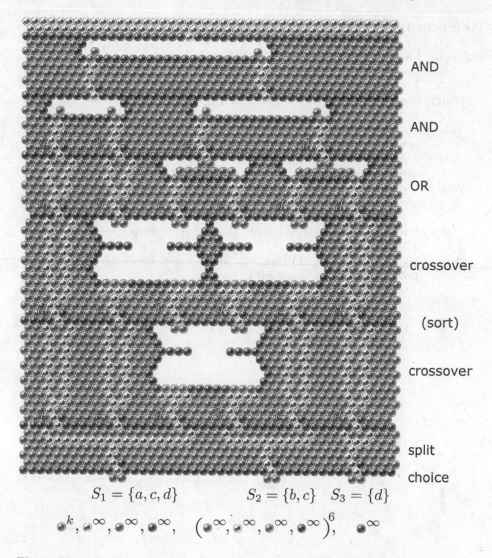

$$S_1 = \{a, c, d\} \qquad S_2 = \{b, c\} \quad S_3 = \{d\}$$

Fig. 8. Figure 7 using actual Puzzle Bobble sprites, thanks to The Spriters Resource.

3   Open Problems

We have proved NP-hardness for just three colors. What about just two colors? Or even one color?

Acknowledgments. We thank Giovanni Viglietta for helpful discussions, in particular for pointing out bugs in earlier versions of this proof.

References

[BDH+04] Breukelaar, R., Demaine, E.D., Hohenberger, S., Jan Hoogeboom, H., Kosters, W.A., Liben-Nowell, D.: Tetris is hard, even to approximate. Int. J. Comput. Geom. Appl. **14**(1–2), 41–68 (2004)

[Hun11] Hunt, S.: Bubble memories: 25 years of bubble bobble. Retro Gamer **95**, 26–35 (2011)

[Thea] The international arcade museum: bubble bobble. http://www.arcade-museum.com/game_detail.php?game_id=7222

[Theb] The international arcade museum: puzzle bobble. http://www.arcade-museum.com/game_detail.php?game_id=9169

[Vig12] Viglietta, G.: Gaming is a hard job, but someone has to do it! In: Proceedings of the 6th International conference on Fun with Algorithms, pp. 357–367 (2012)

[Wik] Wikipedia, the free encyclopedia: puzzle bobble. http://en.wikipedia.org/wiki/Puzzle_Bobble

[YDD89] Young, M.C., Dike, M., Doss, M.: Bust a move. In: Proceedings of Stone Cold Rhymin'. Delicious Vinyl (1989)

Minimum Rectilinear Polygons
for Given Angle Sequences

William S. Evans[1], Krzysztof Fleszar[2], Philipp Kindermann[2,3],
Noushin Saeedi[1], Chan-Su Shin[4], and Alexander Wolff[2(✉)]

[1] Department of Computer Science, University of British Columbia,
Vancouver, Canada
[2] Lehrstuhl für Informatik I, Universität Würzburg, Würzburg, Germany
[3] LG Theoretische Informatik, FernUniversität in Hagen, Hagen, Germany
[4] Division of Computer and Electronic Systems,
Hankuk University of Foreign Studies, Yongin, South Korea
http://www1.informatik.uni-wuerzburg.de/en/staff/wolff_alexander

Abstract. A *rectilinear* polygon is a polygon whose edges are axis-aligned. Walking counterclockwise on the boundary of such a polygon yields a sequence of left turns and right turns. The number of left turns always equals the number of right turns plus 4. It is known that any such sequence can be realized by a rectilinear polygon. In this paper, we consider the problem of finding realizations that minimize the perimeter or the area of the polygon or the area of the bounding box of the polygon. We show that all three problems are NP-hard in general. Then we consider the special cases of x-monotone and xy-monotone rectilinear polygons. For these, we can optimize the three objectives efficiently.

1 Introduction

In this paper, we consider the problem of computing, for a given rectilinear angle sequence, a "small" rectilinear polygon that realizes the sequence. A *rectilinear angle sequence* S is a sequence of left ($+90°$) turns and right ($-90°$) turns, that is, $S = (s_1, \ldots, s_n) \in \{L, R\}^n$, where n is the *length* of S. As we consider only rectilinear angle sequences, we usually drop the term "rectilinear." A polygon P *realizes* an angle sequence S if there is a counterclockwise (*ccw*) walk along the boundary of P such that the turns at the vertices of P, encountered during the walk, form the sequence S. The turn at a vertex v of P is a left or right turn if the interior angle at v is $90°$ (v is convex) or, respectively, $270°$ (v is reflex).

In order to measure the size of a polygon, we only consider polygons that lie on the integer grid. Then, the *area* of a polygon P corresponds to the number of grid cells that lie in the interior of P. The *bounding box* of P is the smallest axis-parallel enclosing rectangle of P. The *perimeter* of P is the sum of the lengths of the edges of P. The task is, for a given angle sequence S, to find a polygon that realizes S and minimizes (i) (the area of) its bounding box, (ii) its area, or (iii) its perimeter. Figure 1 shows that, in general, the three criteria cannot be minimized simultaneously.

J. Akiyama et al. (Eds.): JCDCGG 2015, LNCS 9943, pp. 105–119, 2016.
DOI: 10.1007/978-3-319-48532-4_10

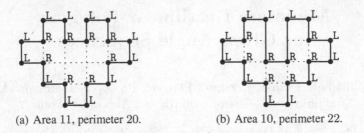

(a) Area 11, perimeter 20.   (b) Area 10, perimeter 22.

Fig. 1. Two polygons realizing the same angle sequence. The bounding box of both polygons has area 20, but (a) has minimum perimeter and (b) has minimum area.

Obviously, the angle sequence of a polygon is unique (up to rotation), but the number of polygons that realize a given angle sequence is unbounded. The formula for the angle sum of a polygon implies that, in any angle sequence, $n = 2r + 4$, where r is the number of right turns, in other words, the number of right turns is exactly four less than the number of left turns.

Related Work. Bae et al. [1] considered, for a given angle sequence S, the polygon $P(S)$ that realizes S and minimizes its area. They studied the following question: Given a number n, find an angle sequence S of length n such that the area of $P(S)$ is minimized (and let $\delta(n)$ be this minimum area), or maximized (and let $\Delta(n)$ be this maximum area). They showed that (i) $\delta(n) = n/2 - 1$ if $n \equiv 4 \bmod 8$, $\delta(n) = n/2$ otherwise, and (ii) $\Delta(n) = (n - 2)(n + 4)/8$ for any $n \geq 4$. The result for $\Delta(n)$ tells us that any angle sequence S of length n can be realized by a polygon with area at most $(n - 2)(n + 4)/8$.

Several authors have explored the problem of realizing a turn sequence. Culberson and Rawlins [4] and Hartley [8] described algorithms that, given a sequence of exterior angles summing to 2π, construct a simple polygon realizing that angle sequence. Culberson and Rawlins' algorithm, when constrained to $\pm 90°$ angles, produces polygons with no colinear edges, implying that any n-vertex polygon can be drawn with area approximately $(n/2 - 1)^2$. However, as Bae et al. [1] showed, the bound is not tight.

In his PhD thesis, Sack [10] introduced label sequences (which are equivalent to turn sequences) and, among others, developed a grammar for label sequences that can be realized as simple rectilinear polygons.

Vijayan and Wigderson [12] considered the problem of efficiently embedding *rectilinear graphs*, of which rectilinear polygons are a special case, using an edge labeling that is equivalent to a turn sequence in the case of paths and cycles.

In graph drawing, the standard approach to drawing a graph of maximum degree 4 orthogonally (that is, with rectilinear edges) is the topology–shape–metrics approach of Tamassia [11]: (1) Compute a planar(ized) embedding; (2) compute an *orthogonal representation*, that is, an angle sequence for each edge and an angle for each vertex; (3) *compact* the graph, that is, draw it inside a bounding box of minimum area. Step (3) has been shown to be NP-complete by Patrignani [9]. Note that an orthogonal representation computed in step (2) is

Table 1. Summary of our results.

Type of sequences	Minimum area	Min. bounding box	Minimum perimeter
General	NP-hard	NP-hard	NP-hard
x-monotone	$O(n^4)$	$O(n^3)$	$O(n^2)$
xy-monotone	$O(n)$	$O(n)$	$O(n)$

essentially an angle sequence for each face of the planarized embedding, so our problem corresponds to step (3) in the special case that the input graph is a simple cycle.

Another related work contains the reconstruction of a simple (non-rectilinear) polygon from partial geometric information. Disser et al. [5] constructed a simple polygon in $O(n^3 \log n)$ time from an ordered sequence of angles measured at the vertices visible from each vertex. The running time was improved to $O(n^2)$, which is the worst-case optimal [3]. Biedl et al. [2] considered polygon reconstruction from points (instead of angles) captured by laser scanning devices.

Our Contribution. First, we show that finding a minimum polygon that realizes a given angle sequence is NP-hard for any of the three measures: bounding box area, polygon area, and polygon perimeter; see Sect. 2. This extends the result of Patrignani [9] and settles an open question that he posed. We also give efficient algorithms for special types of angle sequences, namely xy- and x-*monotone sequences*, which are realized by xy-monotone and x-monotone polygons, respectively. (For example, LLRRLLRLLRLRLLRLRLLR is an x-monotone sequence, see Fig. 1). Our algorithms minimize area (Sect. 3) and perimeter (Sect. 4). For an overview of our results, see Table 1.

2   NP-Hardness of the General Case

In this section we show the NP-hardness of our problem for all three objectives: for minimizing the perimeter of the polygon, the area of the polygon, and the size of the bounding box. We first consider the following special problem from whose NP-hardness we then derive the three desired proofs.

FITUPPERRIGHT: Given an angle sequence S and positive integers W and H, is there a polygon realizing S within an axis-parallel rectangle R of width W and height H such that the first vertex of S lies in the upper right corner of R?

Theorem 1. FITUPPERRIGHT *is NP-hard.*

Proof. Our proof is by reduction from 3-PARTITION: Given a multiset A of $n = 3m$ integers with $\sum_{a \in A} a = mB$, is there a partition of A into m subsets $A_1, \ldots, A_m$ such that $\sum_{a \in A_i} a = B$ for each i? It is known that 3-PARTITION is NP-hard even if B is polynomially bounded in n and, for every $a \in A$, we have $B/4 < a < B/2$, which implies that each of the subsets $A_1, \ldots, A_m$ must contain exactly three elements [7].

For the idea of our reduction, see Fig. 2. For an instance $A = \{a_1, \ldots, a_{3m}\}$ of 3-PARTITION, we construct an LR-sequence S that can be drawn inside an $(W \times H)$-box R if and only if A is a yes-instance. The sequence S consists of a *wall*, and for each number $a_i \in A$, a *snail*, which in turn consists of a *connector* and a *spiral*.

The wall is a box (LLLL) whose top right corner corresponds to the start of S. The connectors are attached to the left side of the wall by introducing two R-vertices. A connector is a thin x-monotone polygon going to the left that can change its y-position $m - 1$ times.

In detail, the LR-sequence S is defined as follows where $\rho = Bm^3$ is the number of windings of the spirals:

$$S = \text{LL } snail_1 snail_2 \ldots snail_{3m} \text{ LL},$$
$$snail_i = \text{R(LRRL)}^{m-1} spiral_i \text{(RLLR)}^{m-1}\text{R},$$
$$spiral_i = \text{(LLLL)}^\rho ladder_i \text{(RRRR)}^{\rho-1}\text{RR},$$
$$ladder_i = \text{(RRLL)}^{(a_i - 1) \cdot m^2}.$$

We choose W and H such that the spirals have to be arranged in m columns of three spirals each. Note that for any order of the numbers in A, we can route the connectors in a planar way such that the triplets of spirals that we desire end up in the same column. Additionally, in each column there must be enough space for the at most $3m$ connectors that go from the wall to spirals further left; see Fig. 2.

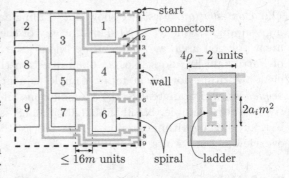

Fig. 2. Spiral i has ρ windings; its height depends on the number a_i from the 3-PARTITION instance.

We set $W = 4\rho m + 16m^2 = \Theta(Bm^4)$ and $H = 12\rho + 2Bm^2 + 6m = \Theta(Bm^3)$. If all spirals are tightly wound, their bounding boxes need total area $(4\rho - 2) \cdot (2Bm^3 + 12\rho m) = \Theta(B^2 m^7)$. The idea of our proof is to show that if a spiral is not tightly wound, we need too much space. The space that is not occupied by spirals is $O(Bm^5)$ in any drawing inside R.

It is clear that our construction is polynomial. By construction, there is a polygon realizing S that fits into R if A is a yes-instance of 3-PARTITION. It remains to show that if S fits into R, then A is a yes-instance of 3-PARTITION.

Fix any feasible drawing of S and a spiral $spiral_i$. Since the first vertex of S has to lie in the upper right corner of R, observe that the 5th L of $spiral_i$ has to lie in the interior of the bounding box of the first four Ls of $spiral_i$. Inductively it follows that, for $5 \leq j \leq 4\rho$, the jth L of $spiral_i$ lies in the interior of the bounding box of the last four Ls of $spiral_i$. Hence, the drawing of $ladder_i$ lies in the bounding box of the last four Ls of the $(\text{LLLL})^\rho$ sequence of $spiral_i$.

By repeating a similar argument for the R vertices, we can observe that every RR edge in $spiral_i$ is lying opposite to a longer LL-edge such that the bounding box spanned by both edges is interiorly empty and completely contained in the polygon. Thus, we can move the RR edge towards the LL edge and assume that the bounding box has width 1. For the last $4\rho - 1$ RR edges in $spiral_i$, we call the bounding box an *arm*.

Hence, any drawing of a spiral consists of a drawing of the ladder and $4\rho - 1$ arms around it. We group the arms into four groups; top, bottom, left, right, depending to which side of the ladder they are lying. Recall that each arm is represented by a pair of LL and RR-edges. We order the arms in each group from the outside to the inside, that is, by the order of their LL edges in S, and define the *level* of an arm as its position in this ordering. We say that *level i is wound tightly* if the distance of all arms of level i to the arms of level $i + 1$ is 1.

Observation 2. *If the first outer i levels are not wound tightly, then the spiral occupies $\Omega(i^2)$ more grid cells than in a tight winding.*

Proof. We consider only the length increase of the top arms. Since the spiral is not wound tightly, the horizontal distance between two consecutive left arms of the first outer i levels is at least two, one more than in a tightly wound spiral. The same is true for the right arms. Hence, the length of the level-i top arm increases at least by 2, that of the level-$(i-1)$ top arm at least by 4, and that of the level-1 top arm at least by $2i$; see Fig. 3. Summing up the increases yields $\Omega(i^2)$.   △

Now, consider any feasible drawing. Recall that the space that is not occupied by spirals is $O(Bm^5)$. Hence, it follows by Observation 2 that at most the first $\lambda := O(\sqrt{B}m^{2.5})$ levels of any spiral are not wound tightly. We simplify the drawing by removing the wall, the connectors and the first λ levels of every spiral. We obtain a set of $3m$ disjoint rectangles, one for each snail. The rectangle for snail i is the bounding box of the inner $O(Bm^3 - \lambda) = O(Bm^3)$ levels of

Fig. 3. A spiral that is not wound tightly in the outer i levels occupies $\Omega(i^2)$ more area.

the snail's spiral, namely, those that must be wound tightly. Rectangle i has width $w := 4Bm^3 - 2 - 4\lambda$ and height $h_i := 4Bm^3 + 2a_im^2 - 4\lambda$. Note that $h' := 4Bm^3 + Bm^2/2 - 4\lambda < h_i < 4Bm^3 + Bm^2 - 4\lambda$. If three rectangles share an x-coordinate, then the remaining height at this coordinate is at most $H - 3h' = 12Bm^3 + 2Bm^2 + 6m - 3(4Bm^3 + 1/2Bm^2 - 4\lambda) = Bm^2/2 + 6m - 4\lambda < h'$; hence, no four rectangles can be drawn at a common x-coordinate. Further, if m rectangles share a y-coordinate, then the remaining width at this coordinate is $W - mw = 4Bm^4 + 16m^2 - m(4Bm^3 - 2 - 4\lambda) = 16m^2 - 2m - 4m\lambda < w$; hence, no $m + 1$ rectangles can be drawn at a common y-coordinate.

These two facts combined imply an assignment of the rectangles to three rows of m rectangles each. To see this, consider three rectangles lying above each other. Then, since there is only $Bm^2/2 + 6m - 4\lambda < h'$ free vertical space, any rectangle has to be intersected by at least one of the three horizontal lines at y-coordinates $Bm^2/2 + 6m - 4\lambda + ih'$ with $i \in \{0, 1, 2\}$. No rectangle can intersect two lines, otherwise at most two rectangles would fit vertically and the third rectangle could not be squeezed in anywhere else. Analogously, we can assign the rectangles to one of the m columns by intersecting them with m vertical lines of distance w.

This assignment of rectangles to lines tells us the solution for the given instance of 3-PARTITION: for $i = 1, \ldots, m$, we put into the set A_i the numbers $a_{i,1}, a_{i,2}, a_{i,3}$ represented by the three rectangles in column i. We claim that $a_{i,1} + a_{i,2} + a_{i,3} \leq B$, which would complete our proof.

In order to see the claim, note that the λ removed levels of each spiral have to be wound completely around the corresponding rectangle. Thus, they also intersect the vertical line that goes through the rectangles in column i. Therefore, the height at this x-coordinate is at least $3 \cdot 4\rho + 2(a_{i,1} + a_{i,2} + a_{i,3})m^2$. The height and, hence, this expression is upperbounded by $H(= 12\rho + 2Bm^2 + 6m)$ since we assumed that the drawing fits into R. This yields $a_{i,1} + a_{i,2} + a_{i,3} \leq B + 3/m$. Exploiting that the $a_{i,j}$'s are integers shows that our above claim holds. $\square$

In order to show the NP-hardness of our three objectives, we adjust the above proof by attaching a very long spiral (with $\omega(WH)$, say $(WH)^2$, windings) to the wall such that it wraps around our construction above. Let T be the resulting LR-sequence. We will provide an upper bound for the objective value of T that holds if and only if the corresponding LR-sequence S is a yes-instance of 3-PARTITION. For this, we will use that any realization of S that is a no-instance causes the very long spiral to stretch by at least one unit horizontally or vertically, which makes the value of the objective increase above the mentioned upper bound.

In more detail, we construct the angle sequence T as follows (see Fig. 4): We tightly draw a spiral around a rectangle of size $(W + 1) \times (H + 3)$ with $\omega(WH)$ windings. By adding the ladder $(\text{LLRR})^{W/2}$ to the innermost horizontal arm and the ladder $(\text{LLRR})^{H/2}$ to the innermost vertical arm of the spiral, we ensure that in any tight drawing with the two ladders being in the inside, the spiral goes around a rectangle of size exactly $W \times (H + 2)$. Further, we add the ladder $(\text{LLRR})^{(4\omega(WH)+W)/2}$ to the outermost horizontal and the ladder $(\text{LLRR})^{(4\omega(WH)+H)/2}$ to the outermost vertical arm of the spiral. Finally, we add S to the spiral by using the appropriate one of the inner-most arms of the spiral as the wall of S. Note that as long as S fits into a bounding box of size $W \times H$ it does not stretch the spiral around it. Hence, if and only if S is a yes-instance, we can draw S inside the spiral without stretching the spiral.

Consider any one of the two objectives: minimizing the inner area and minimizing the perimeter. Observe that in any drawing of S that fits inside the $(W \times H)$-box, the value of the objective is bounded by $3WH$. Let t be the value of the objective of the spiral and its ladders when drawn tightly around a rectangle of size $W \times (H + 2)$. Then $t' := t + 3WH$ is an upper bound of the value of the objective of T in the case that S is a yes-instance.

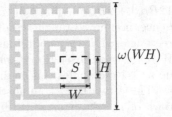

Fig. 4. T containing S inside a long spiral.

Now assume that S is a no-instance. If the spiral is not winding around S, that is, if the bounding box of the first three arms of the spiral (starting with the arms with the attached $(\texttt{LLRR})^{WH}$-ladders) does not contain S, then the other arms of the spiral have to be drawn outside the bounding box of the two arms. Hence, this increases the total length of the other arms by at least $\omega(WH)$, thus leading to a value of the objective greater than $t + \omega(WH) > t'$. If the spiral is winding around S, then, given that S is a no-instance, we have to stretch the spiral as argued above. Stretching the spiral by one unit in any direction, say in the horizontal direction, causes all $\omega(WH)$ many horizontal arms to increase by at least one unit. Hence, the value of the objective is at least $t + \omega(WH) > t'$.

The case of minimizing the bounding box is simpler: Let $W' \times H'$ be the size of the bounding box when the spiral and its ladders are drawn tightly around a rectangle of size $W \times (H+2)$. We claim that T can be drawn inside an $(W' \times H')$-bounding box if and only if S is a yes-instance. If S is not drawn inside the spiral, then the ladders $(\texttt{LLRR})^{WH}$ lie on the innermost arms of the spiral and the claim follows immediately. If S is drawn inside the spiral, we recall that S stretches the spiral (and thus the bounding box of T) if and only if it is a no-instance. This concludes the proof.   □

3   The Monotone Case: Minimum Area

In this section, we show how to compute, for a monotone angle sequence, a polygon of minimum area. We start with the simple xy-monotone case and then consider the more general x-monotone case.

3.1   The xy-monotone Case

An xy-monotone polygon has four *extreme edges*; its leftmost, rightmost, topmost, and bottommost edge. Two consecutive edges are connected by a (possible empty) xy-monotone chain that we will call a *stair*. Starting at the topmost edge, we denote the four stairs in counterclockwise order TL, BL, BR, and TR; see Fig. 5(a). We say that an angle sequence consists of k nonempty *stair sequences* if any xy-monotone polygon that realizes it consists of k nonempty stairs; we also call it a *k-stair sequence*. The extreme edges correspond to the exactly four LL-sequences in an xy-monotone angle sequence and are unique up to rotation. Any xy-monotone angle sequence is of the form $[\texttt{L}(\texttt{LR})^*]^4$, where the single L describes the turn before an extreme edge and $(\texttt{LR})^*$ describes a stair sequence. W.l.o.g., we assume that an xy-monotone sequence always begins with LL and that we always draw the first LL as the topmost edge (the top extreme edge). Then, we can use TL, BL, BR, TR also for the corresponding stair sequences, namely the first, second third and forth $(\texttt{LR})^*$ subsequence after the first LL in cyclic order. Let T be the concatenation of TL, the top extreme edge, and TR; let L, B, and R be defined analogously following Fig. 5(a). For a chain C, let the R-length $r(C)$ be the number of reflex vertices on C.

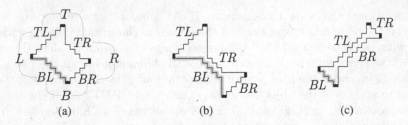

Fig. 5. Extreme edges are bold. Stair BL is highlighted. (a) Notation: The four stairs TL, TR, BR, and BL of an xy-monotone polygon. The sequences T, R, B, and L are unions of neighboring stairs. (b) & (c) Two possible optimum configurations of the polygon.

Theorem 3. *Given an xy-monotone angle sequence S of length n, we can find a polygon P that realizes S and minimizes its (i) bounding box or (ii) area in $O(n)$ time, and in constant time we can find the optimum value of the objective if the R-lengths of the stair sequences are given.*

Part (i) of Theorem 3 follows from the following observation: The bounding box of every polygon that realizes S has width at least $\max\{r(T), r(B)\}+1$ and height at least $\max\{r(L), r(R)\} + 1$. By drawing three stairs with edges of unit length, we can meet these lower bounds.

For part (ii), we first consider angle sequences with at most two nonempty stairs. Here, the only non-trivial case is when the angle sequence consists of two opposite stair sequences, that is, TL and BR, or BL and TR. W.l.o.g., consider the second case.

Lemma 1. *Let S be an xy-monotone angle sequence of length n consisting of two nonempty opposite stair sequences BL and TR. We can find a minimum-area polygon that realizes S in $O(n)$ time. If $r(BL)$ and $r(TR)$ are given, we can compute the area of such a polygon in $O(1)$ time.*

The proof of this lemma is given in the full version of the paper [6] and leads to the following observation.

Observation 4. *In any polygon P of minimum area consisting of two nonempty opposite stairs BL and TR with $b := r(BL) \geq a := r(TR)$, BL consists of only unit-length segments and TR only of segments of lengths $\lfloor b/(a + 1) \rfloor$ and $\lceil b/(a + 1) \rceil$ (in any order).*

We now consider the case of four nonempty stairs. (The case of three nonempty stairs can be solved analogously.) An xy-monotone polygon P with four nonempty stairs TL, TR, BL, and BR is *canonical* if (C1) P has two non-adjacent nonempty stairs, say TL and BR, such that the bounding box B_{TL} of TL and its adjacent extreme edges and the bounding box B_{BR} of BR and its adjacent extreme edges intersect in at most one point, and (C2) the bottom-right corner of B_{TL} as well as the top-left corner of B_{BR} coincides with an endpoint of TR or BL. The proof of the following lemma is given in the full version of the paper [6].

Lemma 2. *For every 4-stair sequence S with |S| > 36, there exists a polygon of minimum area realizing S that is canonical.*

Consider the line segment of TR and the line segment of BL whose endpoints lie on B_{TL} in a canonical polygon. These line segments are (a) both horizontal, (b) both vertical, or (c) perpendicular to each other. Consequently, there is only a constant number of ways in which the stairs outside the two bounding boxes are connected to them.

Consider a (canonical) optimum polygon. We cut the polygon along the edge of B_{TL} that contains an endpoint of both BL and TR. We also cut along the respective edge of B_{BR}. We get three polygons. The polygons on the outside realize the 1-stair sequence defined by TL and BR (including their adjacent extreme edges), respectively, whereas the middle polygon realizes the 2-stair sequence defined by the concatenation of BL, TR, and the edge segments of B_{TL} and B_{BR} that connect them.

This observation leads to the following algorithm: For S with $|S| \leq 36$, we find a solution in constant time by exhaustive search. For larger $|S|$, we guess the partition of the extreme edges whose bounding boxes do not intersect in the (canonical) optimum polygon that we want to compute. W.l.o.g., we guessed B_{TL} and B_{BR} (the other case is symmetric). Then, we guess how TR and BL are connected to the two bounding boxes (see (a)–(c)). This gives us two 1-stair instances and a 2-stair instance. We solve the instances independently and then put the solutions together to form a solution to the whole instance. By Lemma 1 and Observation 4, we solve the middle instance such that the left extreme edge of our solution is of minimum length, and, if possible, also the top extreme edge. The detailed algorithm to prove Theorem 3 the full version of the paper [6].

3.2   The x-monotone Case

For the x-monotone case, we first give an algorithm that minimizes the bounding box of the polygon, and then an algorithm that minimizes the area.

An x-monotone polygon consists of two *vertical extreme* edges, i.e., the leftmost and the rightmost edge, and at least two *horizontal extreme* edges, which are defined to be the horizontal edges of locally maximum or minimum height. The vertical extreme edges divide the polygon into an upper and a lower hull, each of which consists of xy-monotone chains that are connected by the horizontal extreme edges. We call a horizontal extreme edge of type RR an *inner extreme edge*, and a horizontal extreme edge of type LL an *outer extreme edge*; see Fig. 6(a). Similar to the xy-monotone case, we consider a *stair* to be an xy-monotone chain between any two consecutive extreme edges (outer and inner extreme edges as well as vertical extreme edges) and we denote by *stair sequence* the corresponding angle subsequence (LR)*. W.l.o.g., at least one inner extreme edge exists, otherwise the polygon is xy-monotone and we refer to Sect. 3.1. Given an x-monotone sequence, we always draw the first RR-subsequence as the leftmost inner extreme edge of the lower hull. By this, the correspondence between the angle subsequences and the stairs and extreme edges is unique.

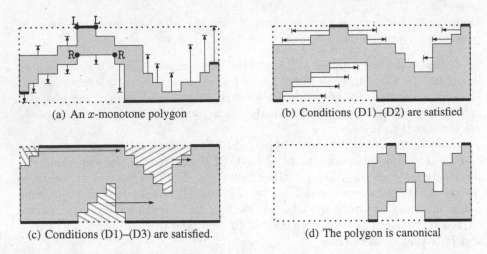

(a) An x-monotone polygon

(b) Conditions (D1)–(D2) are satisfied

(c) Conditions (D1)–(D3) are satisfied.

(d) The polygon is canonical

Fig. 6. Illustration of how to make a polygon canonical. The thick horizontal edges are outer extreme edges, the tiling patterns mark double stairs.

An x-monotone polygon is *canonical* if (D1) all outer extreme edges are lying on the border of the bounding box, (D2) each vertical non-extreme edge that is not incident to an inner extreme edge has length 1, and (D3) each horizontal edge that is not an outer extreme edge has length 1.

In the full version of the paper [6], we show that it suffices to find a canonical x-monotone polygon of minimum bounding box; see Fig. 6 for an illustration.

Lemma 3. *Any x-monotone polygon can be transformed into a canonical x-monotone polygon without increasing the area of its bounding box.*

We observe that the length of the vertical extreme edges depends on the height of the bounding box, while the length of all other vertical edges is fixed by the angle sequence. Thus, a canonical x-monotone polygon is fully described by the height of its bounding box and the length of its outer extreme edges. Furthermore, the y-coordinate of each vertex depends solely on the height of the bounding box.

We use a dynamic program that constructs a canonical polygon of minimum bounding box in time $O(n^3)$. For each possible height h of the bounding box, the dynamic program populates a table that contains an entry for any pair of an extreme vertex p (that is, an endpoint of an outer extreme edge) and a horizontal edge e of the opposite hull. The value of the entry $T[p, e]$ is the minimum width w such that the part of the polygon left of p can be drawn in a bounding box of height h and width w in such a way that the edge e is intersecting the interior of the grid column left of p. The algorithm is given the full version of the paper [6].

Theorem 5. *Given an x-monotone angle sequence S of length n, we can find a polygon P that realizes S and minimizes the area of its bounding box in $O(n^3)$ time.*

For the area minimization, we make two key observations. First, since the polygon is x-monotone, each grid column (properly) intersects either no or exactly two horizontal edges: one edge from the upper hull and one edge from the lower hull. Second, a pair of horizontal edges share at most one column; otherwise, the polygon could be drawn with less area by shortening both edges. With the same argument as for the bounding box, the height of any minimum-area polygon is at most n.

We use a dynamic program to solve the problem. To this end, we fill a three-dimensional table T as follows. Let e be a horizontal edge on the upper hull, let f be a horizontal edge of the lower hull, and let $1 \leq h \leq n$. Then, the entry $T[e, f, h]$ specifies the minimum area required to draw the part of the polygon to the left of (and including) the unique common column of e and f under the condition that e and f share a column and have vertical distance h.

Let $e_1, \ldots, e_k$ be the horizontal edges on the upper hull from left to right and let $f_1, \ldots, f_m$ be the horizontal edges on the lower hull from left to right. We initialize the table with $T[e_1, f_1, h] = h$ for each $1 \leq h \leq n$. To compute any other entry $T[e_i, f_j, h']$, we need to find the correct entry from the column left of the column shared by e_i and f_j. There are three possibilities: this column either intersects e_{i-1} and f_{j-1}, it intersects e_i and f_{j-1}, or it intersects e_{i-1} and f_j. For each of these possibilities, we check which height can be realized if e_i and f_j have vertical distance h' and search for the entry of minimum value. We set

$$T[e_i, f_j, h'] = \min_{h'' \text{ valid}} \{T[e_{i-1}, f_{j-1}, h''], T[e_i, f_{j-1}, h''], T[e_{i-1}, f_j, h'']\} + h'.$$

Finally, we can find the optimum solution by finding $\min_{1 \leq h \leq n}\{T[e_k, f_m, h]\}$. Since the table has $O(n^3)$ entries each of which we can compute in $O(n)$ time, the algorithm runs in $O(n^4)$ time. This proves the following theorem.

Theorem 6. *Given an x-monotone angle sequence S of length n, we can find a minimum-area polygon that realizes S in $O(n^4)$ time.*

4 The Monotone Case: Minimum Perimeter

In this section, we show how to compute a polygon of minimum perimeter for an xy-monotone or x-monotone angle sequence S of length n.

Let P be an x-monotone polygon realizing S. Let e_L be the leftmost vertical edge and let e_R be the rightmost vertical edge of P. Recall that P consists of two x-monotone chains; an upper chain T and a lower chain B connected by e_L and e_R. Without loss of generality, we assume that $r(T) \geq r(B)$.

An x-monotone polygon is *perimeter-canonical* if (P1) every vertical edge except e_R and e_L has unit length, and (P2) every horizontal edge of T has unit length. We show that it suffices to find a perimeter-canonical polygon in the full version of the paper [6].

Lemma 4. *Any x-monotone polygon can be transformed into a perimeter-canonical x-monotone polygon without increasing its perimeter.*

Suppose that P is a minimum-perimeter canonical polygon that realizes S with $r(T) \geq r(B)$, and peri(P) denotes its perimeter. By condition (P2), every edge in T is of unit length, so the length of T is $2r(T)+1$. This implies the width of B should be $r(T)+1$. By condition (P1), the length of the vertical edges in B is $r(B)$, so the total length of B is $r(T)+r(B)+1$. Thus we can observe the following property.

Lemma 5. *Given an x-monotone angle sequence S, there is a canonical minimum-perimeter polygon P realizing S with $r(T) \geq r(B)$ such that* peri$(P) =$ $3r(T) + r(B) + 2 + |e_L| + |e_R|$.

The first three terms of peri(P) in Lemma 5 are constant, so we need to minimize the sum of the last two terms, $|e_L|$ and $|e_R|$, to get a minimum perimeter. However, once one of them is fixed, the other is automatically determined by the fact that all vertical edges in B are unit segments. In other words, minimizing one of them is equivalent to minimizing their sum, consequently minimizing the perimeter. We call the length of the leftmost extreme edge of a polygon the *height* of the polygon.

4.1　The xy-monotone Case

Let P be a minimum-perimeter canonical xy-monotone polygon that realizes an xy-monotone angle sequence S of length n. When $n = 4$, i.e., $r = 0$, a unit square P achieves the minimum perimeter, so we assume here that $r > 0$. Recall that the boundary of P consists of four stairs, TR, TL, BL, and BR. Let (r_1, r_2, r_3, r_4) be a quadruple of the numbers of reflex vertices of TR, TL, BL, and BR, respectively. Then $r = r_1 + r_2 + r_3 + r_4$, where $r_i \geq 0$ for each i. We further assume that (i) r_1 is the largest one among the four r_i's, thus $r_1 > 0$, and (ii) $r_2 \geq r_4$, which directly means $r(T) \geq r(B)$; if not, we can rotate or mirror P so that the assumption holds.

To get peri(P), we have to minimize either $|e_L|$ or $|e_R|$. This implies that if we can draw P such that $|e_L| = 1$ or $|e_R| = 1$, then P has minimum perimeter. We have two cases depending on whether $r_2 = 0$ or not.

We first consider the case when $r_2 = 0$. By assumption (ii), $r_4 = 0$. We then have a quadruple $(r_1, 0, r_3, 0)$. If $r_1 = r_3$, then P satisfying Lemma 4 is uniquely defined as in Fig. 7(a), in which $|e_L| = |e_R| = 2$. So peri$(P) = 3r(T) + r(B) + 6 = 4r_1 + 6$. If $r_1 > r_3$, then we can draw P such that $|e_R| = 1$ and $|e_L| = r_1 - r_3 + 1$ as in Fig. 7(b), so peri$(P) = 3r_1 + r_3 + 2 + (r_1 - r_3 + 1) + 1 = 4r_1 + 4$. This is a minimum because $|e_R| = 1$. This can be rephrased as a general form, peri$(P) =$ $3(r_1 + r_2) + (r_3 + r_4) + |r_3 - (r_1 - r_2 + r_4)| + 4$ under the assumption that $r_2 = r_4 = 0$ and $r_1 > r_3$.

Now we consider the other case when $r_2 > 0$; see Figs. 7(c) and (d). Let h and h' be horizontal lines one unit below the upper end vertices of e_L and e_R, respectively. The bottommost edge e_B must be on or below h'. Since BL and BR share the edge e_B, if $r_3 < (r_1 - r_2 + r_4)$, then e_L should be stretched so that BL can share e_B with BR as in Fig. 7(c), so $|e_L| = (r_1 - r_2 + r_4) - r_3 + 1$ and $|e_R| = 1$. If $r_3 > (r_1 - r_2 + r_4)$ as in Fig. 7(d), then e_R should be stretched, so $|e_L| = 1$ and

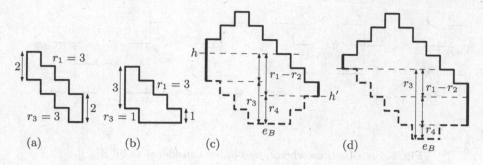

Fig. 7. (a)–(b) The case when $r_2 = 0$. (c)–(d) The case when $r_2 > 0$.

$|e_R| = r_3 - (r_1 - r_2 + r_4) + 1$. Finally, if $r_3 = (r_1 - r_2 + r_4)$, then $|e_L| = |e_R| = 1$. We can express three situations as one equation, $|e_L| + |e_R| = |r_3 - (r_1 - r_2 + r_4)| + 2$. Therefore, $\mathrm{peri}(P) = 3(r_1 + r_2) + (r_3 + r_4) + |r_3 - (r_1 - r_2 + r_4)| + 4$. The minimum perimeter for this case is clearly guaranteed since $|e_L| = 1$ or $|e_R| = 1$.

Theorem 7. *Given an xy-monotone angle sequence S of length n, we can find a polygon P that realizes S and minimizes its perimeter in $O(n)$ time. Furthermore, if the lengths of the stair sequences (r_1, r_2, r_3, r_4) are given as above, then $\mathrm{peri}(P)$ can be expressed as:*

$$\mathrm{peri}(P) = \begin{cases} 4r_1 + 6 & \text{if } (r_1, 0, r_1, 0), \\ 3(r_1 + r_2) + (r_3 + r_4) + |r_3 - (r_1 - r_2 + r_4)| + 4 & \text{otherwise.} \end{cases}$$

4.2 The x-monotone Case

A minimum height polygon P that realizes S can be computed in $O(n^2)$ time using dynamic programming.

From right to left, let $t_1, t_2, \ldots, t_{r(T)}$ be the horizontal edges in T and $b_1, b_2, \ldots, b_{r(B)}$ be the horizontal edges in B. Recall that $r(T) \geq r(B)$. Let $A[i, j]$ be the minimum height of the subpolygon formed with the first i horizontal edges from T and the first j horizontal edges from B. Note that the leftmost vertical edge of the subpolygon whose minimum height is stored in $A[i, j]$ joins the left endpoints of t_i and b_j. To compute $A[i, j]$, we attach the edges t_i and b_j to the upper and lower chains of the subpolygon constructed so far. Since t_i is unit length, t_i and b_j are attached either to the subpolygon with height of $A[i - 1, j - 1]$ or to the subpolygon with height of $A[i - 1, j]$. As in Fig. 8, there are four cases (a)–(d) for the first attachment and two cases (e)–(f) for the second attachment, according to the turns formed at the attachments.

Let u and v be the left end vertex of t_{i-1} and the right end vertex of t_i, respectively. Let u' and v' be the right end vertex of b_j and the left end vertex of b_{j-1}, respectively. Notice that both vertical edges (u, v) and (u', v') are unit-length. For example, let us explain how to calculate $A[i, j]$ when $uv = \mathsf{LR}$ and $u'v' = \mathsf{LR}$, which corresponds to Figs. 8(b) and (f). $A[i, j]$ is the minimum height

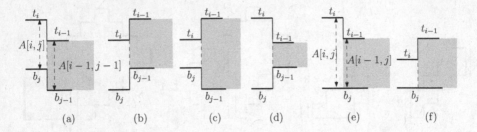

Fig. 8. Six situations when t_i and b_j are considered to fill $A[i,j]$.

of two possible attachments (b) and (f). The height for (b) should be at least 2 to realize $uv = \text{LR}$ and $u'v' = \text{LR}$. If $A[i-1,j-1] > 1$, then t_i and b_j are attached to the subpolygon as illustrated in Fig. 8(b), thus $A[i,j]$ is the same as $A[i-1,j-1]$. Otherwise, if $A[i-1,j-1] = 1$, then we can move the upper chain of the subpolygon one unit upward without intersection so that t_i and b_j are safely attached to the subpolygon with $A[i,j] = 2$. Thus $A[i,j] = \max(A[i-1,j-1],2)$. The height for (f) should be at least 1, so it is expressed as $\max(A[i-1,j]-1,1)$. Therefore,

$$A[i,j] = \min(\max(A[i-1,j-1],2), \max(A[i-1,j]-1,1)).$$

For other turns at uv and $u'v'$, we can similarly define the equations as follows:

$$A[i,j] = \begin{cases} \text{undefined} & \text{if } i=0, j=0 \text{ or } i<j \\ 1 & \text{if } i=1, j=1 \\ A[i-1,j]+1 & \text{if } uv = \text{RL}, j=1 \\ \max(A[i-1,j]-1,1) & \text{if } uv = \text{LR}, j=1 \\ \min(\max(A[i-1,j-1],2), A[i-1,j]+1) & \text{if } uv = \text{RL}, u'v' = \text{RL} \\ \min(\max(A[i-1,j-1],2), \max(A[i-1,j]-1,1)) & \text{if } uv = \text{LR}, u'v' = \text{LR} \\ \min(A[i-1,j-1]+2, A[i-1,j]+1) & \text{if } uv = \text{RL}, u'v' = \text{LR} \\ \min(\max(A[i-1,j-1]-2,1), \\ \quad \max(A[i-1,j]-1,1)) & \text{if } uv = \text{LR}, u'v' = \text{RL} \end{cases}$$

Evaluating each entry takes constant time, so the total time to fill A is $O(n^2)$. Using A, a minimum-perimeter polygon can be reconstructed within the same time bound.

Theorem 8. *Given an x-monotone angle sequence S of length n, we can find a polygon P that realizes S and minimizes its perimeter in $O(n^2)$ time.*

References

1. Bae, S.W., Okamoto, Y., Shin, C.-S.: Area bounds of rectilinear polygons realized by angle sequences. In: Chao, K.-M., Hsu, T., Lee, D.-T. (eds.) ISAAC 2012. LNCS, vol. 7676, pp. 629–638. Springer, Heidelberg (2012). doi:10.1007/978-3-642-35261-4_65

2. Biedl, T.C., Durocher, S., Snoeyink, J.: Reconstructing polygons from scanner data. Theor. Comput. Sci. **412**(32), 4161–4172 (2011)
3. Chen, D.Z., Wang, H.: An improved algorithm for reconstructing a simple polygon from its visibility angles. Comput. Geom. **45**(5–6), 254–257 (2012)
4. Culberson, J.C., Rawlins, G.J.E.: Turtlegons: generating simple polygons from sequences of angles. In: Proceedings of 1st Annual ACM Symposium on Computational Geometry (SoCG 1985), pp. 305–310 (1985)
5. Disser, Y., Mihalák, M., Widmayer, P.: A polygon is determined by its angles. Comput. Geom. **44**(8), 418–426 (2011)
6. Evans, W.S., Fleszar, K., Kindermann, P., Saeedi, N., Shin, C.S., Wolff, A.: Minimum rectilinear polygons for given angle sequences. Arxiv report (2016). https://arxiv.org/abs/1606.06940
7. Garey, M.R., Johnson, D.S.: Computers and Intractability: A Guide to the Theory of NP-Completeness. W.H. Freeman & Co., New York (1979)
8. Hartley, R.I.: Drawing polygons given angle sequences. Inform. Process. Lett. **31**(1), 31–33 (1989)
9. Patrignani, M.: On the complexity of orthogonal compaction. Comput. Geom. **19**(1), 47–67 (2001)
10. Sack, J.R.: Rectilinear computational geometry. Ph.D. thesis, School of Computer Science, McGill University (1984). http://digitool.library.mcgill.ca/R/?func=dbin-jump-full&object_id=71872&local_base=GEN01-MCG02
11. Tamassia, R.: On embedding a graph in the grid with the minimum number of bends. SIAM J. Comput. **16**(3), 421–444 (1987)
12. Vijayan, G., Wigderson, A.: Rectilinear graphs and their embeddings. SIAM J. Comput. **14**(2), 355–372 (1985)

Continuous Folding of Regular Dodecahedra

Takashi Horiyama[1], Jin-ichi Itoh[2], Naoki Katoh[3,4], Yuki Kobayashi[4,5], and Chie Nara[6(✉)]

[1] Graduate School of Science and Engineering, Saitama University, Saitama, Japan
horiyama@al.ics.saitama-u.ac.jp
[2] Faculty of Education, Kumamoto University, Kumamoto 860-8555, Japan
j-itoh@kumamoto-u.ac.jp
[3] Department of Inforatics, Faculty of Science and Technology,
Kwansei Gakuin University, Nishinomiya, Japan
naoki.katoh@gmail.com
[4] JST, CREST, Tokyo, Japan
[5] Graduate School of Engineering, Tokyo Institute of Technology,
2-12-1 Ookayama, Meguro-ku, Tokyo 152-8550, Japan
kobayashi.y.bv@m.titech.ac.jp
[6] Meiji Institute for Advanced Study of Mathematical Sciences,
Meiji University, Nakano, Tokyo 164-8525, Japan
cnara@jeans.ocn.ne.jp

Abstract. Itoh and Nara [3] discussed with Kobayashi the continuous flattening of all Platonic polyhedra; however, a problem was encountered in the case of the dodecahedron. To complete the study, we explicitly show, in this paper, a continuous folding of a regular dodecahedron following the ideas in [3].

Keywords: Dodecahedron · Regular polyhedra · Folding · Continuous flattening

1 Introduction

In this paper we continue the discussion on the continuous flattening of Platonic polyhedra found in [3] and provide a continuous folding of a regular dodecahedron following the approach presented in [3], and also by using the method discussed by Itoh and Nara in [5].

We use the terminology *polyhedron* for a closed polyhedral surface which is permitted to touch itself but not self-intersect. A *flat folding* of a polyhedron is a

---

J.-i. Itoh—Supported by Grant-in-Aid for Scientific Research(B) (15KT0020) and Scientific Research(C)(26400072).

N. Katoh—Supported by JSPS Grant-in-Aid for Scientific Research(A) (25240004).

Y. Kobayashi—Supported by JSPS Grant-in-Aid for Scientific Research(A) (25240004).

C. Nara—Supported by Grant-in-Aid for Scientific Research(C) (16K05258).

© Springer International Publishing AG 2016
J. Akiyama et al. (Eds.): JCDCGG 2015, LNCS 9943, pp. 120–131, 2016.
DOI: 10.1007/978-3-319-48532-4_11

folding by creases into a multilayered planar shape. There are several strategies employed for the continuous flattening of polyhedra; in [3] rhombi are used, in [7] kites, in [5] one and two moving edges, in [6] cut loci and Alexandrov's gluing theorem, in [1] skeleton methods. Approaches for flattening polyhedra are found in [2]. It is proved that any convex polyhedron can be continuously flat folded in [1,6]. However, the moving creases cover the almost all surface during the continuous motions used in [1,6]. We employ a different method so that the moving creases cover a small portion of the surface and that two parallel faces of the regular dodecahedron have no creases, that is, they can be rigid.

We review the definition of continuous flattening.

Definition 1. *Let P be a polyhedron in the Euclidean space $\mathbb{R}^3$. We say that a family of polyhedra $\{P_t : 0 \leq t \leq 1\}$ is a continuous (flat) folding process from $P = P_0$ to P_1 if it satisfies the following conditions:*

(1) for each $0 \leq t \leq 1$, there is an intrinsic isometry from P_t onto P,
(2) the mapping $[0,1] \ni t \longmapsto P_t \in \{P_t : 0 \leq t \leq 1\}$ is continuous,
(3) $P_0 = P$ and P_1 is a (flat) folded state of P.

2   A Folded Rhombus with Wing-Type

We denote by uv the line segment joining u and v for points u and v in $\mathbb{R}^3$, and by $dist(u, v)$ (or simply $|uv|$) the Euclidean distance between u and v. Let h be the center of a rhombus $\mathcal{R} = abcd$ in $\mathbb{R}^3$. Fold $\mathcal{R}$ into halves by a valley crease on bd and fold the resulting figure into halves again to get a four layered right triangle shape. In this motion $dist(a, c)$ decreases to zero first, and then $dist(b, d)$ decreases to zero. We can decrease two distances $dist(a, c)$ and $dist(b, d)$ simultaneously with any desired speed for each. This fact was proved in [3] and extended in [7]. In the proof a special folding of $\mathcal{R}$ plays a key role, in this paper we refer to such a folding of $\mathcal{R}$ as a *folded rhombus with wing-type*.

Let q be any point on hc. Fold $\mathcal{R}$ with a mountain fold on ah, bh, ch, qb, qc, qd and with a valley fold on hq so that $\triangle bhq$ and $\triangle dhq$ overlap on $\triangle abh$ and $\triangle dah$ respectively (see Fig. 1 (a), (b)). We call such figure a *folded rhombus with wing-type* of a rhombus $\mathcal{R} = abcd$ for q. Note that the resulting figure is flexible, that is, distances $dist(a', c')$ and $dist(b', d')$ are not fixed, where the notation u' refers to the point of the resulting figure corresponding to point u of $\mathcal{R}$. However, if we fix those distances, then there is a unique point q on hc which satisfies such distance conditions.

Lemma 1 ([3]). *Let $\mathcal{R} = abcd$ be a rhombus. For any pair of real numbers $l \, (0 \leq l \leq |bd|)$ and $m \, (0 \leq m \leq |ac|)$ there is a unique point q on the line segment hc where h is the center of $\mathcal{R}$, such that a folded rhombus with wing-type of $\mathcal{R}$ satisfies $dist(b', d') = l$ and $dist(a', c') = m$.*

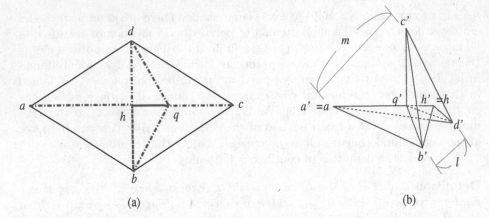

Fig. 1. (a) A rhombus $\mathcal{R} = abcd$ with mountain crease on long-short dotted line segments and valley crease on a bold line segment; (b) the folded rhombus with wing-type such that $dist(b', d') = l$ and $dist(a', c') = m$.

3   A Regular Dodecahedron

Let P be a regular dodecahedron in $\mathbb{R}^3$ with the origin O and denote its twelve regular pentagonal faces by $F_i(1 \leq i \leq 12)$, and its twenty vertices by v_{5i+j} ($0 \leq i \leq 3, 1 \leq j \leq 5$) so that F_1 and F_{12} are parallel to the xy-plane and the center of F_1 is the origin. We can assume without loss of generality that the radius of the circumcircle of F_1 is one. The vertex set is divided into four subsets $V_i = \{v_{5i+j} : 1 \leq j \leq 5\}$ for $0 \leq i \leq 3$ each of which comprises a regular pentagon parallel to the xy-plane, in particular, V_1 and V_4 are vertex sets of F_1 and F_{12}, respectively (Fig. 2(a)). We assume that $F_i(2 \leq i \leq 6)$ and $F_i(7 \leq i \leq 11)$ have common edges with F_1 and F_{12}, respectively such that $F_1 \cap F_4 = v_3v_4$ and $F_{12} \cap F_9 = v_{18}v_{19}$, and that $F_4 \cap F_9 = v_{13}v_9$ (see Fig. 2(a)). Note that $\{V_1, V_4\}$ and $\{V_2, V_3\}$ are located on cylinders about the z-axis such that their radii satisfy the golden ratio $\tau = (\sqrt{5} + 1)/2$.

The final flat folded state P_1 of P is as follows:

(1) P_1 is intrinsically isometric to P,
(2) F_1 and F_{12} have no creases and F_{12} is rotated by $2\pi/5$ onto F_1,
(3) $F_i(2 \leq i \leq 6)$ are folded into halves with valley folds so that v_{i+4} for $2 \leq i \leq 5$ is on v_i, and v_{10} is on v_1,
(4) F_9 is folded with valley folds on line segments $\{v_{13}g_9, v_9g_9, v_{19}g_9\}$ and a mountain fold on the line segment g_9h where g_9 is a center of F_9 and h is the midpoint of v_{13} and v_9 (see Figs. 2(b), (d) and 3), and all five faces F_{6+i} ($1 \leq i \leq 5$) are folded similarly.

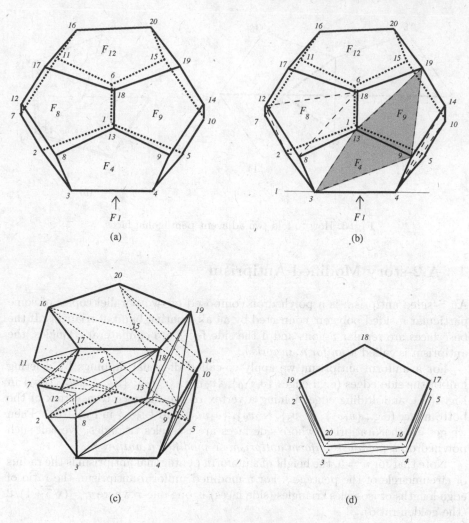

Fig. 2. A regular dodecahedron.

By rotating and pushing down the face F_{12} toward the face F_1, while keeping the two faces parallel, P can be continuously flattened onto the face F_{12}. We revise the continuous motion proposed in [3] for such flattening process because we found that the original motion has self-intersection after reaching some figure Q whose surface will be called a *2-story modified antiprism*. We investigate Q in the next section, and we propose a new continuous motion from Q to a flat folded state of P which can avoid any self-intersection, in the Sect. 6.

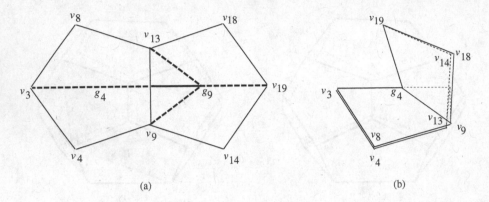

(a)   (b)

Fig. 3. How to fold two adjacent pentagonal faces.

4   A 2-story Modified Antiprism

An n-sided antiprism is a polyhedron composed of two parallel copies of some particular n-sided polygon, connected by an alternating band of triangles. If the base faces are regular n-gons and if the side faces are equilateral triangles, the antiprism is called a *uniform antiprism*.

For a uniform antiprism, we apply so-called diagonal flippings by deleting half of the side edges (e.g., edges $\{v_1v_{10}\}$, $\{v_2v_6\}$, $\{v_3v_7\}$, $\{v_4v_8\}$ and $\{v_5v_9\}$ in Fig. 4(a)), and adding edges joining a vertex of the top face to a vertex of the bottom face (e.g., $\{v_1v_7\}$, $\{v_2v_8\}$, $\{v_3v_9\}$, $\{v_4v_{10}\}$ and $\{v_5v_6\}$ in Fig. 4(b)). Then we get another antiprism whose side faces are isosceles triangles. We call such polyhedron a *modified uniform antiprism* or *modified n-antiprism*.

Note that for $n = 5$, the height of a uniform pentagonal antiprism is the radius of circumcircle of the pentagon. For a modified uniform antiprism, the ratio of edge lengths of isosceles triangles (side faces) is one-one-τ where $\tau = (\sqrt{5}+1)/2$ (the golden ratio).

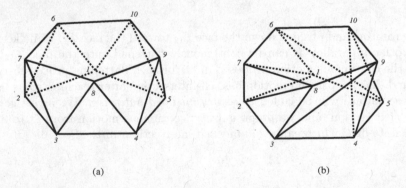

(a)   (b)

Fig. 4. (a) A uniform pentagonal antiprism; (b) a modified 5-antiprism.

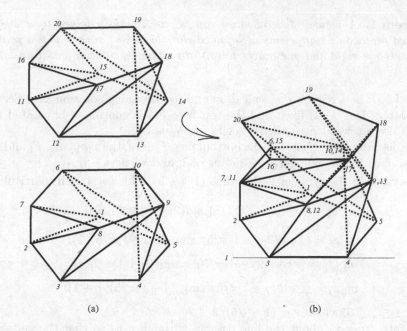

(a)   (b)

Fig. 5. (a) Two congruent copies of a modified pentagonal antiprism; (b) a 2-story modified 5-antiprism

Take two congruent copies of a modified 5-antiprism and put one on top of the other without one pentagonal face, and join them by touching pentagonal edges so that the resulting polyhedron has two parallel pentagonal faces. We call such polyhedron a *2-story modified 5-antiprism* (see Fig. 5).

5   From a Regular Dodecahedron to a 2-story Modified Pentagonal Antiprism

We show that a 2-story modified 5-antiprism is intrinsically isometric to a subset of a regular dodecahedron. Let P be the regular dodecahedron defined in Sect. 3. We call F_1 and F_{12} a bottom face and a top face respectively, and other faces are called side faces and paired as follows, $\{F_2, F_7\}$, $\{F_3, F_8\}$, $\{F_4, F_9\}$, $\{F_5, F_{10}\}$, and $\{F_6, F_{11}\}$. We divide each of those five pairs into triangles. For example, $\{F_4, F_9\}$ is divided by line segments v_3v_{13}, v_3v_9, $v_{19}v_{13}$ and $v_{19}v_9$ (see Fig. 2(b)). Other pairs are divided similarly. If the rhombus of two triangles $\triangle v_3v_{13}v_9$ and $\triangle v_{19}v_{13}v_9$ in the pair $\{F_4, F_9\}$ can be flat folded so that the boundary edges v_3v_{13} and $v_{19}v_{13}$ meet with v_3v_9 and $v_{19}v_9$ respectively, and if we can do a similar operation on each of other pairs, the resulting figure is a 2-story modified pentagonal antiprism Q together with folded rhombi inside.

We show that there is a continuous motion from P to the polyhedron Q, which is a minor revision of a continuous motion proposed in [3].

Theorem 1. *A regular dodecahedron can be continuously folded to a 2-story modified pentagonal antiprisms with folded rhombi inside, such that two pentagonal faces are rigid and one moves toward the other by rotation and translation only.*

Proof. Let P be a regular dodecahedron and use the same notation as in Sect. 3. The bottom face F_1 is fixed and the top face F_{12} is continuously rotated and pushed down toward F_1 until the height of F_{12} is two.

We describe P in $\mathbb{R}^3$ so that the bottom face F_1 is on the xy-plane, F_1 and F_{12} are inscribed in the cylinder with radius one, and vertices $v_{5(i-1)+j}(1 \le j \le 5)$ are inscribed in the cylinder with radius $(1 + \sqrt{5})/2$ for $i = 1, 2$. In particular,

$$v_j = (\cos(2(j-1)\pi/5), \sin(2(j-1)\pi/5), 0),$$

$$v_{5+j} = (\tau \cos(2(j-1)\pi/5), \tau \sin(2(j-1)\pi/5), 1),$$

$$v_{10+j} = (\tau \cos((2j+1)\pi/5), \tau \sin((2j+1)\pi/5, \tau),$$

$$v_{15+j} = (\cos(2(j+1)\pi/5), \sin(2(j+1)\pi/5), \tau + 1),$$

for $1 \le j \le 5$, where $\tau = (1 + \sqrt{5})/2$.

Rotate $\triangle v_3 v_4 v_9$ about the line passing through $v_3 v_4$ so that the vertex v_9 moves along the circular arc with the radius $r = \sqrt{5}/2$ from the point v_9 toward the point v_5. The dihedral angle of $\triangle v_3 v_4 v_9$ and the xy-plane, denoted by θ, decreases from $\pi - \theta_0$ to θ_0 where $\theta_0 = \cos^{-1}(1/\sqrt{5})$. Then, the trace of v_9 for θ, denoted by v_9^θ is

$$v_9^\theta = (r \cos\theta - \cos(\pi/5), -\sin(2\pi/5), r \sin\theta) \quad (0 \le \theta \le \pi - \theta_0).$$

The orthogonal projection of the trace of v_9 to the xy-plane is a line segment of the line passing through v_5 and orthogonal to $v_3 v_4$ in the xy-plane, and it intersects the circumcircle of F_1 (Fig. 6) at the points v_5 and w, where w comprises $\angle wOv_4 = \pi/5$ (see Fig. 6(b)).

Since the angle $\angle wOv_5 = \pi/5$, the motion is divided into two parts; one is the motion until the trace of v_9' reaches w, and the other is from that figure to the flat folded state P_1. When the projection of the trace of v_9 reaches w, the z-coordinate of the trace of v_9 reaches one, and P can be folded into a 2-story modified 5-antiprisms with folded rhombi inside (denoted be $P_{1/2}$).

We define a continuous motion for other vertices of P to obtain a 2-story modified 5-antiprisms as its surface. The set $V_2 = \{v_i : 6 \le i \le 10\}$ is rotated and moves toward the bottom face F_1, while it comprises a regular pentagon with center on z-axis and parallel to F_1.

The motion for the vertex set V_3 is similar to V_2, so that the distance $v_{5+i}v_{10+i}$ $(1 \le i \le 5)$ is preserved. The motion for the vertex set V_4 is determined so that the relative motion of V_3 to V_4 is similar to the relative motion of V_2 to V_1.

When the z-coordinate of the trace of v_9 is one, all vertices V_i $(1 \le i \le 4)$ are on the cylinder with radius one, and v_i and v_{i+4} $(6 \le i \le 10)$ meet. Denote

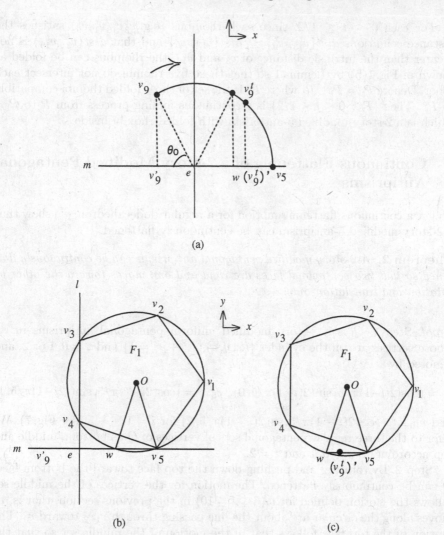

Fig. 6. (a) The trace of the vertex v_9 on the plane parallel to the xz-plane where $\theta_0 = \cos^{(-1)}(1/\sqrt{5})$; (b) the orthogonal projection of the trace of v_9 to the xy-plane where w is the first intersecting point with F_1; (c) $(v_9^t)'$ is the projection of v_9^t for some t.

by P^o the subset of P obtained by removing five (folded) rhombi $v_1v_7v_{17}v_{11}$, $v_2v_8v_{18}v_{12}$, $v_3v_9v_{19}v_{13}$ (shaded in Fig. 2(b)), $v_4v_{10}v_{20}v_{14}$ and $v_5v_6v_{16}v_{15}$. Then the resulting figure of P^o, after such motion, is a 2-story of modified 5-antiprism Q^o. We have a continuous folding process $\{P_t^o : 0 \le t \le 1\}$ from P^o to Q^o.

We show that the five rhombi of P can be continuously flat folded and inserted inside Q^o so that the motion is compatible with the continuous motion from P^o to Q^o. We denote by v^t the vertex in P_t^r corresponding to a vertex v in P^o.

For each $0 \leq t \leq 1/2$, since each rhombus (e.g.,$v_3^t v_{13}^t v_{19}^t v_9^t$) satisfies the distance conditions $dist\{v_{13}^t, v_9^t\} \leq dist\{v_{13}, v_9\}$ and that $dist\{v_3^t, v_{19}^t\}$ is not greater than the intrinsic distance of v_3 and v_{19}, the rhombus can be folded as shown in Fig. 1(b) by Lemma 1 so that those five rhombi do not intersect each other. Denote $P_t = P_t^o \cup R_t$ where R_t is the set of those folded rhombi compatible to P_t^o. Then $\{P_t : 0 \leq t \leq 1/2\}$ is a continuous folding process from P to $P_{1/2}$ which is a 2-story modified 5-antiprism with folded rhombi inside. $\qquad\square$

6   Continuous Flattening of a 2-story Modified Pentagonal Antiprisms

To get a continuous flattening motion for a regular dodecahedron, we show that a 2-story modified 5-antiprism can be continuously flattened.

Theorem 2. *A 2-story modified pentagonal antiprisms can be continuously flattened so that two pentagonal faces are rigid and one moves toward the other by rotation and translation only.*

Proof. Step 1. Q is a 2-story modified uniform pentagonal antiprisms in $\mathbb{R}^3$ whose vertices are on the cylinder $\{(x, y, z) : x^2 + y^2 = 1\}$ and $z = 0, 1$ or 2, and denoted by

$$v_i = (\cos(2(i-1)\pi/5, \sin(2(i-1)\pi/5, 0), \quad v_{5+i} = (\cos(2i-1)\pi/5, \sin((2i-1)\pi/5, 1)$$

and $v_{10+i} = (\cos 2(i - 1)\pi/5, \sin 2(i - 1)\pi/5, 2)$ for $i = 1, \cdots, 5$ (see Fig. 7). We refer to the three regular pentagonal sets of vertices of Q as bottom, middle and top according to $z = 0, 1$ and $z = 2$.

Step 2. By rotating and pushing down the top face toward the bottom face, Q can be continuously flattened. The motion for the vertices of the middle set follows the motion defined for $v_i (6 \leq i \leq 10)$ in the previous section, that is, v_7 moves along the circular arc about the line passing through $v_1 v_2$ toward v_3. The motion of the top face follows that of the motion of the middle set so that the rotated angle and height of the top face is twice those of the middle set.

Step 3. We define a continuous motion $\{Q_t : 0 \leq t \leq 1\}$ of Q. We denote by u^t the trace of $u \in Q$ for t.

$$v_{5+i}^t = (r_t \cos(2i - 1 + t)\pi/5, r_t \sin(2i - 1 + t)\pi/5, s_t),$$

$$v_{10+i}^t = (\cos(2i - 2 + 2t)\pi/5, \sin(2i - 2 + 2t)\pi/5, 2s_t)$$

for $i = 1, 2, \cdots 5$, where

$$r_t = (\cos \pi/10)/\cos(\pi/10 - \pi/5 \cdot t),$$

which is calculated from the fact that the distance of v_9^t from the z-axis equals $|O(v_9^t)'|$ where $(v_9^t)'$ is the projection of v_9^t to the xy-plane (see Fig. 6(a), (c)).

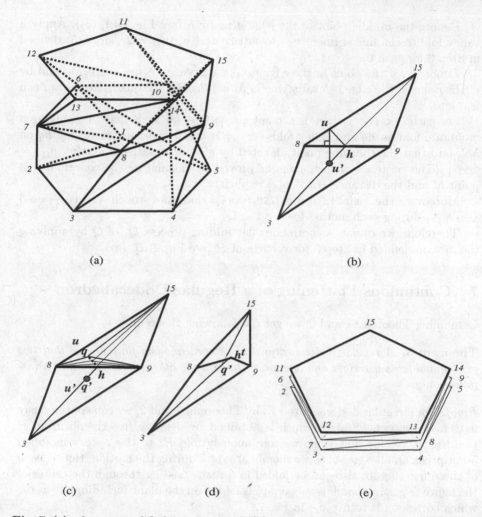

Fig. 7. (a) a 2-story modified 5-antiprism Q; (b) two isosceles triangular faces described in a common plane where uu' is orthogonal to v_8v_9; (c) the faces are described in a common plane with creases; (d) an example of a folded state; (e) a flat folded state of Q.

for $u'_t = v_9^t$ in Fig. 6(b).

Since $|v_1v_6^t| = |v_3v_9| = 2\sin\pi/5$, the z-coordinate s_t of v_9^t is

$$s_t = \sqrt{(3-\sqrt{5})/2 - (r_t)^2 + 2r_t\cos(\pi(1+t)/5)}.$$

Step 4. The ten isosceles faces of Q are divided into five pairs so that each pair has a common edge parallel to the bottom face and comprises a (folded) parallelogram. Those five pairs have similar motions with rotations about the z-axis. We define a motion for a pair $\{\triangle v_8v_9v_3, \triangle v_8v_9v_{15}\}$.

Denote the middle point of the edge v_8v_9 by h (see Fig. 7(b), (c)). Apply a valley fold to the line segment v_3h to satisfy $dist(v_8^t, v_9^t) = 2r_t \sin(\pi/5)$ defined in Step 3 for each $0 \le t \le 1$.

Denote by u the point in the edge v_8v_{15} satisfying $\angle v_8 h v_3 = \angle v_8 h u$, and by u' the point in the edge v_3h satisfying $|u'h| = |uh|$. So, the quadrilateral $v_8u'v_9u$ is a kite.

For each $0 < t < 1$ there is a point $q = q(t) \in hu$ such that by applying a mountain fold on hq and valley folds on v_8q, v_9q and $v_{15}q$, and attaching hq on hu', and that the touching point, denoted by q', on hu' satisfies $dist\{q', v_{15}^t\} = |qv_{15}|$. (The existence of such q can be proved by calculating the coordinates of point h^t and the distance $\{(u')^t, v_{15}^t\}$ explicitly.)

Moreover, the pair $\{\triangle v_8v_9v_3, \triangle v_8v_9v_{15}\}$ does not touch $\triangle v_3v_8v_{15}$ and $\triangle v_4v_9v_{11}$ during such motion for $0 < t < 1$.

Therefore, we obtain a continuous flat folding process Q_t of Q by applying the motion defined in Step 3 for vertices of P (see Fig. 7(d), (e)).   □

7   Continuous Flattening of a Regular Dodecahedron

Combining Theorems 1 and 2, we get the following theorem.

Theorem 3. *A regular dodecahedron can be continuously flattened so that two pentagonal faces are rigid and one moves toward the other by rotation and translation only.*

Proof. For a regular dodecahedron P, by Theorems 1 and 2, we can continuously flatten the subset P^o of P, which is obtained by deleting five rhombi from P (see Fig. 2). Notice that when we continuously fold P^o to the 2-story modified 5-antiprism Q, all edges of those rhombi are rigid during the motion. Hence, each of those five rhombi also can be folded in a plane passing through the center of the figure (e.g., the rhombus $v_3v_{13}v_{19}v_9$ is folded on the plane including $\triangle v_3v_9v_{19}$ which corresponds to $v_3v_9v_{11}$ in Fig. 7.

Therefore, during the continuous motion from the 2-story modified 5-antiprism to the flat folded state, those rhombi also can be folded to avoid any self-intersection.   □

References

1. Abel, Z., Demaine, E., Demaine, M., Itoh, J., Lebiw, A., Nara, C., O'Rourke, J.: Continuously flattening polyhedra using straight skeletons. In: Proceedings of 30th Annual Symposium on Computational Geometry (SoCG), pp. 396–405 (2014)
2. Demaine, E.D., O'Rourke, J.: Geometric Folding Algorithms, Linkages; Origami, Polyhedra. Cambridge University Press, Cambridge (2007)
3. Itoh, J., Nara, C.: Continuous flattening of platonic polyhedra. In: Akiyama, J., Bo, J., Kano, M., Tan, X. (eds.) CGGA 2010. LNCS, vol. 7033, pp. 108–121. Springer, Heidelberg (2011). doi:10.1007/978-3-642-24983-9_11

4. Itoh, J., Nara, C.: Continuous flattening of a regular tetrahedron with explicit mappings. Model. Anal. Inf. Syst. **19**(6), 127–136 (2012)
5. Itoh, J., Nara, C.: Continuous flattening of truncated tetrahedral. J. Geom. **107**(1), 61–75 (2016)
6. Itoh, J., Nara, C., Vîlcu, C.: Continuous flattening of convex polyhedra. In: Márquez, A., Ramos, P., Urrutia, J. (eds.) EGC 2011. LNCS, vol. 7579, pp. 85–97. Springer, Heidelberg (2012). doi:10.1007/978-3-642-34191-5_8
7. Nara, C.: Continuous flattening of some pyramids. Elem. Math. **69**(2), 45–56 (2014)

Escher-like Tilings with Weights

Shinji Imahori[1]([⊠]), Shizuka Kawade[2], and Yoko Yamakata[3]

[1] Department of Information and System Engineering,
Faculty of Science and Engineering, Chuo University, Tokyo, Japan
imahori@ise.chuo-u.ac.jp
[2] Graduate School of Engineering, Nagoya University, Nagoya, Japan
[3] Graduate School of Information Science and Technology,
The University of Tokyo, Tokyo, Japan

Abstract. A tiling of the plane is a set of figures, called tiles, that cover the plane without gaps or overlaps. On tiling we consider "Escherization problem": Given a closed figure in the plane, find a new closed figure that is similar to the original and can tile the plane. In this study, we give a new formulation of the problem with the weighted Procrustes distance and an algorithm to solve the problem optimally. We conduct computational experiments with animal shape tiles to confirm the effectiveness of the proposed method.

1 Introduction

A tiling is a set of figures that cover the plane without gaps or overlaps. Tilings have been used in decoration of walls and floors since ancient times. Tilings have not only artistic but also mathematical aspects, and hence many scientists studied placement rules, properties and varieties of tilings [3,5]. Escher [1,2] is one of the artists who made artistic tilings. He studied tilings from a mathematical viewpoint and made many artistic tilings with one or more kinds of tiles. In this paper, we consider the regular tiling with identical tiles. On this kind of tilings, we can consider *the Escherization problem* named after Escher. This problem is, given a line figure S, to find a new line figure T such that:

1. T resembles S as much as possible,
2. copies of T can cover the plane without gaps or overlaps.

Kaplan and Salesin [6] introduced this problem and proposed a method for it, but the method was inefficient for non-convex input figures. Koizumi and Sugihara [8] approximated the input figure with an n-cornered polygon and reformulated the Escherization problem as an optimization problem. In order to evaluate the similarity of two figures, they use *the Procrustes distance*. They solved this optimization problem in a polynomial time of the input size and showed the effectiveness of their method for not only convex but also non-convex input figures by numerical experiments.

   In this study, we introduce *weights* on the points (heavy weights for important points) of the input figure. We give a new formulation of the problem with *the*

J. Akiyama et al. (Eds.): JCDCGG 2015, LNCS 9943, pp. 132–142, 2016.
DOI: 10.1007/978-3-319-48532-4_12

weighted Procrustes distance and propose a polynomial time algorithm to solve the problem optimally. This new formulation enables us to output a tile along our preference. We conduct computational experiments with animal shape tiles to confirm the effectiveness of the proposed method.

2   Escherization Problem

In this section, we briefly explain the formulation of the Escherization problem given by Koizumi and Sugihara [8] and how to solve it.

2.1   Model of Escherization Problem

The Escherization problem [6] is, given a line figure S, to find a new line figure T such that:

1. T resembles S as much as possible,
2. copies of T can cover the plane without gaps or overlaps.

The first condition is restated as "minimization of the distance $d(S, T)$ between input and output figures." Koizumi and Sugihara [8] used the Procrustes distance to measure the similarity of two figures. Among shapes that can tile the plane, Koizumi and Sugihara only considered the isohedral tilings. It is known that isohedral tilings are easy to be treated mathematically and have enough flexibility to represent various shapes. The details of the objective function and the constraint conditions are explained in the next subsections.

2.2   Objective Function

An input line figure S is approximated with a counterclockwise sequence of n points on the boundary. We put the center of gravity at the origin and choose the first point. Let W (resp., U) be the $2 \times n$ matrix to express the input figure (resp., the output figure that can tile the plane). We use the Procrustes distance $d(U, W)$ for comparing two figures [7,12]; it is defined with matrices U and W with the center of gravity at the origin such that

$$d^2(U, W) = \min_{s, \theta} \left\| sR(\theta)\frac{U}{\|U\|} - \frac{W}{\|W\|} \right\|^2$$
$$= 1 - \frac{\|UW^\top\|^2 + 2\det(UW^\top)}{\|U\|^2\|W\|^2}, \tag{1}$$

where $\|X\|$ is the Frobenius norm of a matrix X, s is a scalar expressing expansion and contraction, and $R(\theta)$ is the matrix of rotation by θ. From the definition, the Procrustes distance is rotation, expansion, contraction and translation invariant. The objective of the Escherization problem is to minimize the distance

$d(U, W)$ between two matrices U and W. Minimizing the Procrustes distance $d(U, W)$ is equivalent to maximizing

$$\frac{\|UW^\top\|^2 + 2\det(UW^\top)}{\|U\|^2}.$$ (2)

Let matrices U, W and vectors u, w be

$$U = \begin{pmatrix} u_x^\top \\ u_y^\top \end{pmatrix}, W = \begin{pmatrix} w_x^\top \\ w_y^\top \end{pmatrix}, u = \begin{pmatrix} u_x \\ u_y \end{pmatrix}, w = \begin{pmatrix} w_x \\ w_y \end{pmatrix},$$

where u_x, u_y, w_x, w_y are n-dimensional column vectors. Then Eq. (2) can be rewritten as

$$\frac{u^\top V u}{u^\top u},$$ (3)

where V is the following $2n \times 2n$ symmetric matrix

$$V = \begin{pmatrix} w_x w_x^\top + w_y w_y^\top & w_x w_y^\top - w_y w_x^\top \\ w_y w_x^\top - w_x w_y^\top & w_x w_x^\top + w_y w_y^\top \end{pmatrix}.$$ (4)

2.3   Constraint Conditions

An isohedral tiling is a tiling in which any two tiles can be transformed to each other by an isometry that leaves the whole tiling unchanged. It is known that the isohedral tilings are classified into 93 types (named IH01–IH93). Constraint conditions to tile the plane depend on the types of the isohedral tilings. Here we consider the IH07 tiling with a hexagonal tile as an example. The incidence symbols of the IH07 tiling are shown in Fig. 1, and the constraint conditions are obtained from these incidence symbols. Let N be $n/6$ and we put n points as Fig. 2; i.e., this figure is represented as $U = (P_N, \ldots, P_1, P_0, P_1' \ldots, Q_N, \ldots, R_{N-1}')$. In Fig. 1, three arrows with the same labels meet at P_0, Q_0, R_0, and hence angles between these arrows ($\angle P_0, \angle Q_0, \angle R_0$) must be $120°$ for symmetry. With the

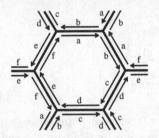

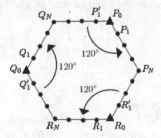

Fig. 1. Incidence symbols of IH07.   **Fig. 2.** Relationship between points.

matrix S of rotation by $120°$, the constraints "edges with the same label must be the same shape" can be formulated as

$$\begin{cases} S(P_i' - P_0) = P_i - P_0 \ (i = 1, \ldots, N) \\ S(Q_i' - Q_0) = Q_i - Q_0 \ (i = 1, \ldots, N) \\ S(R_i' - R_0) = R_i - R_0 \ (i = 1, \ldots, N), \end{cases} \tag{5}$$

where $P_N' = Q_N$, $Q_N' = R_N$ and $R_N' = P_N$ hold. Equations (5) are represented with linear combination of variables without constant terms. We note that other constraint conditions (e.g., translation) appeared in other tiling types can be written in a similar manner. Thus, with a matrix A (whose size is $n \times 2n$ for IH07), we can represent the constraint conditions for the isohedral tilings as

$$Au = \mathbf{0}. \tag{6}$$

Furthermore, with a matrix B composed of the orthonormal basis of $\mathrm{Ker}\,A$ and an arbitrary vector $\boldsymbol{\xi}$, Eq. (6) is rewritten as

$$u = B\boldsymbol{\xi}. \tag{7}$$

2.4   Eigenvalue Problem for Escherization Problem

With Eqs. (3) and (7), the Escherization problem can be rewritten as the following optimization problem without constraints:

$$\text{maximize} \quad \frac{\boldsymbol{\xi}^\top B^\top V B \boldsymbol{\xi}}{\boldsymbol{\xi}^\top \boldsymbol{\xi}}. \tag{8}$$

This is the Rayleigh quotient, and the optimization problem (8) is equivalent to the problem of calculating the maximum eigenvalue of a symmetric matrix $B^\top V B$. A standard method for this purpose is the power method, but it takes much time for some distributions of eigenvalues. Koizumi and Sugihara [8] proposed an explicit method for the problem, but it does not work for tiling types in which a flexible edge coincides with another edge by a glide reflection. Thus, we use a projection method to compute the maximal eigenvalue and its corresponding eigenvector for matrix $B^\top V B$. We use a fact that matrix $B^\top V B$ is a symmetric semidefinite matrix whose rank is at most 2. By using this property, we can compute the maximum eigenvalue and its corresponding eigenvector in $O(n^2)$ time.

We need to give attention to the following two things:

1. the matrix B is determined on the type of tilings,
2. the matrix V is changed by the choice of the first point.

Consequently, we need to solve the following optimization problem

$$\max_{i,j} \quad \frac{\boldsymbol{\xi}^\top B_i^\top V_j B_i \boldsymbol{\xi}}{\boldsymbol{\xi}^\top \boldsymbol{\xi}}, \tag{9}$$

where B_i ($i = 1, \ldots, 93$) is related to the type of the isohedral tiling IHi and V_j ($j = 1, \ldots, n$) depends on the first point. We note that isohedral tilings are categorized into 93 types, but it is enough to consider only 28 types for the Escherization problem. For example, any shape that can tile the plane with IH10 rule can also tile the plane with IH07 rule, hence we need not to consider IH10. We note that, when we treat tiling type IH21, we need to consider two cases for the original input figure and its mirrored figure.

We now evaluate the time complexity of Koizumi and Sugihara's algorithm. For computing B_i, it takes $O(n^3)$ time. It is noted that a matrix B_i depends only on the number n of points and the tiling type IHi. Thus it is possible to compute them in advance for a fixed number n of points. When i and j are fixed, the maximum eigenvalue and its corresponding eigenvector can be computed in $O(n^2)$ time by the projection method. In total, the algorithm runs in $O(n^3)$ time and outputs the figure that is closest to the input figure with the Procrustes distance.

3 Escherization Problem with Weights

By solving the problem stated in the previous section, good tiles are often obtained in a short computation time. However, we still have a question: Whether the Procrustes distance is a truly appropriate criterion to evaluate Escher-like tiles or not. See Fig. 3 for an example. We are given an input figure of "Pegasus" as Fig. 3(a). Figures 3(b) and (c) are shapes that can tile the plane. Which is the better tile for the input figure?

Figure 3(b) is better than Fig. 3(c) if you evaluate solutions with the Procrustes distance (the Procrustes distances to the input figure are 0.1103 and 0.1815, respectively). Figure 3(b) looks better than Fig. 3(c) if you focus on wings or legs. If you focus on the head, however, you may choose Fig. 3(c) as the better output tile for the input figure.

In this section, we introduce *weights* on the points of the input figure. We give a new formulation of the problem with *the weighted Procrustes distance*

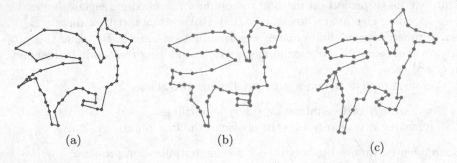

(a)   (b)   (c)

Fig. 3. (a) Input figure, (b) a tile (distance 0.1103), (c) another tile (distance 0.1815).

and propose an algorithm to solve the problem optimally. This new formulation enables us to output a tile along our preference.

3.1   Weighted Procrustes Distance

The weighted Procrustes distance $d_w(U, W, K)$ is used for evaluating the similarity of two figures with weights [9,10], which is defined as follows:

$$
\begin{aligned}
d_w^2(U, W, K) &= \min_{s,\theta} \left\| \left(sR(\theta)U - \frac{W}{\|WK\|} \right) K \right\|^2 \\
&= 1 - \frac{\|UK^2W^\top\|^2 + 2\det(UK^2W^\top)}{\|UK\|^2\|WK\|^2},
\end{aligned}
\tag{10}
$$

where k_i is the positive weight for point i and K is the $n \times n$ diagonal matrix whose diagonal elements are $k_1, k_2, \ldots, k_n$. We assume that the input and output figures are translated such that the matrices W, U satisfy the following equations

$$
\sum_{i=1}^{n} w_{xi}k_i^2 = 0, \quad \sum_{i=1}^{n} w_{yi}k_i^2 = 0,
\tag{11}
$$

$$
\sum_{i=1}^{n} u_{xi}k_i^2 = 0, \quad \sum_{i=1}^{n} u_{yi}k_i^2 = 0.
\tag{12}
$$

3.2   Formulation of Escherization Problem with Weights

We are given a $2 \times n$ matrix W that represents the input figure. We are also given an $n \times n$ diagonal matrix K whose diagonal elements $k_1, k_2, \ldots, k_n$ denote the positive weights for vertices i. The objective is to find a $2 \times n$ matrix U that represents the output figure, where the weighted Procrustes distance $d_w(U, W, K)$ is minimized and U can cover the plane without gaps or overlaps. This problem is summarized as follows:

Input: figure W and positive weights $k_1, k_2, \ldots, k_n$,
Output: figure U,
minimize   the weighted Procrustes distance $d_w(U, W, K)$,
subject to   figure U satisfies the conditions of an isohedral tiling,

where matrices W and U satisfy Eqs. (11) and (12).

Constraint conditions are very similar to the case without weights. The difference only appears in equations representing translation; that is, we use Eq. (12) instead of $\sum_{i=1}^{n} u_{xi} = 0$ and $\sum_{i=1}^{n} u_{yi} = 0$.

We then rewrite the objective function. Minimizing the weighted Procrustes distance can be represented as maximizing

$$
\frac{\|UK^2W^\top\|^2 + 2\det(UK^2W^\top)}{\|UK\|^2},
\tag{13}
$$

and it is also expressed as

$$\frac{u^\top G^2 V G^2 u}{u^\top G^2 u},$$ (14)

where V is the symmetric matrix defined in Eq. (4) and G is the $2n \times 2n$ diagonal matrix whose diagonal elements are $k_1, k_2, \ldots, k_n, k_1, k_2, \ldots, k_n$.

3.3 Eigenvalue Problem for Escherization Problem with Weights

We now have the following optimization problem to compute a tile whose weighted Procrustes distance to the input figure is minimized:

maximize $\dfrac{u^\top G^2 V G^2 u}{u^\top G^2 u}$,   (15)

subject to $Au = 0$.

As explained in Sect. 2, Koizumi and Sugihara [8] rewrote the constraints to $u = B\xi$ with a matrix B composed of *the orthonormal basis* of $\mathrm{Ker}A$ and an arbitrary vector ξ. We, however, rewrite the constraints to

$$u = B'\xi$$ (16)

with a matrix B' composed of *the basis* of $\mathrm{Ker}A$ and an arbitrary vector ξ. We note that the orthonormality is not required for tiling, and we explain how to choose a matrix B' later.

The optimization problem (15) can be rewritten as

maximize $\dfrac{\xi^\top B'^\top G^2 V G^2 B' \xi}{\xi^\top B'^\top G^2 B' \xi}$ (17)

with Eq. (16). If we consider *the generalized Rayleigh Quotient* and solve *a generalized eigenvalue problem*, it is possible to solve this optimization problem. However we consider another approach to solve the problem; if we choose a matrix B' that satisfies $B'^\top G^2 B' = I$ (where I is the identity matrix), then the problem (15) becomes the following optimization problem without constraints:

maximize $\dfrac{\xi^\top B'^\top G^2 V G^2 B' \xi}{\xi^\top \xi}$.   (18)

This is the (normal) Rayleigh quotient, and the optimization problem (18) is equivalent to the problem of calculating the maximum eigenvalue of the symmetric matrix $B'^\top G^2 V G^2 B'$. Let ξ' be an eigenvector corresponding to the maximum eigenvalue, then the output figure becomes $u = B'\xi'$.

We now explain how to compute a $2n \times m$ matrix B' that is composed of the basis of $\mathrm{Ker}A$ and satisfies $B'^\top G^2 B' = I$. Let $a_i\,(i = 1, 2, \ldots, 2n - m)$ be the set of vectors of A and $b'_i\,(i = 1, 2, \ldots, m)$ be the set of vectors of B' (where m depends on tiling types and is about n). Each vector b'_i must satisfy $(b'_i, a_l) = 0$ for $1 \le l \le 2n - m$ and $(Gb'_i, Gb'_l) = \delta_{il}$ for $1 \le l \le m$, where δ_{il}

is the Kronecker delta. We first apply the Gram-Schmidt orthonormalization to vectors $a_1, a_2, \ldots, a_{2n-m}$:

$$a_i^* = a_i - \sum_{k=1}^{i-1} (a_i, \hat{a}_k)\hat{a}_k, \tag{19}$$

$$\hat{a}_i = \frac{a_i^*}{\|a_i^*\|}. \tag{20}$$

We then compute b_i' with the following equations:

$$b_i^* = r_i - \sum_{k=1}^{2n-m} (r_i, \hat{a}_k)\hat{a}_k, \tag{21}$$

$$b_i^{**} = b_i^* - \sum_{k=1}^{i-1} (Gb_i^*, Gb_k')b_k', \tag{22}$$

$$b_i' = \frac{b_i^{**}}{\|Gb_i^{**}\|}. \tag{23}$$

It is easy to see that the resulting matrix B' satisfies the above two conditions.

We note that vectors $b_1', b_2', \ldots, b_m'$ are linearly independent under the assumption that every weight k_i must be positive.

3.4   Algorithm for Escherization Problem with Weights

We summarize the proposed algorithm and evaluate the time complexity. For a given input figure W with n points and weights $k_1, k_2, \ldots, k_n$, we solve the following optimization problem

$$\max_{i,j} \frac{\xi^\top B_{ij}'^\top G^2 V_j G^2 B_{ij}' \xi}{\xi^\top \xi}, \tag{24}$$

where i $(=1, 2, \ldots, 28)$ denotes the meaningful tiling types IHi and j $(=1, 2, \ldots, n)$ designates the first point. For computing B_{ij}', it takes $O(n^3)$ time. It should be noted that a matrix B_{ij}' depends on the number n of points, the tiling type IHi and given weights K, and hence we need to compute it for every input figure. (Given weights K on points will be different for different input figures.) On the other hand, when we consider the case without weights, we can use the same matrix B_i for a fixed number n of points and tiling type IHi.

When i and j are fixed, this optimization problem is equivalent to the maximum eigenvalue problem. By using the projection method in the previous section (where we use the characteristics of matrix $B_{ij}'^\top G^2 V_j G^2 B_{ij}'$), the problem can be solved in $O(n^2)$ time. In total, our algorithm runs in $O(n^4)$ time and outputs the figure that is closest to the input figure with the weighted Procrustes distance.

4 Computational Experiments

In this section, we show some computational results. The proposed algorithm was implemented in Java and ran on an ordinary personal computer with Intel Core i7-4770 (3.40 GHz) and 8 GB RAM. In our implementation, 28 types of isohedral tilings are considered. For each tiling type IHi, we compute the best tiling and finally output the best of them.

We are given an input figure with 60 points ($n = 60$) and their weights. In Fig. 4 we show results of our method: (a) represents the input figure "Nessie" with 60 points. In this figure, colored points have heavier weights ($k_i = 2$ for colored points and $k_i = 1$ for white points). Figure 4(b) is an optimal tile for Fig. 4(a) if all points have identical weights. In other words, this is the output tile by Koizumi and Sugihara's method [8]. Figure 4(c) is an optimal tile when we evaluate the similarity by the weighted Procrustes distance. Figure 4(d) is a tiling generated by Fig. 4(c). We also conduct experiments with a "pig" and results are represented in Fig. 5. These results show that the proposed method can keep the shapes of a set of heavy weight points. We also report the Procrustes distance and the weighted Procrustes distance between input figures and output tiles in Table 1. The row of "Procrustes" shows the Procrustes distance between the input figure and output tiles for "Nessie" and "pig" instances ignoring the weights on points. The row of "weighted" reports the weighted

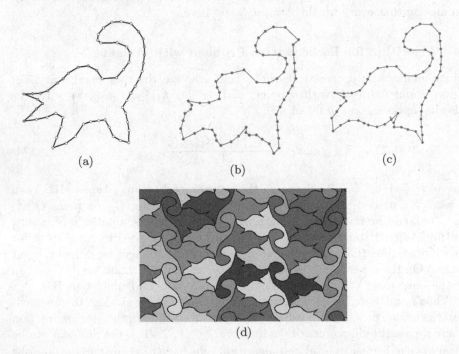

(a)   (b)   (c)

(d)

Fig. 4. Escher-like tiling of a Nessie; (a) input figure, (b) optimal solution without weights, (c) optimal solution with weights, (d) Nessie tiling generated by (c).

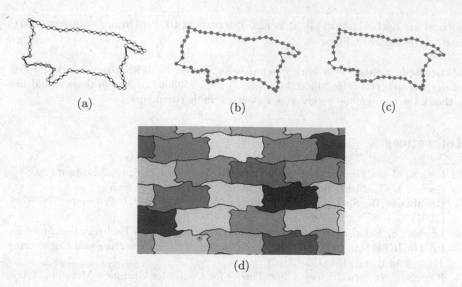

(a)

(b)

(c)

(d)

Fig. 5. Escher-like tiling of a pig; (a) input figure, (b) optimal solution without weights, (c) optimal solution with weights, (d) tiling of pigs generated by (c).

Table 1. Distance between input figure and output tile.

	Nessie		Pig	
	(b)	(c)	(b)	(c)
Procrustes	0.141	0.160	0.078	0.081
Weighted	0.149	0.122	0.077	0.072

Procrustes distance between them. It takes about 0.8 s to compute a tile minimizing the Procrustes distance and about 4.5 s for the weighted Procrustes distance.

5 Conclusions

For the Escherization problem, we introduced a new formulation with weights. By putting heavy weights for a part of the input figure, we can keep the shape of the region as much as possible. We applied the weighted Procrustes distance to evaluate the similarity of figures, and proposed a method to compute an optimal tile along the weights efficiently. We conducted computational experiments with complex input figures and confirmed the effectiveness of the proposed method.

One of our next goal is to compute appropriate weights on points automatically. Some computational geometry concepts including the local feature size [11] will be useful for this question. Another goal is to improve the quality of output tiles by combining with some other techniques such as a local-search based method for Escherization [4]. It is also interesting to treat more complex tilings than the isohedral tilings.

Acknowledgments. This work was partly supported by JSPS Grant-in-Aid for Scientific Research (B) (No. 24360039) and (C) (No. 25330024). The authors would like to thank the anonymous reviewers for their valuable comments.

References

1. Escher, M.C.: The Graphic Work, Taschen America Llc, Special Edition (2008)
2. Escher, M.C.: The Official Website. http://www.mcescher.com/
3. Grünbaum, B., Shephard, G.C.: Tiling and Patterns. W. H. Freeman, New York (1987)
4. Imahori, S., Sakai, S.: A local-search based algorithm for the Escherization problem. In: The IEEE International Conference on Industrial Engineering and Engineering Management, pp. 151–155 (2012)
5. Kaplan, C.S.: Introductory Tiling Theory for Computer Graphics. Morgan & Claypool Publishers, San Rafael (2009)
6. Kaplan, C.S., Salesin, D.H.: Escherization. In: Proceedings of SIGGRAPH, pp. 499–510 (2000)
7. Kendall, D.G.: Shape manifolds, procrustean metrics, and complex projective spaces. Bull. Lond. Math. Soc. **16**, 81–121 (1984)
8. Koizumi, H., Sugihara, K.: Maximum eigenvalue problem for Escherization. Graphs Comb. **27**, 431–439 (2011)
9. Koschat, M.A., Swayne, D.F.: A weighted Procrustes criterion. Psychometrika **56**, 229–239 (1991)
10. Mooijaart, A., Commandeur, J.J.F.: A general solution of the weighted orthonormal Procrustes problem. Psychometrika **55**, 657–663 (1990)
11. Ruppert, J.: A Delaunay refinement algorithm for quality 2-dimensional mesh generation. J. Algorithms **18**, 548–585 (1995)
12. Werman, M., Weinshall, D.: Similarity and affine invariant distances between 2D point sets. IEEE Trans. Pattern Anal. Mach. Intell. **17**, 810–814 (1995)

Number of Ties and Undefeated Signs in a Generalized Janken

Hiro Ito[1,2](✉) and Yoshinao Shiono[1]

[1] The University of Electro-Communications (UEC), Tokyo, Japan
itohiro@uec.ac.jp
[2] CREST, JST, Tokyo, Japan

Abstract. Janken, which is a very simple game and is usually used as a coin-toss in Japan, originated in China, and many variants are seen throughout the world. A variant of janken can be represented by a tournament (a complete asymmetric digraph), where a vertex corresponds to a sign and an arc (x, y) indicates that sign x defeats sign y. However, not all tournaments define useful janken variants, i.e., some janken variants may include a useless sign, which is strictly inferior to any other sign in any case. In a previous paper by one of the authors, a variant of janken (or simply janken) was said to be *efficient* if it contains no such useless signs, and some properties of efficient jankens were presented. The jankens considered in the above research had no tie between different signs. However, some actual jankens do include such ties. In the present paper, we investigate jankens that are allowed to have a tie between different signs. That is, a janken can be represented as an asymmetric digraph, where no edge between two vertices x and y indicates a tie between x and y. We first show the tight upper and lower bounds of the number of ties in an efficient janken with n-vertices. Moreover, it is shown that for any integer t between the upper and lower bounds, there is an efficient janken having just t ties. We next consider *undefeated* vertices, which are vertices that are not defeated by any sign. We show that there is an efficient janken with n vertices such that the number of vertices that are *not* undefeated is $o(n)$, i.e., almost all vertices are undefeated.

1 Introduction

1.1 Background

Janken is a simple game to decide a winner by simultaneously holding out one hand in one of three gestures (signs) to signify *rock* (closed hand), *paper* (open hand), or *scissors* (closed hand with index and middle fingers extended). As such, janken is also called *rock-paper-scissors*. Rock defeats scissors, scissors defeats paper, and paper defeats rock. These relations can be represented by an asymmetric complete digraph (a *tournament*), where an arc (x, y) indicates that x defeats y. See Fig. 1(a), for example.

© Springer International Publishing AG 2016
J. Akiyama et al. (Eds.): JCDCGG 2015, LNCS 9943, pp. 143–154, 2016.
DOI: 10.1007/978-3-319-48532-4_13

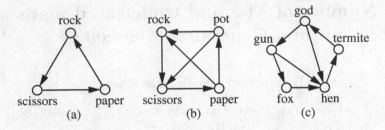

Fig. 1. Janken variants represented by digraphs

Janken has many variants throughout the world [4,6]. For example, in a part of France, a local variant includes pot^1 (forming a hole) as an additional sign, and hence four signs are used. Pot defeats rock and scissors (since they are sunk) but is defeated by paper (since it covers the mouth of the pot). This variant of janken can also be represented by a corresponding tournament, which has four vertices (see Fig. 1(b)), referred to herein as *pot-janken*. Pot-janken has a curious feature, namely, throwing pot is always better than throwing rock, since both defeat scissors and are defeated by paper, but pot defeats rock. Thus, rock is never used, and so, in a practical sense, pot-janken is essentially identical to the original version of janken.

A janken variant (or simply a janken) is said to be *efficient* if it contains no useless signs, and exhibits the properties of an efficient janken [2]. In the abovementioned study, however, we considered jankens that have no ties between different signs. In fact, there are jankens that have such ties, e.g., a janken in a part of Guangdong, China, uses five signs, god, hen, gun, fox, and termite; and "god and fox," "gun and termite," and "fox and termite" result in ties (see Fig. 1(c)) [4].

Thus, the previous study cannot be applied to such jankens directly.

1.2   Contribution of the Present Study

In the present paper, we investigate jankens that are allowed to have ties between different signs, such as "god and fox" in the Guangdong janken. A janken can be represented by an asymmetric digraph $G = (V, A)$, $(x, x) \notin A$ for all $x \in V$ and $|\{(x, y), (y, x)\} \cap A| \leq 1$ for all $x, y \in V$.

Definition 1. *A* janken *is an asymmetric digraph. A janken with n vertices is called an n-*janken. *For a janken* $G = (V, A)$ *and a distinct vertex pair* $x, y \in V$, *if* $(x, y) \in A$, *then we say that* x *defeats* y. *If* $\{(x, y), (y, x)\} \cap A = \emptyset$, *then the pair* $\{x, y\}$ *is called a* tie. *The set of ties is denoted by* T_G, *i.e.,* $T_G = \{\{x, y\} \mid x, y \in V, x \neq y, \{(x, y), (y, x)\} \cap A = \emptyset\}$. *Here,* T_G *may be written as* T *if* G *is clear.*

Definition 2. *For a janken* $G = (V, A)$, *if a pair of vertices* x *and* y *satisfies the following conditions, then* x *is* superior *to* y, *and* y *is* useless:

- $(y, x) \notin A$.
- *for any vertex* $z \in V$;

---

^{1} Sometimes "well" is used in place of pot.

- *if $(y, z) \in A$, then $(x, z) \in A$, and*
- *if $(z, y) \notin A$, then $(z, x) \notin A$.*

If a janken has no useless vertices, then the janken is said to be efficient.

Definition 3. *For a janken $G = (V, A)$, let $t(G)$ be the number of ties, i.e., $t(G) = |T_G|$. Clearly, $t(G) = \binom{n}{2} - |A|$, where $n = |V|$. Let $t_{\max}(n)$ (resp., $t_{\min}(n)$) be the maximum (resp., minimum) number of $t(G)$ among efficient n-jankens G.*

We have the following theorem.

Theorem 1. *For $n \geq 3$, the following equations hold:*

$$t_{\max}(n) = \begin{cases} 0, & \text{if } n = 3, \\ \binom{n}{2} - n + 1, & \text{if } n \geq 4. \end{cases}$$

$$t_{\min}(n) = \begin{cases} 1, & \text{if } n = 4, \\ 0, & \text{otherwise.} \end{cases}$$

Moreover, for any integer $t_{\min}(n) \leq t \leq t_{\max}(n)$, there is an efficient n-janken G with $t(G) = t$. □

Note that there are no efficient 2-jankens, and, for $n = 1$, only the trivial 1-janken, which consists of one vertex and no arcs, is efficient.

Next, we consider undefeated vertices.

Definition 4. *A vertex $x \in V$ is said to be* undefeated *if $(y, x) \notin A$ for all $y \in V$. If x is not undefeated, then it is said to be* ordinary. *For an n-janken G, let $\nu_0(G)$ and $\nu_1(G)$ be the numbers of undefeated vertices and ordinary vertices, respectively, in G*

Note that $\nu_0(G) + \nu_1(G) = n$. We then obtain the following result.

Theorem 2. *For any positive integer m, there exists an efficient $\binom{2m+1}{m}$-janken G with $\nu_1(G) = 2m + 1$.* □

This means that, for any integer n_0, there is an integer $n \geq n_0$ and an efficient n-janken such that the number of vertices that are *not* undefeated is $o(n)$, i.e., almost all vertices are undefeated.

1.3   Definitions

The following definitions are used in the present paper. For a janken $G = (V, A)$, let $U(G) = (V, E(A))$ be the *corresponding undirected graph* of G such that $E(A) := \{(x, y) \mid (x, y) \in A\}$. A vertex subset $W \subseteq V$ is called a *component* of G if it is a connected component of $U(G)$.

For any integer $n \geq 2$, an *n-cycle* is a digraph isomorphic to $(\{0, 1, \ldots, n-1\}, \{(0, 1), (1, 2), \ldots, (n-2, n-1), (n-1, 0)\})$ and it is represented by C_n. (In the present paper, we consider only asymmetric digraphs, and thus C_2 never

appears.) For any integer $n \geq 2$, an n-path is a digraph isomorphic to
$(\{0, 1, \ldots, n-1\}, \{(0,1), (1,2), \ldots, (n-2, n-1)\})$ and is represented by P_n.
A digraph isomorphic to $(\{0\}, \emptyset)$ is represented by K_1. If a digraph includes no
n-cycles as subgraphs for all $n \geq 2$, then it is said to be *acyclic*, and otherwise is
said to be *cyclic*. For two digraphs $G_1 = (V_1, A_1)$ and $G_2 = (V_2, A_2)$, a digraph
$G_1 + G_2$ is defined as $(V_1 \cup V_2, A_1 \cup A_2)$. For other basic terms, see [1,3,5].

2   Number of Ties

2.1   Upper and Lower Bounds

In this section, we consider jankens that have the maximum number of ties. Having the maximum number of ties is equivalent to having the minimum number
of arcs. Although it may appear that such a janken must be a cycle C_n (see
Fig. 2(a)), this is not correct. An n-janken having the maximum number of ties
is $C_{n-1} + K_1$ (see Fig. 2(b)). Surprisingly, the janken includes an isolated vertex,
which will never win or lose! However, we can easily confirm that no other vertex
is superior to the isolated vertex, and vice versa. Here, C_{n-1} can be separated
into a number of cycles (see Fig. 2(c)).

The values of $t_{\min}(n)$ for all $n \geq 3$, except for $n = 4$ and $t_{\max}(3)$ in Theorem 1,
have already been obtained in [2]. Then, $t_{\min}(4) = 1$ is obtained from janken
$(\{0, 1, 2, 3\}, \{(0, 1), (0, 2), (1, 2), (2, 3), (3, 4)\})$ and is efficient. This janken is the
right-most janken in Fig. 3, which appears in Sect. 2.2.

In order to show $t_{\max}(n)$ for $n \geq 4$, we introduce the complementary value
of $t_{\max}(n)$, as follows:

$$m_{\min}(n) := \binom{n}{2} - t_{\max}(n).$$

We first show $m_{\min}(n) = n - 1$ for $n \geq 4$. We previously defined undefeated
vertices, and we next introduce a complementary concept as follows.

Definition 5. *A vertex $x \in V$ is said to be* zero-defeating *if $(x, y) \notin A$ for all
$y \in V$. If a vertex x is both undefeated and zero-defeating, then it is said to be*
isolated.

We first describe a simple observation. The proof is directly obtained from
the definition and so is omitted.

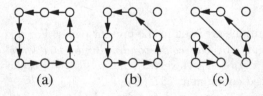

(a)   (b)   (c)

Fig. 2. Efficient 8-jankens with large ties (= small arcs): (a) eight arcs, (b) and (c)
seven arcs (minimum).

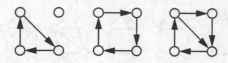

Fig. 3. Examples of the continuous existence of the number of ties for $n = 4$.

Lemma 1. *If a janken $G = (V, A)$ includes two distinct vertices $x \in V$ that is undefeated and $y \in V$ that is zero-defeating, then x is superior to y, and thus G is not efficient.*

The following corollary also follows directly.

Corollary 1. *An efficient janken $G = (V, A)$ includes at most one isolated vertex.*

The following lemma is also easily obtained from Lemma 1.

Lemma 2. *If a janken $G = (V, A)$ includes an acyclic component consisting of at least two vertices, then it is not efficient.*

Proof. An acyclic component consisting of at least two vertices includes two distinct vertices x and y such that x has no edge entering it, and y has no edge leaving it. In other words, x is an undefeated vertex, and y is a zero-defeating vertex. Then, the result follows from Lemma 1.   □

Next, we prove the first half of Theorem 1 using the following lemma:

Lemma 3. *All values of $t_{\max}(n)$ and $t_{\min}(n)$ in Theorem 1 are correct.*

Proof. Assume $n \geq 4$. Since it is clear that $(C_{n-1} + K_1)$ is efficient, $m_{\min}(n) \leq n - 1$ is obtained. Thus, we show that $m_{\min}(n) \geq n - 1$. Let G be an efficient n-janken. From Corollary 1, G includes at most one isomorphic vertex. In other words, any component other than the isomorphic vertex consists of at least two vertices. From Lemma 2, any such component is cyclic. Any cyclic component with k vertices has at least k arcs. It follows that G has at least $n - 1$ arcs, i.e., $m_{\min}(n) \geq n - 1$. Therefore, $t_{\max}(n) = \binom{n}{2} - m_{\min}(n) = \binom{n}{2} - n + 1$ is proven.

For the remaining values, $t_{\min}(4) = 1$ is shown by the fact that janken $(\{0, 1, 2, 3\}, \{(0, 1), (0, 2), (1, 2), (2, 3), (3, 4)\})$ is efficient. In [2], $t_{\max}(3) = t_{\min}(3) = 0$ and $t_{\min}(n) = 0$ for $n \geq 5$ have been shown. Therefore, we have proven all of these bounds.   □

2.2 Algorithm for Constructing a Series of Jankens Having a Continuous Number of Ties

The last half of Theorem 1 is proven by the following lemma.

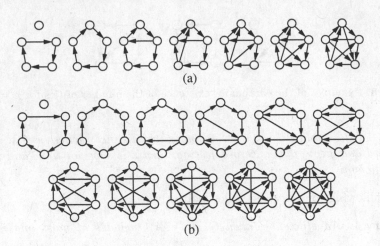

Fig. 4. Examples of the continuous existence of the number of ties for (a) $n = 5$ and (b) $n = 6$.

Lemma 4. *For any integers $n \geq 3$ and $t_{\min}(n) \leq t \leq t_{\max}(n)$, there is an efficient n-janken G with $t(G) = t$.*

For small n, we can prove this by presenting concrete examples. Figures 3 and 4(a), (b) show the cases of $n = 4, 5$, and 6, respectively. We can check the jankens one by one to confirm that all of them are efficient. (Using Lemmas 5 and 7, which are presented later herein, makes this task easier.)

For the general $n \geq 4$, we present an algorithm that constructs $G_n^i = (V_n, A_n^i)$, denoting the efficient n-janken with i ties for every integer $0 \leq i \leq \binom{n}{2} - n + 1$. For the case of $n \leq 6$, examples are shown in Figs. 3 and 4.

Next, we consider the case of $n \geq 7$. From the previous discussion, we know that $G_n^{\binom{n}{2}-n+1} = C_{n-1} + K_1$ and $G_n^{\binom{n}{2}-n} = C_n$. We represent the latter relation as

$$V_n = \{0, 1, \ldots, n-1\}.$$

$$A_n^{\binom{n}{2}-n} = \{(0,1), (1,2), \ldots, (n-2, n-1), (n-1, 0)\}.$$

For $i = \binom{n}{2} - n - 1, \binom{n}{2} - n - 2, \ldots, 0$, the algorithm generates G_n^i by simply adding an arc a_i to G_n^{i+1}. For this purpose, the algorithm calls CYCLEADD(n), which outputs a series of arcs $a_{\binom{n}{2}-n-1}, a_{\binom{n}{2}-n-2}, \ldots, a_0$ in this order.

Each a_i for $i = \binom{n}{2} - n - 1, \ldots, \binom{n}{2} - 3n - 7$ is defined as follows (see Fig. 5):

$$a_i = \begin{cases} (n-3, 0), & \text{if } i = \binom{n}{2} - n - 1, \\ (n-1, j), & \text{if } i = \binom{n}{2} - n - 1 - j \text{ for } j = 1, \ldots, n-3, \\ (j, n-2), & \text{if } i = \binom{n}{2} - 2n + 1 - j \text{ for } j = 0, \ldots, n-4. \end{cases} \tag{1}$$

When i becomes $\binom{n}{2} - 3n + 5$, i.e., we obtain $G_n^{\binom{n}{2}-3n+5}$, then vertices $n-2$ and $n-1$ are fully connected to the other vertices. On the other hand, $G_n^{\binom{n}{2}-3n+5}$

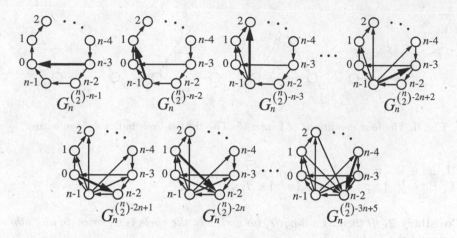

Fig. 5. Addition of arcs by CYCLEADD(n): each bold arc is a_i.

contains a chord-free cycle C_{n-2}, i.e., the subdigraph of $G_n^{\binom{n}{2}-3n+5}$ induced by $V_{n-2} = \{0, 1, \ldots, n-3\}$ is C_{n-2}. (Note that $\binom{n}{2} - 3n + 5 = \binom{n-2}{2} - (n-2)$.)

Here, CYCLEADD(n) recursively calls CYCLEADD($n-2$) and adds arcs to C_{n-2} one by one, in the same manner as with C_n if $n \geq 7$, or in the manner shown in Fig. 4 if $n = 5$ or 6.

Following this algorithm, we finally obtain G_n^0, which is a complete asymmetric digraph. A formal expression of CYCLEADD(n) for $n \geq 4$ is shown as follows.

Procedure CYCLEADD(n)
begin
if $n \leq 6$ **then**
    output $a_{\binom{n}{2}-n-1}, \ldots, a_0$ according to Fig. 3 or 4;
else
    output $a_{\binom{n}{2}-n-1}, \ldots, a_{\binom{n}{2}-3n-5}$ according to (1);
    call CYCLEADD($n-2$)
stop
end.

2.3 Correctness of the Algorithm

In this subsection, we prove the correctness of CYCLEADD(n). First, we introduce a condition for determining whether a vertex is useless. The proof is omitted, since it follows from the definitions.

Lemma 5. *Let $G = (V, A)$ be an n-janken. Let $x, y \in V$ be a pair of vertices. A necessary and sufficient condition for x to not be superior to y is that at least one of the following conditions holds (see Fig. 6):*

I. $(y, x) \in A$,
II. $\exists z \in V, (y, z), (z, z) \in A$,

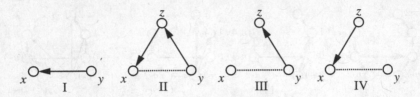

Fig. 6. The four conditions of Lemma 5: The dotted lines indicate "don't care."

III. $\exists z \in V$, $(y, z) \in A$ and $\{x, z\} \in T_G$,
IV. $\exists z \in V$, $(z, x) \in A$ and $\{y, z\} \in T_G$.

Corollary 2. *If there is a 3-cycle, no vertex in the cycle is superior to any other vertex in the cycle.*

We introduce an operation for adding two new vertices to a janken, as follows.

Definition 6. *For an n-janken $G = (V, A)$, let $G^+ = (V^+, A^+)$ be an $(n + 2)$-janken constructed as follows, where $u, v \notin V$:*

$$V^+ = V \cup \{u, v\}, \quad A^+ = A \cup \{(u, w), (w, v) \mid \forall w \in V\} \cup \{(v, u)\}.$$

Lemma 6. *If G is efficient, then G^+ is also efficient.*

Proof. For $x, y \in V$, x is not superior to y in G. From Lemma 5, x and y satisfy one of conditions I through IV. Since adding a new vertex never causes a condition not to be satisfied, x is not superior to y in G^+. For u, v, and $\forall w \in V$, based on Corollary 2, since there is a 3-cycle ($\{u, w, v\}, \{(u, w), (w, v), (v, u)\}$), no vertex in the cycle is superior to any other vertex in the cycle.   □

Lemma 7. *Let G be an efficient n-janken, and let $\{v, w\} \in T_G$ be a tie pair. Let $G' = (V, A')$ be the graph made by adding an arc (v, w) to G, i.e., $A' = A \cup (v, w)$. If there is a pair of vertices $x, y \in V$ such that x is superior to y in G', then at least one of the following two conditions holds:*

i. $x = v$ and $(y, w) \in A$,
ii. $y = w$ and $(v, x) \in A$.

Proof. Although x is not superior to y in G, x is superior to y in G'. From Lemma 5, it follows that one of conditions I through IV is not satisfied by adding arc (v, w). This only occurs by adding arc (x, z) to condition III or by adding arc (z, y) to condition IV. (See Fig. 7.) The former is case i, and the latter is case ii.   □

We define $C'_n := G_n^{\binom{n}{2} - n - 1} = (V_n, A_n^{\binom{n}{2} - n - 1})$ and the following arc subsets:

$$A_1 := \{(n - 1, i) \mid i \in \{1, \ldots, n - 3\}\},$$
$$A_2 := \{(i, n - 2) \mid i \in \{0, \ldots, n - 4\}\}.$$

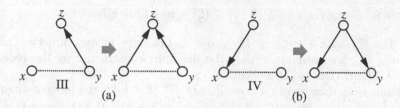

Fig. 7. (a) Adding arc (x, z) to III, and (b) adding arc (z, y) to IV.

Lemma 8. *For any subset of arcs $A \subseteq A_1 \cup A_2$, $G_A = (V_n, A_n^{\binom{n}{2} - n - 1} \cup A)$ is efficient.*

Proof. An n-cycle is clearly efficient. Assume that the arcs in A are added one by one. If G_A is not efficient, there is an arc, say $a \in A \cup \{(0, n - 3)\}$, such that the janken changes from efficient to inefficient when a is added. We separately prove the following three cases:

- If $a = (n - 3, 0)$:
  - If Fig. 7(a) (case i of Lemma 7) occurs: $x = n - 3$, $z = 0$, and thus y must be $n - 1$, i.e., the only possibility is that $n - 3$ is superior to $n - 1$. However, by the existence of vertex $n - 4$, Case IV of Lemma 5 occurs, and thus $n - 3$ is not superior to $n - 1$.
  - If Fig. 7(b) (case ii of Lemma 7) occurs: $z = n - 3$, $y = 0$, and thus x must be $n - 2$, i.e., the only possibility is that $n - 2$ is superior to 0. However, by the existence of vertex 1, Case III of Lemma 5 occurs, and thus $n - 2$ is not superior to 0.
- If $a = (n - 1, i) \in A_1$:
  - If Fig. 7(a) occurs: $x = n - 1$, $z = i$, and thus y must be $i - 1$, i.e., the only possibility is that $n - 1$ is superior to i. However, by the existence of vertex $n - 2$, Case II or IV of Lemma 5 occurs, and thus $n - 1$ is not superior to $i - 1$.
  - If Fig. 7(b) occurs: $z = n - 1$, $y = i$, and thus x must be 0, i.e., the only possibility is that 0 is superior to i. However, by the existence of vertex $n - 3$, Case II or III of Lemma 5 occurs, and thus 0 is not superior to i.
- If $a = (i, n - 2) \in A_2$:
  - If Fig. 7(a) occurs: $x = i$, $z = n - 2$, and thus y must be $n - 3$, i.e., the only possibility is that i is superior to $n - 3$. However, by the existence of vertex 0, Case II or III of Lemma 5 occurs, and thus i is not superior to $n - 3$.
  - If Fig. 7(b) occurs: $z = i$, $y = n - 2$, and thus x must be $i + 1$, i.e., the only possibility is that $i + 1$ is superior to $n - 2$. However, by the existence of vertex $n - 1$, Case II or III of Lemma 5 occurs, and thus $i + 1$ is not superior to $n - 2$.

From the above discussions, in no case is one vertex superior to any other vertex in G_A, i.e., G_A is efficient. □

Lemma 9. *Every janken G_n^i ($0 \leq i \leq \binom{n}{2} - n + 1$) is efficient.*

Proof. For $\binom{n}{2} - 3n + 5 \leq i \leq \binom{n}{2} - n + 1$, Lemma 8 follows directly. For the remaining G_n^i, we must prove that the digraphs constructed by the recursive calls of CYCLEADD are also effective.

Assume that there exists a non-efficient G_n^i. Let k be the integer satisfying $\binom{n-2k}{2} - (n - 2k) \leq i < \binom{n-2(k+1)}{2} - (n - 2(k + 1))$. In other words, G_n^i is constructed in calling CYCLEADD($n - 2k$). From the above discussions, $k \geq 1$. Let F_n^i be the subdigraph of G_n^i induced by $\{0, 1, \ldots, n - 2k\}$. From Lemma 8, F_n^i is efficient. G_n^i is constructed by applying the operation $^+$ (of Definitions 6) k times to F_n^i. Since F_n^i is efficient, from Lemma 6, G_n^i is efficient, which is a contradiction.   □

Next, we finish the proof of Theorem 1.

Proof of Theorem 1: This theorem is easily proven by combining Lemmas 3 and 9.   □

3   Number of Undefeated Vertices and Zero-Defeating Vertices

In this section, we investigate the number of undefeated or zero-defeating vertices that can be included in an efficient janken. From the symmetry of the undefeated and zero-defeating vertices, we assume that there is at least one undefeated vertex, say z, in an efficient n-janken $G = (V, A)$.

If z is an isolated vertex, from Lemma 1, there are no other undefeated or zero-defeating vertices in G. Thus, we assume that z is not zero-defeating. Again, from Lemma 1, there are no zero-defeating vertices in G. Therefore, it is sufficient to consider the case in which G includes no zero-defeating vertices, and we investigate the number of undefeated vertices that G can include.

We will show that, for any positive integer m, there exists an efficient $\binom{2m+1}{m}$-janken G with $\nu_1(G) = 2m + 1$ (Theorem 2). This means that there is an efficient n-janken with $\nu_1(G) = o(n)$.

Proof of Theorem 2: This theorem is proven by constructing such a graph.

Let $m \geq 1$ be a positive integer, and let $n = 2m + 1$. $F_n = (X_n, B_n)$ denotes an n-janken such that

$$X_n = \{0, 1, \ldots, n - 1\},$$
$$B_n = \{(i, i + j) \mid i \in \{0, \ldots, n - 1\}, j \in \{1, \ldots, m\}\},$$

where the indices are taken in cyclic order, with 0 following $n - 1$. For example, see Fig. 8(a). Let $\mathcal{S}_n \subset 2^{\{0, \ldots, n-1\}}$ be the family of subsets $S \subset \{0, \ldots, n-1\}$ such that $|S| = m$, where S does not contain m consecutive integers. For examples,

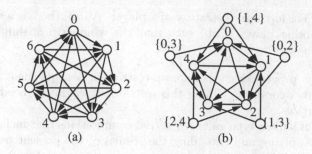

Fig. 8. (a) F_7, (b) G_5.

$$\mathcal{S}_5 = \{\{0,2\},\{0,3\},\{1,3\},\{1,4\},\{2,4\}\},$$
$$\mathcal{S}_7 = \{\{0,1,3\},\{0,1,4\},\{0,1,5\},\{0,2,3\},\{0,2,4\},\{0,2,5\},\{0,2,6\},\{0,3,4\},$$
$$\{0,3,5\},\{0,3,6\},\{0,4,5\},\{0,4,6\},\{1,2,4\},\{1,2,5\},\{1,2,6\},\{1,3,4\},$$
$$\{1,3,5\},\{1,3,6\},\{1,4,5\},\{1,4,6\},\{1,5,6\},\{2,3,5\},\{2,3,6\},\{2,4,5\},$$
$$\{2,4,6\},\{2,5,6\},\{3,4,6\},\{3,5,6\}\}.$$

We construct the janken $G_n = (V_n = X_n \cup \mathcal{S}_n, A_n = B_n \cup E_n)$ such that

$$E_n = \{(S,i) \mid S \in \mathcal{S}, i \in S\}.$$

For example, see Fig. 8(b). All vertices in $\mathcal{S}_n$ are undefeated.

Next, we confirm that G_n is efficient. For a vertex $x \in V_n$, let $N(x)$ be the subset of vertices y such that $(x,y) \in A_n$, i.e., $N(x) = \{y \in V_n \mid (x,y) \in A_n\}$. For any pair of distinct vertices $x, y \in V_n$ $(x \neq y)$, $N(x) - N(y) \neq \emptyset$ and $N(y) - N(x) \neq \emptyset$. From the definitions, no pair is superior to the others. Thus, G_n is efficient. $|V_n| = \binom{2m+1}{m}$, and $\nu_1(G_n) = |X_n| = 2m + 1$. $\qquad\square$

4 Summary and Future Work

In the present paper, we extended the results presented in [2] to jankens that allow ties. We presented two theorems, one regarding the number of ties and one regarding the number of undefeated vertices. Since undefeated vertices are invincible in a sense, a janken may never finish if both players use these vertices. Thus, the existence of undefeated vertices may make a janken meaningless. For a janken that has no ties, even though "efficient (no useless signs)" would seem to be a sufficient restriction to make the janken useful, for a janken with ties, the property may not be enough. We can introduce a stronger restriction, as follows:

Restriction: For any mixed strategy of a player (Alice), there is a mixed strategy of the other player (Bob) such that the winning probability of Bob is greater than zero.

It is easily proven that the above restriction is equivalent to the no-undefeated-vertex condition. Under this restriction, we do not need to consider undefeated vertices.

A number of janken-type games or actual competitions that include ties have applications. Applying and extending the results of the present paper to such games and competitions is an interesting area for future work.

Acknowledgements. The present study was supported in part by the Algorithms on Big Data Project (ABD14) of CREST, JST, the ELC project (MEXT KAKENHI Grant Number 24106003), and JSPS KAKENHI (Grant Numbers 24650006 and 15K11985).

References

1. Chartrand, G., Lesniak, L., Zhang, P.: Graphs & Digraphs, 5th edn. CRC Press, Boca Raton (2011)
2. Ito, H.: How to generalize janken – rock-paper-scissors-king-flea. In: Akiyama, J., Kano, M., Sakai, T. (eds.) TJJCCGG 2012. LNCS, vol. 8296, pp. 85–94. Springer, Heidelberg (2013)
3. Moon, J.W.: Topics on Tournaments. Holt, Rinehart and Winston, New York (1968)
4. Ohbayashi, T., Kishino, U., Sougawa, T., Yamashita, S. (eds.): Encyclopedia of Ethnic Play and Games, Taishukan Shoten (1998). (in Japanese)
5. Nisan, N., Roughgarden, T., Tardos, E., Vazirani, V. (eds.): Algorithmic Game Theory. Cambridge University Press, Cambridge (2007)
6. Rock-paper-scissors. http://en.wikipedia.org/wiki/Rock-paper-scissors

γ-Labeling of a Cycle with One Chord

Supaporn Saduakdee[✉] and Varanoot Khemmani

Department of Mathematics, Srinakharinwirot University,
Sukhumvit 23, Bangkok 10110, Thailand
aa_o_rr@hotmail.com, varanoot@g.swu.ac.th

Abstract. Let G be a graph of order n and size m. A γ-labeling of G is a one-to-one function $f : V(G) \to \{0, 1, 2, \ldots, m\}$ that induces an edge-labeling $f' : E(G) \to \{1, 2, \ldots, m\}$ on G defined by

$$f'(e) = |f(u) - f(v)|, \quad \text{for each edge } e = uv \text{ in } E(G).$$

The value of f is defined as

$$\text{val}(f) = \sum_{e \in E(G)} f'(e).$$

The maximum value of a γ-labeling of G is defined as

$$\text{val}_{\max}(G) = \max\{\text{val}(f) : f \text{ is a } \gamma\text{-labeling of } G\};$$

while the minimum value of a γ-labeling of G is

$$\text{val}_{\min}(G) = \min\{\text{val}(f) : f \text{ is a } \gamma\text{-labeling of } G\}.$$

In this paper, we determine the maximum and minimum values of a γ-labeling of a graph derived from cycle with adding one chord.

Keywords: γ-Labeling · Value of a γ-labeling · Cycle with one chord

1 Introduction

Let G be a graph of order n and size m. A γ-*labeling* of G is defined in [2] as a one-to-one function $f : V(G) \to \{0, 1, \ldots, m\}$ that induces an *edge-labeling* $f' : E(G) \to \{1, \ldots, m\}$ on G defined by $f'(e) = |f(u) - f(v)|$ for each edge $e = uv$ of G. The *value* of f is defined by

$$\text{val}(f) = \sum_{e \in E(G)} f'(e).$$

If the edge-labeling f' of a γ-labeling f of a graph is also one-to-one, then f is a *graceful labeling*. Among all labelings of graphs, graceful labelings are probably the best known and most studied. Graceful labelings originated with a paper

V. Khemmani—Research supported by Srinakharinwirot University, Year 2015.

J. Akiyama et al. (Eds.): JCDCGG 2015, LNCS 9943, pp. 155–166, 2016.
DOI: 10.1007/978-3-319-48532-4_14

of Rosa [11], who used the term β-valuations. A few years later, Golomb [10] called these labelings "graceful" and this is the terminology that has been used since then. Gallian [9] has written an extensive survey on labelings of graphs. The subject of γ-labelings of graphs was studied in [1–3,5–8].

Obviously, since f is one-to-one, it follows that $f'(e) \geq 1$, for any edge e, and therefore, $\mathrm{val}(f) \geq m$. Moreover, G has a γ-labeling if and only if $m \geq n - 1$ and every connected graph has a γ-labeling.

The *maximum value* and the *minimum value* of a γ-labeling of G are defined in [2] as

$$\mathrm{val_{max}}(G) = \max\{\mathrm{val}(f)\colon f \text{ is a } \gamma\text{-labeling of } G\}$$

and

$$\mathrm{val_{min}}(G) = \min\{\mathrm{val}(f)\colon f \text{ is a } \gamma\text{-labeling of } G\},$$

respectively. A γ-labeling g of G is a γ-max *labeling* if $\mathrm{val}(g) = \mathrm{val_{max}}(G)$ and a γ-labeling h is a γ-min *labeling* if $\mathrm{val}(h) = \mathrm{val_{min}}(G)$. In [1–3], the maximum and minimum values of a γ-labeling of a path P_n of order n, a cycle C_n of order n, the complete graph K_n, a double star $S_{p,q}$ and the complete bipartite graph $K_{r,s}$ are determined. Next, we recall the extremal values of a γ-labeling of a cycle of order n.

Theorem 1 [2]. *If C_n is a cycle of order $n \geq 3$, then*

$$\mathrm{val_{min}}(C_n) = 2(n - 1)$$

and

$$\mathrm{val_{max}}(C_n) = \begin{cases} \frac{(n-1)(n+3)}{2} & \text{if } n \text{ is odd} \\ \frac{n(n+2)}{2} & \text{if } n \text{ is even.} \end{cases}$$

The extremal values of a cycle with a triangle $C_n^{\triangle}$, i.e., a cycle with a chord joining two nonadjacent vertices but adjacent to some vertex in the cycle C_n, are determined in [6], as we state next.

Theorem 2 [6]. *If $C_n^{\triangle}$ is a cycle with a triangle of order $n \geq 5$, then*

$$\mathrm{val_{min}}(C_n^{\triangle}) = 2n - 1$$

and

$$\mathrm{val_{max}}(C_n^{\triangle}) = \begin{cases} 32 & \text{if } n = 6 \\ \frac{n^2+6n-10}{2} & \text{if } n \text{ is even and } n \geq 8 \\ \frac{n^2+6n-3}{2} & \text{if } n \text{ is odd.} \end{cases}$$

In [2] a simple and useful connection between minimum and maximum values of a connected graph and that of a proper connected subgraph was found.

Proposition 1 [2]. *If H is a proper connected subgraph of a connected graph G, then*

$$\mathrm{val_{min}}(H) < \mathrm{val_{min}}(G) \text{ and } \mathrm{val_{max}}(H) < \mathrm{val_{max}}(G).$$

The following result appeared in [7] will be useful to us.

Theorem 3 [7]. *If f is a γ-max labeling of a nontrivial graph G of order n and size m, then $\{0, m\} \subseteq f(V(G))$.*

For a nonnegative integer δ and a γ-labeling of a connected graph G of order n and size m, the *extension labeling* γ^δ-*labeling* of G is defined in [6] as a one-to-one function $f : V(G) \to \{0, 1, \ldots, m+\delta-1, m+\delta\}$ that induces an edge-labeling $f' : E(G) \to \{1, \ldots, m+\delta\}$ on G defined by

$$f'(uv) = |f(u) - f(v)|, \quad \text{for each edge } uv \text{ in } E(G).$$

The *value* of a γ^δ-labeling f is defined as $\mathrm{val}(f) = \sum_{e \in E(G)} f'(e)$. The *maximum value* and the *minimum value* of a γ^δ-labeling of G are defined as

$$\mathrm{val}^\delta_{\max}(G) = \max\{\mathrm{val}(f) : f \text{ is a } \gamma^\delta\text{-labeling of } G\}$$

and

$$\mathrm{val}^\delta_{\min}(G) = \min\{\mathrm{val}(f) : f \text{ is a } \gamma^\delta\text{-labeling of } G\},$$

respectively. The other definitions are similar to the γ-labeling case.

The maximum value of a γ^δ-labeling of a cycle C_n of order $n \geq 4$ are shown in [6] as follows.

Theorem 4 [6]. *For every pair δ, n of nonnegative integers with $n \geq 4$,*

$$\mathrm{val}^\delta_{\max}(C_n) = \begin{cases} \mathrm{val}_{\max}(C_n) + n\delta & = \frac{n(n+2\delta+2)}{2} & \text{if } n \text{ is even} \\ \mathrm{val}_{\max}(C_n) + (n-1)\delta & = \frac{(n-1)(n+2\delta+3)}{2} & \text{if } n \text{ is odd}. \end{cases}$$

Now, we consider a cycle of order $n \geq 4$ with one chord, say $C_n + e$, i.e., a cycle with a chord e joining two nonadjacent vertices in the cycle C_n. Therfore $C_n^\triangle$ is also a cycle with one chord that joins two nonadjacent vertices with distance 2 in the cycle C_n. In this paper, we naturally generalize a cycle C_n with one chord e that joins two nonadjacent vertices with distance r in the cycle C_n where $2 \leq r \leq \lfloor \frac{n}{2} \rfloor$ and determine the maximum and minimum values of a γ-labeling of $C_n + e$.

The reader is referred to Chartrand and Zhang [4] for basic definitions and terminology not mentioned here.

2   The Minimum Value of Cycle with One Chord

In this section we establish $\mathrm{val}_{\min}(C_n + e)$ for a cycle with one chord $C_n + e$.

Theorem 5. *For every integer $n \geq 4$,*

$$\mathrm{val}_{\min}(C_n + e) = 2n - 1.$$

Proof. By Theorem 1 and Proposition 1, we have

$$\mathrm{val}_{\min}(C_n + e) \geq 2(n-1) + 1 = 2n - 1.$$

Hence it remains to show that $\mathrm{val}_{\min}(C_n + e) \leq 2n - 1$.

Suppose that $C_n + e$ is a cycle $C_n : v_1, v_2, \ldots, v_{r-1}, v_r, v_{r+1}, \ldots, v_n, v_1$ with a chord $e = v_1 v_r$ where $3 \leq r \leq n-1$. Consider now the γ-labeling f of $C_n + e$ defined by

$$f(v_i) = \begin{cases} r - 1 & \text{if } i = 1 \\ i - 2 & \text{if } 2 \leq i \leq r \\ n + r - i & \text{if } r + 1 \leq i \leq n. \end{cases}$$

Then

$$\begin{aligned} \mathrm{val}(f) &= \sum_{i=3}^{r} (f(v_i) - f(v_{i-1})) + \sum_{i=r+1}^{n-1} (f(v_i) - f(v_{i+1})) + (f(v_1) - f(v_2)) \\ &\quad + (f(v_n) - f(v_1)) + (f(v_{r+1}) - f(v_r)) + (f(v_1) - f(v_r)) \\ &= 2n - 1. \end{aligned}$$

Therefore, $\mathrm{val}_{\min}(C_n + e) \leq 2n - 1$. $\qquad\square$

3 The Maximum Value of Odd Cycle with One Chord

In order to discuss $\mathrm{val}_{\max}(C_n + e)$, we first consider the maximum value of an odd cycle with one chord.

Theorem 6. *For every odd integer $n \geq 5$,*

$$\mathrm{val}_{\max}(C_n + e) = \frac{n^2 + 6n - 3}{2}.$$

Proof. Let $C_n + e$ be an odd cycle with one chord of order $n = 2k + 1$ with $k \geq 2$, which is obtained from a cycle $C_{2k+1} : x_1, y_1, x_2, y_2, \ldots, x_r, y_r, \ldots, x_k, y_k, x_{k+1}, x_1$ and a chord e. Since for each r with $2 \leq r \leq k$, an odd cycle with one chord $C_n + x_1 y_r$ is isomorphic to $C_n + x_1 x_{k-r+2}$, without loss of generality, we may assume that $e = x_1 y_r$. Define a γ-labeling f of $C_n + e$ by

$$f(x_i) = \quad i - 1 \qquad\qquad\quad \text{if } 1 \leq i \leq k + 1$$
$$f(y_i) = \begin{cases} n + 1 - i & \text{if } 1 \leq i \leq r - 1 \\ n + 1 & \text{if } i = r \\ k - r + 2 + i & \text{if } r + 1 \leq i \leq k. \end{cases}$$

Then

$$\begin{aligned} \mathrm{val}(f) &= 3(f(y_r) - f(x_1)) + 2 \left(\sum_{i=1}^{r-1} f(y_i) + \sum_{i=r+1}^{k} f(y_i) - \sum_{i=2}^{k} f(x_i) \right) \\ &= \tfrac{n^2 + 6n - 3}{2}. \end{aligned}$$

Thus $\mathrm{val}_{\max}(C_n + e) \geq \frac{n^2 + 6n - 3}{2}$.

It remains therefore to show that $\mathrm{val_{max}}(C_n + e) \leq \frac{n^2+6n-3}{2}$. Let g be a γ-max labeling of $C_n + e$. Then

$$
\begin{aligned}
\mathrm{val_{max}}(C_n + e) &= \mathrm{val}(g) \\
&= \sum_{e \in E(C_n)} g'(e) + g'(x_1 y_r) \\
&\leq \mathrm{val_{max}^1}(C_n) + (n+1) \\
&= \frac{(n-1)(n+2\cdot 1+3)}{2} + (n+1) \qquad \text{(by Theorem 4)} \\
&= \frac{n^2+6n-3}{2}.
\end{aligned}
$$

Therefore, $\mathrm{val_{max}}(C_n + e) \leq \frac{n^2+6n-3}{2}$. □

4   The Maximum Value of Even Cycle with One Chord

In Sect. 3, we considered the maximum value of an odd cycle with one chord. We now study maximum value of an even cycle with one chord, $\mathrm{val_{max}}(C_n + e)$ for even integer $n \geq 4$. First, we determine the maximum value of $C_n + e$ where e is a chord joining two vertices with odd distance in the even cycle C_n.

Theorem 7. *For every even integer $n \geq 6$,*

$$
\mathrm{val_{max}}(C_n + e) = \frac{n^2 + 6n + 2}{2}
$$

where e is a chord joining two vertices with odd distance in the even cycle C_n.

Proof. Let $C_n + e$ be an even cycle with one chord of order $n = 2k$ with $k \geq 3$, which is obtained from a cycle $C_{2k} : x_1, y_1, x_2, y_2, \ldots, x_r, y_r, \ldots, x_k, y_k, x_1$ and a chord $e = x_1 y_r$, where $2 \leq r \leq k - 1$. Define a γ-labeling f of $C_n + e$ by

$$
f(x_i) = \quad i - 1 \qquad\qquad\ \text{if } 1 \leq i \leq k
$$
$$
f(y_i) = \begin{cases} n+1-i & \text{if } 1 \leq i \leq r-1 \\ n+1 & \text{if } i = r \\ k-r+1+i & \text{if } r+1 \leq i \leq k. \end{cases}
$$

Then

$$
\begin{aligned}
\mathrm{val}(f) &= 3(f(y_r) - f(x_1)) + 2\left(\sum_{i=1}^{r-1} f(y_i) + \sum_{i=r+1}^{k} f(y_i) - \sum_{i=2}^{k} f(x_i) \right) \\
&= \frac{n^2+6n+2}{2}.
\end{aligned}
$$

Thus $\mathrm{val_{max}}(C_n + e) \geq \frac{n^2+6n+2}{2}$.

In order to show that $\mathrm{val_{max}}(C_n + e) \leq \frac{n^2+6n+2}{2}$, let g be a γ-max labeling of $C_n + e$. Then

$$
\begin{aligned}
\mathrm{val_{max}}(C_n + e) &= \mathrm{val}(g) \\
&= \sum_{e \in E(C_n)} g'(e) + g'(x_1 y_r) \\
&\leq \mathrm{val_{max}^1}(C_n) + (n+1) \\
&= \frac{n(n+2\cdot 1+2)}{2} + (n+1) \qquad \text{(by Theorem 4)} \\
&= \frac{n^2+6n+2}{2}.
\end{aligned}
$$

Therefore, $\mathrm{val_{max}}(C_n + e) \leq \frac{n^2+6n+2}{2}$. □

In the remaining part of this section, we present the maximum value of an even cycle with one chord, $C_n + e$ where e is a chord joining two vertices with even distance in the even cycle C_n. In order to do this, for $k \geq 2$, we let $n = 2k$ and $e = x_1 x_r$ be a chord in the even cycle $C_n : x_1, y_1, x_2, y_2, \ldots, x_r, y_r, \ldots, x_k, y_k, x_1$, where $2 \leq r \leq k$.

Proposition 2

$$\text{val}_{\max}(C_4 + e) = 17$$
$$\text{val}_{\max}(C_6 + e) = 32$$

where e is a chord joining two vertices with even distance in the cycles C_4 and C_6, respectively.

Proof. First, let $C_4 + e = C_4 + x_1 x_2$. The γ-labeling f of $C_4 + e$ is defined by

$$f(x_1) = 0, \quad f(x_2) = 1, \quad f(y_1) = 5 \text{ and } \quad f(y_2) = 4.$$

Then $\text{val}_{\max}(C_4 + e) \geq \text{val}(f) = 17$.

On the other hand, let g be a γ-max labeling of $C_4 + e$. By Theorem 3, we may assume that $g(V(C_4 + e)) = \{0, 5, a, b\}$ where $a, b \in \{1, 2, 3, 4\}$. Notice that $\deg(x_1) = \deg(x_2) = 3$ and $\deg(y_1) = \deg(y_2) = 2$. We consider four cases, according to the vertices of $C_4 + e$ labeled 0 and 5.

> Case 1. $\{0, 5\} = \{g(x_1), g(x_2)\}$. Then $\text{val}(g) = 15$.
> Case 2. $\{0, 5\} = \{g(y_1), g(y_2)\}$. Then $\text{val}(g) = 10 + |a - b| \leq 13$.
> Case 3. $0 \in \{g(x_1), g(x_2)\}$ and $5 \in \{g(y_1), g(y_2)\}$. Then
>
> $$\text{val}(g) = 10 + \max\{a, b\} + |a - b| \leq 17.$$
>
> Case 4. $0 \in \{g(y_1), g(y_2)\}$ and $5 \in \{g(x_1), g(x_2)\}$. Then
>
> $$\text{val}(g) = 15 - \min\{a, b\} + |a - b| \leq 17.$$

Since $\text{val}(g) \leq 17$, it follows that $\text{val}_{\max}(C_4 + e) = \text{val}(g) \leq 17$.

Next, we compute $\text{val}_{\max}(C_6 + e)$. Since $C_6 + e = C_6 + x_1 x_2$ or $C_6 + x_1 x_3$, it then follows that $C_6 + e = C_6^{\triangle}$, and by Theorem 2, $\text{val}_{\max}(C_6 + e) = \text{val}(C_6^{\triangle}) = 32$. $\square$

Next, we establish a lower bound for the maximum values of $C_n + e$ of even order $n \geq 8$, where e is a chord joining two vertices with even distance in the even cycle C_n.

Lemma 1. *For every even integer $n \geq 8$,*

$$\text{val}_{\max}(C_n + e) \geq \frac{n^2 + 6n - 10}{2}$$

where e is a chord joining two vertices with even distance in the even cycle C_n.

Proof. Let f be a γ-labeling of $C_n + e$ defined by

$$f(x_i) = \begin{cases} i - 1 & \text{if } 1 \le i \le r - 1 \\ n + 1 & \text{if } i = r \\ i - 2 & \text{if } r + 1 \le i \le k \end{cases}$$

$$f(y_i) = \begin{cases} n + 1 - i & \text{if } 1 \le i \le r - 2 \\ k - r + 2 + i & \text{if } r - 1 \le i \le k. \end{cases}$$

Then

$$\text{val}(f) = 3\left(f(x_r) - f(x_1)\right) + 2\left(\sum_{i=1}^{r-2} f(y_i) + \sum_{i=r+1}^{k} f(y_i) - \sum_{i=2}^{r-1} f(x_i) - \sum_{i=r+1}^{k} f(x_i)\right)$$
$$= \frac{n^2 + 6n - 10}{2}.$$

Therefore $\text{val}_{\max}(C_n + e) \ge \text{val}(f) = \frac{n^2 + 6n - 10}{2}.$ □

In order to present an upper bound for $\text{val}_{\max}(C_n + e)$, where $e = x_1 x_r$ is a chord in the even cycle $C_n : x_1, y_1, x_2, y_2, \ldots, x_r, y_r, \ldots, x_k, y_k, x_1$ of order $n = 2k$ with $k \ge 4$ and $2 \le r \le k$, we need some additional notation and new definitions. Let f be a γ-max labeling of $C_n + e$. For each integer i, with $1 \le i \le k$, we define the 3-term sequences

$$S_i(f) = (f(x_i), f(y_i), f(x_{i+1})) \text{ and } T_i(f) = (f(y_i), f(x_{i+1}), f(y_{i+1})),$$

where the addition is taken modulo k, and let

$$ST(f) = \{S_1(f), S_2(f), \ldots, S_k(f), T_1(f), T_2(f), \ldots, T_k(f)\}$$

be a set of 3-term sequences $S_i(f)$ and $T_i(f)$, for all i with $1 \le i \le k$. Furthermore, let $C_n(f)$ be an oriented cycle obtained from $(C_n + e) - x_1 x_r$ by assigning to the edge uv the orientation (u, v) if $f(u) < f(v)$.

Theorem 8

$$\text{val}_{\max}(C_8 + e) = 51$$

where e is a chord joining two vertices with even distance in the cycle C_8.

Proof. Let $e = x_1 x_r$ be a chord in the cycle $C_8 : x_1, y_1, x_2, y_2, x_3, y_3, x_4, y_4, x_1$, where $r = 2, 3, 4$. If $r = 2$ or 4, then it follows by Theorem 2, $\text{val}_{\max}(C_8 + e) = \text{val}(C_8^\triangle) = 51$. Assume that $r = 3$. By Lemma 1, we have $\text{val}_{\max}(C_8 + e) \ge \frac{8^2 + 6 \cdot 8 - 10}{2} = 51$.

On the other hand, let f be a γ-max labeling of $C_8 + e$. We consider the two cases according to the set $ST(f)$.

Case 1. No element of $ST(f)$ is monotone. Then, for each i with $1 \le i \le 4$, the vertices x_i and y_i of the oriented cycle $C_8(f)$ have

either $\text{id}(x_i) = 0, \text{id}(y_i) = 2$ **or** $\text{id}(x_i) = 2, \text{id}(y_i) = 0$.

First, assume that $f(x_1) > f(x_r)$. We consider two subcases, according to whether $\mathrm{id}(x_i) = 0, \mathrm{id}(y_i) = 2$ or $\mathrm{id}(x_i) = 2, \mathrm{id}(y_i) = 0$ in the oriented cycle $C_8(f)$ for each i with $1 \leq i \leq 4$.

Case 1.1. For each i with $1 \leq i \leq 4$, $\mathrm{id}(x_i) = 0$, $\mathrm{id}(y_i) = 2$ of the oriented cycle $C_8(f)$. So, $f(x_i) < f(y_i)$ and $f(x_{i+1}) < f(y_i)$ where addition is taken modulo 4. Then

$$\mathrm{val}_{\max}(C_8 + e) = \mathrm{val}(f) = 2\sum_{i=1}^{4} f(y_i) - (f(x_1) + 2f(x_2) + 3f(x_3) + 2f(x_4)).$$

Since the vertices $x_3, x_2(\text{or } x_4), x_1$ can be assigned $0, 1(\text{or } 2), 3$, respectively and the vertices in $\{y_1, y_2, y_3, y_4\}$ can be assigned each of the labels $9, 8, 7, 6$, it follows that

$$2\sum_{i=1}^{4} f(y_i) \leq 60 \quad \text{and} \quad f(x_1) + 2f(x_2) + 3f(x_3) + 2f(x_4) \geq 9.$$

Then $\mathrm{val}_{\max}(C_8 + e) = \mathrm{val}(f) \leq 60 - 9 = 51$.

Case 1.2. For each i with $1 \leq i \leq 4$, $\mathrm{id}(x_i) = 2$, $\mathrm{id}(y_i) = 0$ of the oriented cycle $C_8(f)$. So, $f(x_i) > f(y_i)$ and $f(x_{i+1}) > f(y_i)$ where addition is taken modulo 4. Then

$$\mathrm{val}_{\max}(C_8 + e) = \mathrm{val}(f) = (3f(x_1) + 2f(x_2) + f(x_3) + 2f(x_4)) - 2\sum_{i=1}^{4} f(y_i).$$

Since the vertices $x_1, x_2(\text{or } x_4), x_3$ can be assigned $9, 8(\text{or } 7), 6$, respectively and the vertices in $\{y_1, y_2, y_3, y_4\}$ can be assigned each of the labels $0, 1, 2, 3$, it follows that

$$3f(x_1) + 2f(x_2) + f(x_3) + 2f(x_4) \leq 63 \quad \text{and} \quad 2\sum_{i=1}^{4} f(y_i) \geq 12.$$

Then $\mathrm{val}_{\max}(C_8 + e) = \mathrm{val}(f) \leq 63 - 12 = 51$.

Next, if $f(x_1) < f(x_r)$, with similar argument we can show that $\mathrm{val}_{\max}(C_8 + e) \leq 51$.

Case 2. Some element of $ST(f)$ is monotone. Then the oriented cycle $C_8(f)$ contains a directed path a, b, c of order 3. If we delete the chord $e = x_1 x_r$ and vertex b from $C_8 + e$ and join the vertices a and c, the resulting graph G is isomorphic to C_7 and the restriction g of f to $V((C_8 + e) - x_1 x_r) - \{b\}$ has the same value on G as f on $(C_8 + e) - x_1 x_r$, that is

$$\mathrm{val}(g) = \mathrm{val}(f) - f'(x_1 x_r) \geq \mathrm{val}_{\max}(C_8 + e) - 9.$$

Then $\mathrm{val}_{\max}(C_8 + e) \leq \mathrm{val}(g) + 9$. Moreover, since g is a γ^2-labeling of a graph G that is isomorphic to C_7, it follows that $\mathrm{val}(g) \leq \mathrm{val}^2_{\max}(C_7)$. Therefore, by Theorem 4, $\mathrm{val}_{\max}(C_8 + e) \leq 51$. $\square$

As a consequence of Proposition 2 and Theorem 8, we have the following result.

Corollary 1. *For $n = 4, 6$ and 8,*

$$\text{val}_{\max}(C_n + e) = \frac{n^2 + 5n - 2}{2}$$

where e is a chord joining two vertices with even distance in the even cycle C_n.

Lemma 2. *For every even integer n, with $n \geq 8$, let f be a γ-max labeling of $C_n + e$ where e is a chord joining two vertices with even distance in even cycle C_n. If some element of $ST(f)$ is monotone, then there are exactly two monotone elements of $ST(f)$.*

Proof. Suppose that some element of $ST(f)$ is monotone. Then the oriented cycle $C_n(f)$ contains a directed path of order 3. Since the size of the oriented cycle $C_n(f)$ is even, it follows that there are ℓ directed paths of order 3 in $C_n(f)$, for some even integer l with $2 \leq l \leq n$. Next, we show that there are no more than two directed paths of order 3 in oriented cycle $C_n(f)$.

Assume, to the contrary, that there are at least 4 directed paths of order 3 in the oriented graph $C_n(f)$. For each i with $1 \leq i \leq 4$, let P_i be a directed path of order 3 having internal vertex u_i in the oriented graph $C_n(f)$. Then the cycle $(C_n + e) - x_1 x_r$ is not only isomorphic to C_n but it is also a subdivision of the cycle C_{n-4}.

We can construct a graph G which is isomorphic to C_{n-4} obtained from $C_n + e$ by deleting the chord $e = x_1 x_r$ and the internal vertices u_1, u_2, u_3, u_4 and then adding one or more edges to G. Then the restriction of f to $V((C_n + e) - x_1 x_r) - \{u_1, u_2, u_3, u_4\}$, g, has the same value on G as f does on $(C_n + e) - x_1 x_r$, that is

$$\text{val}(g) = \text{val}(f) - f'(x_1 x_r) \geq \text{val}_{\max}(C_n + e) - (n+1).$$

Then $\text{val}_{\max}(C_n + e) \leq \text{val}(g) + (n+1)$. Moreover, since g is a γ^5-labeling of a graph G which is isomorphic to C_{n-4}, it follows that $\text{val}(g) \leq \text{val}^5_{\max}(C_{n-4})$. Therefore,

$$
\begin{aligned}
\text{val}_{\max}(C_n + e) &\leq \text{val}(g) + (n+1) \\
&\leq \text{val}^5_{\max}(C_{n-4}) + (n+1) \\
&= \frac{(n-4)((n-4)+2\cdot5+2)}{2} + (n+1) \qquad \text{(by Theorem 4)} \\
&< \frac{n^2+6n-10}{2}
\end{aligned}
$$

a contradiction with Lemma 1. Thus, for $n \geq 8$, the oriented cycle $C_n(f)$ contains exactly two directed paths of order 3 and there are also exactly two monotone elements in $ST(f)$. $\square$

By Theorem 8 and Lemma 2, we are able to characterize any γ-max labeling f of $C_8 + e$, in terms of set $ST(f)$ and the oriented cycle $C_8(f)$.

Proposition 3. *Let f be a γ-max labeling of $C_8 + e$ where e is a chord joining two vertices with even distance in cycle C_8. Then either no element of $ST(f)$ is monotone or there are exactly two monotone elements of $ST(f)$.*

Proposition 4. *For every even integer n with $n \geq 10$, let f be a γ-max labeling of $C_n + e$ where e is a chord joining two vertices with even distance in the even cycle C_n. Then $ST(f)$ contains exactly two monotone elements.*

Proof. Assume, to the contrary, that the property does not verify. Then, by Lemma 2, $ST(f)$ contains no monotone element. Hence, for the vertices x_i and y_i of the oriented cycle $C_n(f)$, **either** $\mathrm{id}(x_i) = 0, \mathrm{id}(y_i) = 2$ **or** $\mathrm{id}(x_i) = 2, \mathrm{id}(y_i) = 0$ for each i with $1 \leq i \leq k$. We consider the two cases separately.

Case 1. For each i with $1 \leq i \leq k$, $\mathrm{id}(x_i) = 0$, $\mathrm{id}(y_i) = 2$ of the oriented cycle $C_n(f)$. Thus $f(x_i) < f(y_i)$ and $f(x_{i+1}) < f(y_i)$ where addition is taken modulo k. Then

$$\mathrm{val}_{\max}(C_n + e) = \mathrm{val}(f) = 2\sum_{i=1}^{k} f(y_i) - 2\sum_{i=1}^{k} f(x_i) + |f(x_1) - f(x_r)|.$$

If $f(x_1) > f(x_r)$, then

$$\mathrm{val}_{\max}(C_n + e) = \mathrm{val}(f) = 2\sum_{i=1}^{k} f(y_i) - \left(3f(x_r) + 2\sum_{\substack{2 \leq i \leq k \\ i \neq r}} f(x_i) + f(x_1)\right).$$

The vertices in $\{y_1, y_2, \ldots, y_k\}$ and $\{x_1, x_2, \ldots, x_k\}$ can be assigned labels in the consecutive integers sets $\{k + 2, k + 3, \ldots, n + 1\}$ and $\{0, 1, \ldots, k - 1\}$, respectively. Moreover, x_r and x_1 can be assigned 0 and $k - 1$, respectively. Hence,

$$\mathrm{val}_{\max}(C_n + e) = \mathrm{val}(f) \leq 2\sum_{i=k+2}^{n+1} i - \left(0 + 2\sum_{i=1}^{k-2} i + (k-1)\right)$$
$$= n^2 + 3n - 2k^2 - k - 1$$
$$= \tfrac{n^2 + 5n - 2}{2}.$$

On the another hand, if $f(x_1) < f(x_r)$, with similar argument we can show that $\mathrm{val}_{\max}(C_n + e) \leq \frac{n^2 + 5n - 2}{2}$. However, by Lemma 1, we have $\mathrm{val}_{\max}(C_n + e) \geq \frac{n^2 + 6n - 10}{2}$, which is a contradiction.

Case 2. For each i with $1 \leq i \leq k$, $\mathrm{id}(x_i) = 2$, $\mathrm{id}(y_i) = 0$ of the oriented cycle $C_n(f)$. Thus $f(x_i) > f(y_i)$ and $f(x_{i+1}) > f(y_i)$ where addition is taken modulo k. Then

$$\mathrm{val}_{\max}(C_n + e) = \mathrm{val}(f) = 2\sum_{i=1}^{k} f(x_i) - 2\sum_{i=1}^{k} f(y_i) + |f(x_1) - f(x_r)|.$$

First, if $f(x_1) > f(x_r)$. Then

$$\text{val}_{\max}(C_n + e) = \text{val}(f) = \left(3f(x_1) + 2\sum_{\substack{2 \le i \le k \\ i \ne r}} f(x_i) + f(x_r)\right) - 2\sum_{i=1}^{k} f(y_i).$$

The vertices in $\{x_1, x_2, \ldots, x_k\}$ and $\{y_1, y_2, \ldots, y_k\}$ can be assigned labels in the consecutive integers sets $\{k+2, k+3, \ldots, n+1\}$ and $\{0, 1, \ldots, k-1\}$, respectively. Moreover, x_1 and x_r can be assigned $n+1$ and $k+2$, respectively. Therefore,

$$\text{val}_{\max}(C_n + e) = \text{val}(f) \le \left(3(n+1) + 2\sum_{i=k+3}^{n} i + (k+2)\right) - 2\sum_{i=0}^{k-1} i$$
$$= n^2 + 4n - 2k^2 - 3k - 1$$
$$= \frac{n^2 + 5n - 2}{2}.$$

Next, if $f(x_1) < f(x_r)$, with similar argument we can show that $\text{val}_{\max}(C_n + e) \le \frac{n^2+5n-2}{2}$. However, by Lemma 1, we have $\text{val}_{\max}(C_n + e) \ge \frac{n^2+6n-10}{2}$, which is a contradiction. $\qquad\square$

We are now prepared to establish a general formula for $\text{val}_{\max}(C_n + e)$, when $n \ge 10$.

Theorem 9. *For every even integer $n \ge 10$,*

$$\text{val}_{\max}(C_n + e) = \frac{n^2 + 6n - 10}{2}$$

where e is a chord joining two vertices with even distance in the even cycle C_n.

Proof. By Lemma 1, we have $\text{val}_{\max}(C_n + e) \ge \frac{n^2+6n-10}{2}$. Next, we show that $\text{val}_{\max}(C_n + e) \le \frac{n^2+6n-10}{2}$. Let f be a γ-labeling of $C_n + e$. By Proposition 4, the oriented cycle $C_n(f)$ contains a directed path a, b, c of order 3. If we delete the chord $e = x_1 x_r$ and the vertex b from $C_n + e$ and then join the vertices a and c, the resulting graph G is isomorphic to C_{n-1} and the restriction g of f to $V((C_n + e) - x_1 x_r - b)$, verifies

$$\text{val}(g) = \text{val}(f) - f'(x_1 x_r) \ge \text{val}_{\max}(C_n + e) - (n+1).$$

Therefore,

$$\text{val}_{\max}(C_n + e) \le \text{val}(g) + (n+1).$$

Moreover, since g is a γ^2-labeling of graph G that is isomorphic to C_{n-1}, it follows that $\text{val}(g) \le \text{val}^2_{\max}(C_{n-1})$. Therefore,

$$\text{val}_{\max}(C_n + e) \le \text{val}^2_{\max}(C_{n-1}) + (n+1)$$
$$= \frac{((n-1)-1))((n-1)+2\cdot2+3)}{2} + (n+1)$$
$$= \frac{n^2+6n-10}{2}. \qquad\square$$

As a consequence of Corollary 1 and Theorem 9, we have the following result.

Corollary 2. *For every even integer* $n \geq 4$,

$$\mathrm{val}_{\max}(C_n + e) = \begin{cases} \frac{n^2+5n-2}{2} & if\, n = 4,6,8 \\ \frac{n^2+6n-10}{2} & if\, n \geq 10 \end{cases}$$

where e *is a chord joining two vertices with even distance in the even cycle* C_n.

5   Final Remarks

In [2,8] the γ-spectra of doublestars, paths, cycles, and complete graphs are determined. Later, in [6], the extremal values of a γ-labeling of a cycle with a triangle are established. It would be interesting to determine the γ-spectrum of a cycle with one chord and find whether it is continuous or not. Observe that the value of any γ-labeling of a cycle in [8] is always even.

References

1. Bullington, G.D., Eroh, L.L., Winters, S.J.: γ-labelings of complete bipartite graphs. Discuss. Math. Graph Theor. **30**, 45–54 (2010)
2. Chartrand, G., Erwin, D., VanderJagt, D.W., Zhang, P.: γ-labelings of graphs. Bull. Inst. Combin. Appl. **44**, 51–68 (2005)
3. Chartrand, G., Erwin, D., VanderJagt, D.W., Zhang, P.: On γ-labelings of trees. Discuss. Math. Graph Theor. **25**(3), 363–383 (2005)
4. Chartrand, G., Zhang, P.: Introduction to Graph Theory. The Walter Rudin Student Series in Advanced Mathematics. McGraw-Hill Higher Education, Boston (2005)
5. Crosse, L., Okamoto, F., Saenpholphat, V., Zhang, P.: On γ-labelings of oriented graphs. Math. Bohem. **132**, 185–203 (2007)
6. Fonseca, C.M., Saenpholphat, V., Zhang, P.: Extremal values for a γ-labeling of a cycle with a triangle. Utilitas Math. **92**, 167–185 (2013)
7. Fonseca, C.M., Khemmani, V., Zhang, P.: On γ-labelings of graphs. Utilitas Math. **98**, 33–42 (2015)
8. Fonseca, C.M., Saenpholphat, V., Zhang, P.: The γ-spectrum of a graph. Ars Combin. **101**, 109–127 (2011)
9. Gallian, J.A.: A dynamic survey of graph labeling. Electron. J. Combin. **16**(6), 1–219 (2009)
10. Golomb, S.W.: How to number a graph. In: Graph Theory and Computing, pp. 349–355. Academic Press, New York (1972)
11. Rosa, A.: On certain valuations of the vertices of a graph. In: Theory of Graphs (International Symposium, Rome, 1966), pp. 349–355, Gordon and Breach, New York (1966)

Box Pleating is Hard

Hugo A. Akitaya[1]([⊠]), Kenneth C. Cheung[2], Erik D. Demaine[3],
Takashi Horiyama[4], Thomas C. Hull[5], Jason S. Ku[3], Tomohiro Tachi[6],
and Ryuhei Uehara[7]

[1] Tufts University, Medford, USA
hugo.alves_akitaya@tufts.edu
[2] NASA, Washington, D.C., USA
kenneth.c.cheung@nasa.gov
[3] MIT, Cambridge, USA
{edemaine,jasonku}@mit.edu
[4] Saitama University, Saitama, Japan
horiyama@al.ics.saitama-u.ac.jp
[5] Western New England University, Springfield, USA
thull@wne.edu
[6] The University of Tokyo, Tokyo, Japan
tachi@idea.c.u-tokyo.ac.jp
[7] JAIST, Nomi, Japan
uehara@jaist.ac.jp

Abstract. Flat foldability of general crease patterns was first claimed to be hard for over twenty years. In this paper we prove that deciding flat foldability remains NP-complete even for box pleating, where creases form a subset of a square grid with diagonals. In addition, we provide new terminology to implicitly represent the global layer order of a flat folding, and present a new planar reduction framework for grid-aligned gadgets.

1 Introduction

In their seminal 1996 paper, Bern and Hayes initiated investigation into the computational complexity of origami [BH96]. They claimed that it is NP-hard to determine whether a given general crease pattern can be folded flat, both when the creases have or have not been assigned crease directions (mountain fold or valley fold). Since that time, there has been considerable work in analyzing the computational complexity of other origami related problems. For example, Arkin et al. [ABD+04] proved that deciding foldability is hard even for simple folds, while Demaine et al. [DFL10] proved that optimal circle packing for origami design is also hard.

While the gadgets in the hardness proof presented in [BH96] for unassigned crease patterns are relatively straightforward, their gadgets for assigned crease patterns are considerably more convoluted, and quite difficult to check. In fact, we have found an error in even their unassigned crossover gadget where signals are not guaranteed to transmit correctly for wires that do not cross orthogonally,

© Springer International Publishing AG 2016
J. Akiyama et al. (Eds.): JCDCGG 2015, LNCS 9943, pp. 167–179, 2016.
DOI: 10.1007/978-3-319-48532-4_15

which is required in their construction. Part of the reason no one found this error until now is that there was no formal framework in which to prove statements about flat-folded states. We attempt to provide such a framework.

At the end of their paper, Bern and Hayes pose some interesting open questions to further their work. While most of them have been investigated since, two in particular (problems 2 and 3) have remained untouched until now. First, is there a simpler way to achieve a proof for assigned crease patterns (i.e. "without tabs")? Second, their reductions construct creases at a variety of unconstrained angles. Is deciding flat foldability easy under more restrictive inputs? For example, *box pleating* involves folding creases only along on a subset of a square grid and the diagonals of the squares, a special case of particular interest in transformational robotics and self-assembly, with a universality result constructing arbitrary polycubes using box pleating [BDDO10].

In this paper we address both these questions. We prove that deciding flat foldability of box-pleated crease patterns is NP-hard in both the unassigned and assigned cases, using relatively simple gadgets containing no more than 25 layers at any point.

2   Definitions

In general, we are guided by the terminology laid out in [DO07,Rob77]. An *isometric flat folding* of a paper P is a function $f : P \to \mathbb{R}^2$ such that if γ is a piecewise-geodesic curve on P parameterized with respect to arc-length, then $f(\gamma)$ is also a piecewise-geodesic curve parameterized with respect to arc-length. It is not hard to show that under these conditions f must be continuous and non-expansive. Let X_f be the boundary of a paper P together with the set of points not differentiable under f. Then one can prove that X_f is a straight-line graph embedded in the paper [Rob77], with vertex set V_f and edge set C_f, the *creases* of our folding f. A vertex or crease in V_f or C_f is *external* if it contains a boundary point of P, and *internal* otherwise. Subtracting X_f from P results in a disconnected set of open polygons F_f we call *faces*. For any face $F \in F_f$, $f(F)$ is either an isotopic transformation in $\mathbb{R}^2$, or the transformation involves a reflection and is anisotopic. Define $u_f : P \setminus X_f \to \{-1, 1\}$ such that $u_f(p) = -1$ if the face containing p is reflected under f and $u_f(p) = 1$ otherwise. We call $u_f(p)$ the *orientation* of the face containing p. Every point in P is in exactly one of V_f, C_f, or F_f. We call this partition of P the *isometrically flat foldable crease pattern* $\Sigma_f = (V_f, C_f, F_f)$ induced by f. We call a folding *box pleating* if every vertex lies on two dimensional integer lattice, and the creases are aligned at multiples of $45°$ to each other.

We say two disjoint simply connected subsets of P are *adjacent* to each other if their closures intersect; we call such an intersection the *adjacency* of the adjacent subsets. We say a simply connected subset of P is *uncreased* under f if f is injective when restricted to the subset. We say two simply connected subsets of P *overlap* under f if the interiors of their images under f intersect. We say two simply connected subsets of P *strictly overlap* under f if their images under f

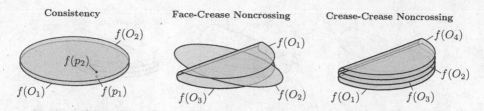

Fig. 1. Topologically different local interactions within an isometric flat folding. Forbidden configurations are shown for Face-Crease and Crease-Crease Non-Crossing.

exactly coincide. It is known that the set of creases adjacent to an internal vertex of a crease pattern obey the so called Kawasaki-Justin Theorem: the alternating sum of angles between consecutive creases when cyclically ordered around the vertex equals zero [DO07]. This condition turns out to be necessary sufficient: given a paper P exhaustively partitioned into a set of isolated points V, open line segments C, and open disks F such that every point in V is adjacent to more than two segments in C, then (V, C, F) is an isometrically flat foldable crease pattern induced by a unique isometric flat folding if and only if (V, C, F) obeys the Kawasaki-Justin Theorem.

Let a function $\lambda_f : P \times P \rightarrow \{-1, 1\}$ be a *global layer ordering* of an isometric flat folding f if it obeys the following six properties.

Existence: λ_f satisfies *existence* if $\lambda_f(p, q)$ is defined for every distinct pair of points p and q that strictly overlap under f and at least one of p or q is not in X_f; otherwise $\lambda_f(p, q)$ is undefined. Informally, order is only defined between a point on a face and another point overlapping it in the folding.

Antisymmetry: λ_f is *antisymmetric* if $\lambda_f(p, q) = -\lambda_f(q, p)$, where λ_f is defined. Informally, if p is above q, then q is below p.

Transitivity: λ_f is *transitive* if $\lambda_f(p, q) = \lambda_f(q, r)$ implies $\lambda_f(p, r) = \lambda_f(p, q)$, where λ_f is defined. Informally, if q is above p and r is above q, then r is above p.

Consistency (Tortilla-Tortilla Property): For any two uncreased simply connected subsets O_1 and O_2 of P that strictly overlap under f, λ_f is *consistent* if $\lambda_f(p_1, p_2)$ has the same value for all $(p_1, p_2) \in O_1 \times O_2$, where λ_f is defined. See Fig. 1. Informally, if two regions completely overlap in the folding, one must be entirely above the other.

Face-Crease Non-crossing (Taco-Tortilla Property): For any three uncreased simply connected subsets O_1, O_2, and O_3 of P such that O_1 and O_3 are adjacent and strictly overlap, and O_2 overlaps the adjacency between O_1 and O_3 under f, λ_f is *face-crease non-crossing* if $\lambda_f(p_1, p_2) = -\lambda_f(p_2, p_3)$ for any points $(p_1, p_2, p_3) \in O_1 \times O_2 \times O_3$, where λ_f is defined. See Fig. 1. Informally, if a region overlaps a nonadjacent internal crease, the region cannot be between the regions adjacent to the crease.

Crease-Crease Non-crossing (Taco-Taco Property): For any two adjacent pairs of uncreased simply connected subsets (O_1, O_2) and (O_3, O_4) of P such that

All same   Half-half   Odd one out
Adjacent   Nested   Intersecting

Fig. 2. Local interaction between overlapping regions around two distinct creases.

every pair of subsets strictly overlap and the adjacency of O_1 and O_2 strictly overlaps the adjacency of O_3 and O_4 under f, λ_f is *crease-crease non-crossing* if either $\{\lambda_f(p_1, p_3), \lambda_f(p_1, p_4), \lambda_f(p_2, p_3), \lambda_f(p_2, p_4)\}$ are all the same or half are $+1$ and half are -1, for any points $(p_1, p_2, p_3, p_4) \in O_1 \times O_2 \times O_3 \times O_4$, where λ_f is defined. See Fig. 2. Informally, if two creases overlap in the folding, either the regions incident to one crease lie entirely above the regions incident to the other (all same), or the regions incident to one crease nest inside the regions incident to the other (half-half).

If there exists a global layer ordering for a given isometrically flat foldable crease pattern, we say the crease pattern is *globally flat foldable*. Consider an isometrically flat foldable crease pattern Σ_f containing two adjacent uncreased simply connected subsets O_1 and O_2 of P that strictly overlap under f, and let p and q be points in O_1 and O_2 respectively that overlap under f. O_1 and O_2 are subsets of disjoint adjacent faces of the crease pattern mutually bounding a crease. If λ_f is a global flat folding of Σ_f, then it induces a *mountain/valley assignment* $\alpha_{\lambda_f}(c) = u(p)\lambda_f(p, q)$ for each crease point c in the adjacency of O_1 and O_2. This assignment is unique by consistency. We call a crease point c a *valley fold* (V) if $\alpha_{\lambda_f}(c) = 1$ and a *mountain fold* (M) if $\alpha_{\lambda_f}(c) = -1$. In the figures, mountain folds are drawn in red while valley folds are drawn in blue. By convention, if $\lambda_f(p, q) = -1$ we say that p is *above* q, and if $\lambda_f(p, q) = 1$ we say that p is *below* q.

Given an isometrically flat foldable crease pattern Σ_f, the UNASSIGNED-FLAT-FOLDABILITY problem asks whether there exists a global layer ordering for f. Alternatively, given an isometrically flat foldable crease pattern Σ_f and an assignment $\alpha : C_f \to \{M, V\}$ mapping creases to either mountain or valley, the ASSIGNED-FLAT-FOLDABILITY problem asks whether there exists a global layer ordering for f whose induced mountain valley assignment is consistent with α.

We now prove the following implied properties of globally flat foldable crease patterns relating the layer order between points contained in multiple overlapping faces. Informally, *Pleat-Consistency* says if a face is adjacent and overlapping two larger faces, then the creases between them must have different M/V assignment, forming a pleat. *Path-Consistency* says that a face overlapping creases connecting an adjacent sequence of faces is either above or below all of them.

Lemma 1 (Pleat-Consistency). *If Σ_f is a globally flat foldable crease pattern containing disjoint uncreased simply connected subsets O_1, O_2, and O_3 of P with*

O_2 adjacent to both O_1 and O_3 such that O_2 strictly overlaps subsets $O_1' \subset O_1$ and $O_3' \subset O_3$, and the interiors of O_1 and O_3 overlap the adjacencies of O_2, O_3 and O_1, O_2 respectively, then $\lambda_f(p_1, p_2) = \lambda_f(p_2, p_3)$ for any pairwise overlapping points $(p_1, p_2, p_3) \in O_1 \times O_2 \times O_3$.

Proof. Taco-Tortilla applied to O_3 which overlaps the adjacency of strictly overlapping sets O_2 and O_1' implies $\lambda_f(p_2, p_3) = -\lambda_f(p_3, p_1)$. Similarly, Taco-Tortilla applied to O_1 which overlaps the adjacency of strictly overlapping sets O_3' and O_2 implies $\lambda_f(p_3, p_1) = -\lambda_f(p_1, p_2)$, so $\lambda_f(p_1, p_2) = \lambda_f(p_2, p_3)$.   □

Lemma 2 (Path-Consistency). *If Σ_f is a globally flat foldable crease pattern containing uncreased simply connected subset T of P and a disjoint sequence of adjacent uncreased simply connected subsets $O_1, \ldots, O_n$ of P such that O_i strictly overlaps some subset T_i of T and the interior of O overlaps the adjacency of each pair O_i and O_{i+1} for $i = \{1, \ldots, n-1\}$, then $\lambda_f(t_j, p_j) = \lambda_f(t_k, p_k)$ for any two pairs of overlapping points $(t_j, p_j) \in T_j \times O_j$ and $(t_k, p_k) \in T_k \times O_k$ for $j, k \in \{1, \ldots, n\}$.*

Proof. If some O_i and O_{i+1} overlap, Taco-Tortilla and Consistency ensure that $\lambda_f(t_i, p_i) = \lambda_f(t_{i+1}, p_{i+1})$ for $(t_i, p_i) \in T_i \times O_i$ and $(t_{i+1}, p_{i+1}) \in T_{i+1} \times O_{i+1}$. Alternatively, O_i and O_{i+1} do not overlap and the closure of $O_i \cup O_{i+1}$ is an uncreased region for which $\lambda_f(t_i, p_i) = \lambda_f(t_{i+1}, p_{i+1})$ by consistency. Applying sequentially to each pair of faces proves the claim.   □

The proofs in Sects. 5 and 6 contain many examples of the application of these properties. When proving the existence of a global layer ordering λ_f, it is often impractical to define λ_f between every pair of points. Frequently λ_f is uniquely induced by a M/V assignment, consistency, and transitivity. When it is not, we will provide λ_f between additional point pairs so that it will be. We present crease patterns with this implicit layer ordering information and encourage readers to fold them to reconstruct the unique layer orderings they induce.

3   Bern and Hayes and k-Layer-Flat-Foldability

Two crossover gadgets are presented in the reduction to UNASSIGNED-FLAT-FOLDABILITY provided in [BH96]. For each, they claim that the M/V assignment of the crease pair intersecting one edge of the gadget deterministically implies the M/V assignment of the crease pair on the opposite side. This claim is true for their perpendicular crossover gadget, but is unfortunately not true for the other for wires meeting at 45°. The gadget as described requires an exterior 45° angle between incoming wires that is the smallest angle at a four-crease vertex, forbidding the wires to be independently assigned by Pleat-Consistency. For completeness, we have also checked the family of possible gadgets of this form, with a rotated internal parallelogram, and no choice of rotation allows the gadget to function correctly as a crossover for the range of widths of wires that appear in the construction. Our proof to follow only uses the perpendicular crossover, avoiding this complication.

Also in [BH96], they define k-LAYER-FLAT-FOLDABILITY to be the same as UNASSIGNED-FLAT-FOLDABILITY or ASSIGNED-FLAT-FOLDABILITY but with the additional constraint that f maps at most k distinct points to the same point. They claim that their reduction implies hardness of UNASSIGNED-k-LAYER-FLAT-FOLDABILITY for $k = 7$. But in fact their perpendicular crossover gadget requires nine points to be mapped to the same point. Our reduction uses the same gadget as a crossover, so we reconfirm that UNASSIGNED-k-LAYER-FLAT-FOLDABILITY is NP-complete for $k \geq 9$, even for box pleated crease patterns. Also, because of the complexity of their assigned crease pattern reduction, they were unable to bound the number of layers in their reduction. We explicitly provide gadgets for the assigned case to prove ASSIGNED-k-LAYER-FLAT-FOLDABILITY is NP-complete for $k \geq 25$, even for box pleated crease patterns.

4   SCN-Satisfiability

Our reductions will be from the following NP-complete problem [Sch78].

Problem 1 (**Not-All-Equal 3-SAT**). Given a collection of clauses each containing three variables, NOT-ALL-EQUAL 3-SAT (NAE3-SAT)[1] asks if variables can be assigned True or False so that no clause contains variables of only one assignment.

We can construct a planar directed graph G embedded in $\mathbb{R}^2$ from an instance $\mathcal{N}$ of NAE3-SAT. For each clause, construct a Complex Clause Gadget as the one shown in Fig. 3. The motivation behind the Complex Clause Gadget is to

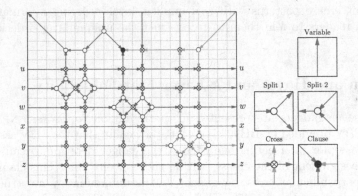

Fig. 3. SCN Gadgets. [Left] A Complex Clause Gadget constructed from the Not-All-Equal clause on variables v, w, and y of a NAE3-SAT instance on six variables. [Right] The five elemental SCN Gadgets.

---

[1] This problem is sometimes called 'positive' as variables cannot appear negated within clauses, however we follow the naming convention from [Sch78].

encode the bipartite graph implicit in $\mathcal{N}$ in a planar grid embedding that can be modularly connected. Each directed edge of the Complex Clause Gadget is associated with a different variable, and we associate a different color with each variable. Some variables do not participate in the clause and simply form a straight chain of directed segments from left to right. However, the three variables participating in the clause are rerouted to intersect at the black dot. We construct a Complex Clause Gadget for each clause in the instance of NAE3-SAT and chain them together side by side, so the arrows exiting the right side of one enter the left side of another. Graph G has vertices that are adjacent to edges associated with exactly one, two, or three variables. We call these vertices *split*, *cross*, and *clause* vertices respectively. In the figures, they are labeled with white circles, crossed circles, and black circles respectively. We call such a directed graph G a *Split-Cross-Not-All-Equal* (SCN) graph.

Problem 2 (**SCN-Satisfiability**). Given a SCN graph, SCN-SATISFIABILITY asks if variables can be assigned True or False so that no clause vertex is adjacent to edges associated with variables of only one assignment.

The authors introduce SCN-Satisfiability as a useful intermediate problem because it is equivalent to NAE3-SAT but its embedding is planar, lies on a grid, and is constructed only by a small number of local elements. SCN-SATISFIABILITY is equivalent to NAE3-SAT because the bipartite graph connecting SCN variables to clause vertices is exactly the bipartite graph representing $\mathcal{N}$ by construction. However, G has useful structure for many problems. It is planar, the embedding contains edges with only four slopes, and the edges are directed meaning that a variable can be represented locally with respect to that direction. Further G is constructed from only a small number of local elements: a variable gadget, two split gadgets, a cross gadget, and a clause (simple) gadget as shown in Fig. 3. We call these the five *elemental* SCN Gadgets. If we can simulate each of these gadgets in another context, proving that edges of the same color in each gadget must all have the same value, and edges adjacent to a clause vertex do not all have equal value, we can prove other problems NP-hard. This will be our strategy in the following sections.

Theorem 1. *If a problem X can simulate the elemental SCN gadgets such that edges of the same color in each gadget have the same value and edges adjacent to a clause vertex do not all have equal value and if the correspondent gadgets in X can be connected consistently, then X is NP-Hard.*

5 Unassigned Crease Patterns

In this section we present gadgets simulating the elemental SCN gadgets with unassigned crease patterns. They are shown in Fig. 4.

We define a variable gadget to be a pair of parallel creases placed close together having an direction as shown in Fig. 4. By pleat-consistency and transitivity, $\lambda_f(a,b) = \lambda_f(b,c) = \lambda_f(a,c)$ so, local to the gadget, it has exactly two

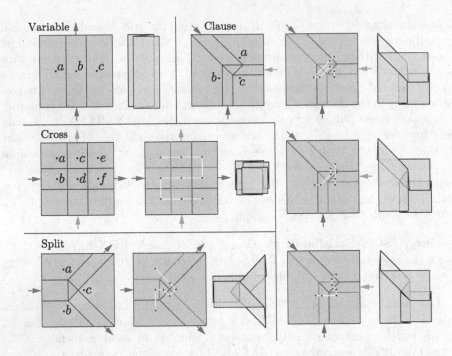

Fig. 4. Elemental SCN Gadgets simulated with unassigned crease patterns.

globally flat foldable states. We say the variable is *True* if the face to the right of the variable direction is above the face to left ($\lambda_f(a,c) = 1$), and *False* otherwise.

Lemma 3. *The unassigned crossover gadget is a globally flat foldable crease pattern if and only if opposite variables are equal.*

Proof. Refer to Fig. 4. Assume global flat foldability. Let A, B, C, D, E, F be the maximal subsets of the faces respectively containing points a, b, c, d, e, f such that every pair strictly overlap. First assume that $\lambda_f(a,b) = \lambda_f(c,d)$. By Taco-Taco with respect to adjacencies A, C and B, D, $\lambda_f(a,d) = \lambda_f(c,b)$. By Taco-Taco with respect to adjacencies A, B and C, D, $\lambda_f(a,c) = -\lambda_f(b,d)$. By Pleat-Consistency on A, C, E, $\lambda_f(a,c) = \lambda_f(c,e)$. By Pleat-Consistency on B, D, F, $\lambda_f(b,d) = \lambda_f(d,f)$. So $\lambda_f(c,e) = -\lambda_f(d,f)$. By Taco-Taco with respect to adjacencies C, D and E, F, $\lambda_f(c,f) = -\lambda_f(d,e)$. By Taco-Taco with respect to adjacencies C, E and D, F, $\lambda_f(c,d) = \lambda_f(e,f)$. Thus because $\lambda_f(a,b) = \lambda_f(e,f)$, the variable on the left has the same value as the one on the right. Alternatively if $\lambda_f(a,b) = -\lambda_f(c,d)$, the same series of arguments yields that $\lambda_f(c,d) = -\lambda_f(e,f)$, so $\lambda_f(a,b) = \lambda_f(e,f)$. Thus if global flat foldability holds, opposite variables are equal. Now assume that opposite variables are equal. The M/V assignment in Fig. 4 completely induces λ_f, along with consistency and transitivity. The path shown is a linear order on the faces satisfying global layer ordering. Further, every other assignment of variables can be represented by a reflection of this crease pattern.   □

Lemma 4. *The unassigned split gadget is a globally flat foldable crease pattern if and only if its three variables are equal.*

Proof. Refer to Fig. 4. Assume global flat foldability. Let A and B be the faces containing points a and b respectively. The region highlighted in the figure and A must satisfy Path-Consistency, so $\lambda_f(a, b) = \lambda_f(a, c)$. Since the crease pattern is symmetric, $\lambda_f(b, a) = \lambda_f(b, c)$. Then, by antisymmetry, $\lambda_f(a, b) = \lambda_f(c, b)$, and therefore all variables are equal. Now assume all variables are equal. The path shown in Fig. 4 is a linear order on the faces satisfying global layer ordering. Further, every other assignment of variables can be represented by a reflection of this crease pattern.   □

Lemma 5. *The clause gadget is a globally flat foldable crease pattern if and only if its three variables are not all equal.*

Proof. Refer to Fig. 4. Assume for contradiction the clause gadget is global flat foldable and all variables are equal. By consistency $\lambda_f(a, b) = \lambda_f(b, c) = \lambda_f(c, a)$. By transitivity, $\lambda_f(a, b) = \lambda_f(a, c)$. By antisymmetry, $\lambda_f(a, b) = -\lambda_f(c, a)$, a contradiction. Thus the variables are not all equal. Now assume all variables are not all equal. The paths shown in Fig. 4 are linear orders on the faces satisfying global layer ordering. Further, every other assignment of variables can be represented by the negation of one of these (M/V) assignments.   □

Theorem 2. Unassigned-Flat-Foldability *is NP-complete, even for box pleated crease patterns.*

Proof. Given λ_f as our certificate, we can check in polynomial time whether it satisfies all conditions for global flat foldability, therefore Unassigned-Flat-Foldability is in NP. By Lemmas 3, 4 and 5, Unassigned-Flat-Foldability can simulate the SCN-Satisfiability gadgets. It remains to check if the gadgets can be consistently connected. Let the width of a variable be the distance between its parallel creases. The crossover gadget connects variables of the same width while the clause and split gadgets both connect variables whose ratios differ by a factor of $\sqrt{2}$. Setting the width of one variable in any gadget induces the width of the other variables in the gadget. Fixing the width of one variable in the Complex Clause Gadget (Fig. 3), a consistent unique width for all other variables is induced, resulting in the same width for each variable intersecting a left or right edge. Therefore, by Theorem 1, Unassigned-Flat-Foldability is NP-Hard.   □

6   Assigned Crease Patterns

In this section we present gadgets simulating the elemental SCN gadgets with assigned crease patterns. They are shown in Fig. 5.

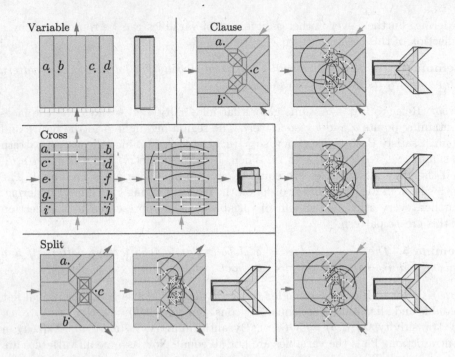

Fig. 5. Elemental SCN Gadgets simulated with assigned crease patterns.

We define a variable gadget as a set of parallel creases placed close together having a direction and a crease assignment as shown in Fig. 5. By Taco-Tortilla, $\lambda_f(a,c) = \lambda_f(b,c) = \lambda_f(a,d) = \lambda_f(b,d)$, so, local to the gadget, it has exactly two globally flat foldable states. We say the variable is *True* if the faces to the right of the variable direction are above the faces to left ($\lambda_f(a,c) = 1$), and *False* otherwise.

Lemma 6. *The assigned crossover gadget is a globally flat foldable crease pattern if and only if opposite variables are equal.*

Proof. Refer to Fig. 5. Assume global flat foldability. Let A, B, C, D be the maximal subsets of the faces respectively containing points a, b, c, d such that every pair strictly overlap. By transitivity on subset of λ_f induced by the M/V assignment shown, $\lambda_f(a,d) = \lambda_f(b,c) = -1$. By Taco-Taco with respect to adjacencies A, C and B, D, $\lambda_f(a,b) = -\lambda_f(c,d)$. Repeating this argument for adjacent rows of faces all the way down implies $\lambda_f(a,b) = -\lambda_f(c,d) = \lambda_f(e,f) = -\lambda_f(g,h) = \lambda_f(i,j)$. Thus, the variable on the top edge of the gadget has the same value as the one on the bottom. First assume $\lambda_f(g,a) = \lambda_f(a,b)$. Then previous implications imply $\lambda_f(g,a) = -\lambda_f(g,h)$. By transitivity and antisymmetry, $\lambda_f(g,a) = \lambda_f(h,b)$. Thus, the variable on the left side of the gadget has the same value as the one on the right. Alternatively, assume $-\lambda_f(g,a) = \lambda_f(a,b)$ so $\lambda_f(c,i) = \lambda_f(d,c)$. Then previous implications

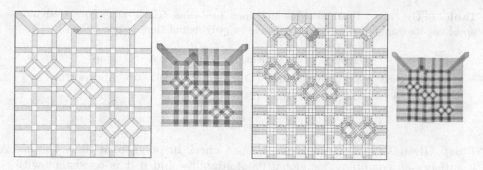

Fig. 6. A folded example of our assigned reduction with two clauses on four variables.

imply $\lambda_f(c, i) = \lambda_f(i, j)$. By transitivity and antisymmetry, $\lambda_f(c, i) = \lambda_f(d, j)$. Thus, the variable on the left side of the gadget has the same value as the one on the right. So, if globally flat foldable, opposite variables are equal. Now assume that opposite variables are equal. One can fix a unique λ_f by choosing a subset of λ_f in addition to the subset induced by the M/V assignment and consistency. The path shown in Fig. 5 is a linear order on the faces satisfying global layer ordering. Further, every other assignment of variables can be represented by a reflection of this crease pattern.   □

Lemma 7. *The assigned split gadget is a globally flat foldable crease pattern if and only if its three variables are equal (Fig. 6).*

Proof. Refer to Fig. 5. Assume global flat foldability. Let A and B be the faces containing points a and b respectively. The region highlighted in the figure and A must satisfy Path-Consistency, so $\lambda_f(a, b) = \lambda_f(a, c)$. Since the crease pattern is symmetric, $\lambda_f(b, a) = \lambda_f(b, c)$. Then, by antisymmetry, $\lambda_f(a, b) = \lambda_f(c, b)$, and therefore all variables are equal. Now assume all variables are equal. The path shown in Fig. 5 is a linear order on the faces satisfying global layer ordering. Further, any other assignment of variables can be attained by a reflection.   □

Lemma 8. *The assigned clause gadget is a globally flat foldable crease pattern if and only if its three variables are not all equal.*

Proof. Refer to Fig. 5. Assume for contradiction the clause gadget is global flat foldable and all variables are equal. By consistency $\lambda_f(a, b) = \lambda_f(b, c) = \lambda_f(c, a)$. By transitivity, $\lambda_f(a, b) = \lambda_f(a, c)$. By antisymmetry, $\lambda_f(a, b) = -\lambda_f(c, a)$, a contradiction. Thus the variables are not all equal. Now assume all variables are not all equal. The paths shown in Fig. 5 are linear orders on the faces satisfying global layer ordering. Further, any other assignment of variables can be attained by reversing the arrows in the figure.   □

Theorem 3. ASSIGNED-FLAT-FOLDABILITY *is NP-complete, even for box pleated crease patterns.*

Table 1. Overview of **our results** and open problems. 'Hard' and 'Poly' designate problems that are NP-complete or solvable in polynomial time respectively.

	General	Box pleating	Orthogonal	Map
Unassigned	Hard [BH96]	**Hard (Ours)**	Poly [ABD+04]	Always true
Assigned	Hard [BH96]	**Hard (Ours)**	Open	Open

Proof. Given λ_f as our certificate, we can check in polynomial time whether it satisfies all conditions for global flat foldability and if it is consistent with the crease assignment, therefore ASSIGNED-FLAT-FOLDABILITY is in NP. By Lemmas 6, 7 and 8, ASSIGNED-FLAT-FOLDABILITY can simulate the SCN-SATISFIABILITY gadgets. It remains to check if the gadgets can be consistently connected. Let the width of a variable be the distance between its two parallel mountain creases. By the same argument as in the proof of Theorem 2, widths of variables can be assigned consistently. Therefore, by Theorem 1, ASSIGNED-FLAT-FOLDABILITY is NP-Hard.   □

7   Conclusion

Table 1 overviews our results and open problems. We proved UNASSIGNED-FLAT-FOLDABILITY and ASSIGNED-FLAT-FOLDABILITY are NP-complete, even for box pleated crease patterns containing no more than 9 and 25 layers respectively. Are these problems still hard for even more restricted inputs? The computational complexity of ASSIGNED-FLAT-FOLDABILITY is still open when the crease pattern is a $m \times n$ square grid called a map [ABD+04]. Orthogonal folding, with crease patterns restricted to orthogonally aligned creases, is also open.

Acknowledgements. This work was begun at the 2015 Bellairs Workshop on Computational Geometry, co-organized by Erik Demaine and Godfried Toussaint. We thank the other participants of the workshop for stimulating discussions, and to Barry Hayes for helpful comments leading to some simplifications. The research of H. Akitaya was supported by NSF grant CCF-1422311 and Science without Borders. The research of T. Hull and T. Tachi were respectively supported by NSF grant EFRI-ODISSEI-1240441 and the JST Presto Program.

References

[ABD+04]   Arkin, E.M., Bender, M.A., Demaine, E.D., Demaine, M.L., Mitchell, J.S.B., Sethia, S., Skiena, S.S.: When can you fold a map? Comput. Geom. Theory Appl. **29**(1), 23–46 (2004)
[BDDO10]   Benbernou, N.M., Demaine, E.D., Demaine, M.L., Ovadya, A.: Universal hinge patterns to fold orthogonal shapes. In: Origami[5], pp. 405–420. A.K Peters, Singapore, July 2010

[BH96] Bern, M., Hayes, B.: The complexity of flat origami. In: Proceedings of the Seventh Annual ACM-SIAM Symposium on Discrete Algorithms SODA 1996, pp. 175–183. Society for Industrial and Applied Mathematics, Philadelphia, PA, USA (1996)

[DFL10] Demaine, E.D., Fekete, S.P., Lang, R.J.: Circle packing for origami design is hard. In: Origami^5, pp. 609–626. A.K Peters, Singapore (2010)

[DO07] Demaine, E.D., O'Rourke, J., Algorithms, G.F.: Linkages, Origami, Polyhedra. Cambridge University Press, Cambridge (2007)

[Rob77] Robertson, S.A.: Isometric folding of Riemannian manifolds. Proc. Roy. Soc. Edinburgh **79**(3–4), 275–284 (1977)

[Sch78] Schaefer, T.J.: The complexity of satisfiability problems. In: Proceedings of the Tenth Annual ACM Symposium on Theory of Computing, pp. 216–226. ACM (1978)

Symmetric Assembly Puzzles are Hard, Beyond a Few Pieces

Erik D. Demaine[1], Matias Korman[2], Jason S. Ku[1(✉)], Joseph S.B. Mitchell[3], Yota Otachi[4], André van Renssen[5,6], Marcel Roeloffzen[5,6], Ryuhei Uehara[4], and Yushi Uno[7]

[1] MIT, Cambridge, USA
{edemaine,jasonku}@mit.edu
[2] Tohoku University, Sendai, Japan
mati@dais.is.tohoku.ac.jp
[3] Stony Brook University, Stony Brook, USA
joseph.mitchell@stonybrook.edu
[4] JAIST, Nomi, Japan
{otachi,uehara}@jaist.ac.jp
[5] National Institute of Informatics, Tokyo, Japan
{andre,marcel}@nii.ac.jp
[6] JST, ERATO, Kawarabayashi Large Graph Project, Tokyo, Japan
[7] Osaka Prefecture University, Sakai, Japan
uno@mi.s.osakafu-u.ac.jp

Abstract. We study the complexity of symmetric assembly puzzles: given a collection of simple polygons, can we translate, rotate, and possibly flip them so that their interior-disjoint union is line symmetric? On the negative side, we show that the problem is strongly NP-complete even if the pieces are all polyominos. On the positive side, we show that the problem can be solved in polynomial time if the number of pieces is a fixed constant.

1 Introduction

The goal of a 2D *assembly puzzle* is to arrange a given set of pieces so that they do not overlap and form a target silhouette. The most famous example is the Tangram puzzle, shown in Fig. 1. Its earliest printed reference is from 1813 in China, but by whom or exactly when it was invented remains a mystery [5]. There are over 2,000 Tangram assembly puzzles [5], and many more similar 2D assembly puzzles [3]. A recent trend in the puzzle world is a relatively new type of 2D assembly puzzle which we call *symmetric assembly puzzles*. In these puzzles the target shape is not specified. Instead, the objective is to arrange the pieces so that they form a symmetric silhouette without overlap.

The first symmetric assembly puzzle, "Symmetrix", was designed in 2003 by Japanese puzzle designer Tadao Kitazawa and was distributed by Naoyuki Iwase as his exchange puzzle at the 2004 International Puzzle Party (IPP) in Tokyo [4]. In this paper, we aim for arrangements that are line symmetric (reflection through a line), but other symmetries such as rotational symmetry could

© Springer International Publishing AG 2016
J. Akiyama et al. (Eds.): JCDCGG 2015, LNCS 9943, pp. 180–192, 2016.
DOI: 10.1007/978-3-319-48532-4_16

Fig. 1. [Left] The seven Tangram pieces (1) can be assembled into non-simple silhouettes (2) and (3). [Right] A symmetric assembly puzzle invented by Hiroshi Yamamoto [7]: given the two black pieces (right) from the classic T puzzle (left), make two different line symmetric shape. (Used with permission.)

also be considered. The lack of a specified target shape makes these puzzles quite difficult to solve.

We study the computational complexity of symmetric assembly puzzles in their general form. We define a *symmetric assembly puzzle* or SAP to be a set of k simple polygons $\mathcal{P} = \{P_1, P_2, \ldots, P_k\}$, with $n = |P_1| + \cdots + |P_k|$ the total number of vertices in all pieces. By *simple polygon* we mean a closed subset of $\mathbb{R}^2$ homeomorphic to a disk bounded by a closed path of straight line segments where nonadjacent edges and vertices do not intersect. A *symmetric assembly* $f : \mathcal{P} \to \mathbb{R}^2$ of a SAP $\mathcal{P}$ is a planar isometric embedding of the pieces so that their mapped interiors are disjoint and their mapped union forms a simple polygon that is line symmetric. We allow pieces to flip over (reflect), but other variants of the puzzle may disallow this. Given that humans have difficulty SAPs with even few low-complexity pieces, we consider two different generalizations: bounded piece complexity ($|P_i| = O(1)$) and bounded piece number ($k = O(1)$). In the former case, we prove strong NP-completeness, while in the latter case, we solve the problem in polynomial time (the exponent is linear in k).

2   Many Pieces

First we show that it is hard to solve symmetric assembly puzzles with a large number of pieces, even if each piece has bounded complexity ($|P_i| = O(1)$).

Theorem 1. *Symmetric assembly puzzles are strongly NP-complete even if each piece is a polyomino with at most six vertices and area upper bounded by a polynomial function of the number of pieces.*

If a SAP has a solution, the location and orientation of each piece within a symmetric assembly is a solution certificate of polynomial size checkable in polynomial time, so symmetric assembly puzzles are in NP. We reduce from the RECTANGLE PACKING PUZZLE problem, known to be strongly NP-hard [2]. Specifically, it is (strongly) NP-complete to decide whether k given rectangular pieces—sized $1 \times x_1, 1 \times x_2, \ldots, 1 \times x_k$, where the x_i's are positive integers bounded above by a polynomial in k—can be exactly packed into a specified rectangular box with given width w and height h and area $x_1 + x_2 + \cdots + x_k = wh$.

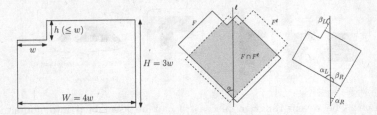

Fig. 2. [Left] The frame piece F. [Middle] If ℓ and ℓ_B form an angle of $\pi/4$, then $F \cap F^\ell$ is contained in a rectangle in an $H \times H$ and thus O^* cannot be line symmetric. [Right] The angles α_L, β_L, α_R, and β_R.

Let $I = (x_1, \ldots, x_k, w, h)$ be a rectangle packing puzzle. Without loss of generality, we assume that $w \geq h$. Now let $I' = (P_1, \ldots, P_k, F)$ be the SAP where P_i is the $1 \times x_i$ rectangle for each $i \in \{1, \ldots, k\}$, and F is the polyomino in Fig. 2. We call F the *frame piece* of I'. We show that I has a rectangle packing if and only if I' has a symmetric assembly.

Clearly, if I has a rectangle packing, then the pieces $P_1, \ldots, P_k$ can be packed into the $w \times h$ hole in the frame piece creating a line symmetric $W \times H$ rectangle, solving the SAP. Now we show the reverse implication. Assume that I' has a symmetric assembly, and let O^* be a line symmetric polygon formed by the pieces $\{P_1, \ldots, P_k, F\}$. We claim that O^* must be a $W \times H$ rectangle, which will imply that I is a yes-instance of RPP. Fix a placement of the pieces of I' that forms O^*, and let ℓ be one of its lines of symmetry. Assume, without loss of generality, that ℓ is a vertical line. Let F^ℓ be the reflection of F about ℓ.

Observation 1. $\mathsf{area}(F \cap F^\ell) \geq WH - 2wh \geq 10w^2$

Proof. Since O^* contains F^ℓ and F, it holds that $\mathsf{area}(F^\ell \setminus F) \leq \mathsf{area}(O^* \setminus F) = wh$. Since $F \cup F^\ell$ is mirror-symmetric, $\mathsf{area}(F^\ell \setminus F) = \mathsf{area}(F \setminus F^\ell)$. Hence, it follows that $\mathsf{area}(F \cap F^\ell) = \mathsf{area}(F) - \mathsf{area}(F \setminus F^\ell) \geq WH - 2wh \geq 10w^2$. □

Observation 1 implies that ℓ passes through an interior point of F. Let ℓ_B be the line containing the segment of F with length $4w$. Let c be the center of the frame piece's bounding box.

Lemma 1. ℓ_B *is either parallel or orthogonal to* ℓ.

Proof. Suppose for contradiction that ℓ_B is neither parallel nor orthogonal to ℓ. Let α be the smaller angle made by ℓ_B and ℓ. We partition the edges of F crossed by ℓ into two at their intersection points. Let F_L and F_R be the sets of segments on the left and right portions of F, respectively. Consider the set of counter-clockwise angles between ℓ and the lines containing segments of F_L. The assumptions that ℓ_B and ℓ are neither parallel nor orthogonal, and that F is a polyomino together imply that the set contains exactly two angles α_L and β_L, where $\alpha_L \leq \beta_L$ and $\alpha_L + \pi/2 = \beta_L$. Similarly, let α_R and β_R be the clockwise angles between ℓ and the lines containing segments of F_R, where $\alpha_R \leq \beta_R$ and

$\alpha_R + \pi/2 = \beta_R$. Since $\alpha_L + \beta_R = \pi$, it holds that $\alpha_L + \alpha_R = \pi/2$. Note that $\alpha \in \{\alpha_L, \alpha_R\}$.

Two distinct pieces of I' are *connected* if the fixed placement of the two pieces to form O^* have a non-degenerate line segment on their edges in common. Let $\mathcal{P}$ be the subset of $\{P_1, \ldots, P_n, F\}$ such that each $P_i \in \mathcal{P}$ can be reached from F by repeatedly following connected pieces in O^*.

As before, consider the angles formed by ℓ and the lines containing segments in the left and right parts of $\mathcal{P}$. Since all pieces are polyominoes, these lines cannot make angles other than α_L, β_L, α_R, and β_R with ℓ. Further note that the subset O' of Q^* covered by $\mathcal{P}$ must be mirror-symmetric with respect to ℓ, or else O^* would not be. This implies that $\alpha_L = \alpha_R$. Since $\alpha_L + \alpha_R = \pi/2$, the only solution in which ℓ is not parallel or orthogonal to ℓ_B is when $\alpha_L = \alpha_R = \pi/4$ and $\alpha = \pi/4$. However, if $\alpha = \pi/4$, then $F \cap F^\ell$ is a subset of an $H \times H$ rectangle (see Fig. 2), whose area is at most $H^2 = 9w^2$, contradicting Observation 1. $\qquad\square$

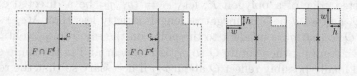

Fig. 3. [Left] When ℓ passes to the left of c, the portion of F to the left of ℓ is too small. If it passes to the right, the right portion would be too small. [Right] If ℓ passes through c, and is either orthogonal or parallel to ℓ_B, the symmetric assembly puzzle can only be completed into a rectangle.

So ℓ is either parallel or orthogonal to ℓ_B. Further, it passes through c (see Fig. 3). In either case, $F \cup F^\ell$ is a $W \times H$ rectangle, and thus $O^* = F \cup F^\ell$. This implies that $O^* \setminus F$ is a $w \times h$ rectangle that must contain the remaining pieces of I'. In particular, we have that this placement packing of $P_1, \ldots, P_n$ gives a solution to the instance I of RPP, completing the proof of Theorem 1.

3   Constant Pieces

Next we analyze symmetric assembly puzzles with a constant number of pieces but many vertices, and show they can be solved in polynomial time.

Theorem 2. *Given a symmetric assembly puzzle with a constant number of pieces k containing at most n vertices in total, deciding whether it has a symmetric assembly can be decided in polynomial time with respect to n.*

To prove this theorem, we present a brute force algorithm for solving a SAP that runs in polynomial time for constant k. We say two pieces in a symmetric assembly are *connected* to each other if their intersection in the symmetric assembly contains a non-degenerate line segment, and let the *connection* between two connected pieces be their intersection not including isolated points. We will call two pieces *fully* connected if their connection is exactly an edge of one of the pieces, and *partially* connected otherwise. Call a piece a *leaf* if it connects to at most one piece, and a *branch* otherwise. Given a leaf, let its *parent* be the piece connected to it (if it exists), and let its *siblings* be all other pieces connected to its parent. An illustration demonstrating these terms can be found in Fig. 4.

We will use a few utility functions in our algorithm. Deciding whether a single simple polygon has a line of symmetry can be done in linear time [6]. We will use isSym(P) to denote this algorithm, returning TRUE if polygo n P has a line of symmetry and FALSE otherwise. In addition, we can test congruence of polygons in linear time using cong(P, Q), returning TRUE if P and Q are congruent polygons, and FALSE otherwise.

In addition, we will need to construct simple polygons from provided simple polygons by laying them next to each other along an edge. Let E_P denote the set of directed edges (p_i, p_j) from a vertex p_i to an adjacent vertex p_j of some simple polygon P. Given an edge $e \in E_P$, we denote its length by $\lambda(e)$. Let $e_P = (p_1, p_2)$ be a directed edge of a polygon P, let $e_Q = (q_1, q_2)$ be a directed edge of a polygon Q, and let d be a nonnegative length strictly less than $\lambda(e_P) + \lambda(e_P)$. Translate Q so that q_1 is incident to the point on the ray from p_1 containing e_P a distance d from p_1; then rotate Q so e_Q is collinear and in the same direction as e_P; and finally possibly reflect Q about e_Q if necessary so that the respective interiors of P and Q incident to e_P and e_Q lie in different half planes. Call these transformations the mapping $g : P \cup Q \to \mathbb{R}^2$. Then we define join e_P, e_Q, d to be either, $g(P) \cup g(Q)$ if it is a simple polygon and the interior of $g(P) \cap g(Q)$ is empty (forms a simple polygon without overlapping pieces), or otherwise the empty set. See Fig. 4.

Fig. 4. [Left] Visualization of a join operation. [Right] Example symmetric assembly $\mathcal{P}$ showing its connection graph. Pieces a and d are fully connected to piece b, with c partially so. Pieces b, c, and d are branches. Piece a is a leaf, with b its parent and c and d the siblings of a.

If a SAP has a symmetric assembly, let its *connection graph* be a graph on the pieces with an edge connecting two pieces if they are connected in the symmetric assembly. Because a symmetric assembly is a simple polygon by definition, its connection graph is connected and has a spanning tree; we can then construct the assembly using a concatenation of join procedures in breadth-first-search order from an arbitrary root. Because parameter d is not discrete, the total solution space of simple polygons that are constructible from the pieces of a SAP may

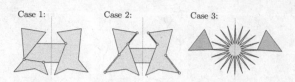

Fig. 5. Examples of symmetric assemblies belonging to each case. Case 1 highlights vertices of connected pieces that intersect. Case 2 highlights `join` operations using lengths of piece edges. Case 3 is constructed from one symmetric piece and a pair of congruent pieces.

be uncountable. However, we can exploit the structure of symmetric assemblies to search only a finite set of configurations.

In order to enumerate possible configurations, we would like to distinguish between three cases of puzzle (see Fig. 5), specifically:

Case 1: the puzzle has a symmetric assembly in which two connected pieces share a vertex on their connection;

Case 2: the puzzle has a symmetric assembly not satisfying Case 1 in which the distance between vertices from the connecting edges between two connected pieces has the same length as an edge from a third piece (we say the connection between two pieces *constructs* the length of another edge); or

Case 3: the puzzle has a symmetric assembly not satisfying Case 1 or Case 2 where a nonempty set of pieces are symmetric about the line of symmetry of the symmetric assembly, and any remaining pieces are pairs of congruent pieces.

Lemma 2. *If a SAP has a symmetric assembly, it can be described by one to the above three cases.*

Proof. Suppose for contradiction we have a symmetric assembly $f : \mathcal{P} \to \mathbb{R}^2$ of a SAP $\mathcal{P}$ that does not satisfy any of the above cases let $s : f(\mathcal{P}) \to f(\mathcal{P})$ be an automorphism reflecting $f(\mathcal{P})$ across a line of symmetry L, and let $\mu = s \circ f$, mapping a point $p \in \mathcal{P}$ to the reflection of $f(p)$ across L.

Consider the connection graph of $f(\mathcal{P})$. Because the symmetric assembly forms a simple polygon and no two connected pieces share a vertex, by exclusion from Case 1 the connection graph is a tree which we call a *connection tree*, or else the symmetric assembly would not be homeomorphic to a disk. Further, all connections are single non-degenerate line segments.

Let P be a leaf in the symmetric assembly, whose siblings include at most one branch. We claim that either P is a line symmetric polygon, or $\mu(P)$ is itself a piece of the SAP congruent to P contradicting exclusion from Case 3. First, if P has no parent and is the only piece in the symmetric assembly, P must be a line symmetric polygon. Otherwise, let Q be the parent of P with edge e_P from E_P touching edge e_Q from E_Q. Let e_{QP} denote the subset of e_Q that maps to the intersection $f(e_P) \cap f(e_Q)$. Segment $f(e_{QP})$ cannot lie along L or else

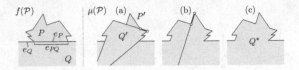

Fig. 6. Possible topological configurations of $\mu(P)$.

one of $f(e_P)$ or $f(e_Q)$ would share a vertex with another piece, contradicting exclusion from Case 1. Alternatively suppose $f(e_{QP})$ and $\mu(e_{QP})$ are the same line segment. As a leaf, P connects to the rest of the symmetric assembly only through $f(e_{QP})$, so for the assembly to be symmetric, $f(P)$ must be the same as $\mu(P)$, and piece P is a line symmetric polygon.

Lastly, suppose $f(e_{QP})$ and $\mu(e_{QP})$ are not the same line segment; we claim $\mu(P)$ is itself a piece of the SAP congruent to P. Suppose for contradiction it were not. Then $\mu(P)$ either (a) contains a piece as a strict subset, (b) does not fully contain a piece but intersects interiors of multiple pieces, or (c) is a strict subset of a single piece (see Fig. 6).

First suppose (a), so $\mu(P)$ contains some piece S as a strict subset. Root the connection tree at a piece R with the shortest graph distance to S in the connection tree for which $f(R) \cap \mu(P) \neq \emptyset$ and $f(R) \setminus \mu(P) \neq \emptyset$ which exists because $\mu(e_{PQ})$ must intersect some piece. Then a leaf P' with a longest root to leaf path that contains S is also fully contained in $\mu(P)$. Let Q' be its parent with edge e'_P from P' touching edge e'_Q from Q'. Because R is the piece crossing the boundary of $\mu(P)$ closest to S in the connection tree and P' has the longest root to leaf path, e'_Q connects to at most one branch piece that intersects $\mu(P)$. Segment $f(e'_P)$ cannot contain an edge of the symmetric assembly or else it would construct a length equal to an edge of P, contradicting exclusion from Case 2. So every leaf fully contained in $\mu(P)$ connected to e'_Q is fully connected to Q'. Each endpoint of the subset of e'_Q in $\mu(P)$ has shortest Euclidean distance to the connection of one leaf intersecting $\mu(P)$ connected to e'_Q. But at least one of these leaves is fully contained in $\mu(P)$ which that would construct a length equal to an edge of P, contradicting exclusion from Case 2. So $\mu(P)$ does not fully contain a leaf, contradicting case (a).

Now suppose (b), and suppose two connected pieces intersect $\mu(P)$. The edges connecting these two pieces must overlap in $\mu(P)$ to construct a length equal to an edge of P, contradicting exclusion from Case 2. So $\mu(P)$ does not intersect the interior of multiple branch pieces.

Finally suppose (c), and let $\mu(P)$ be the strict subset of some piece Q^*. Segment $f(e_P)$ cannot contain an edge of the symmetric assembly or else it would create a length equal to an edge of Q^*, contradicting exclusion from Case 2. So P is fully connected. A useful corollary of the preceding three arguments is that the reflection of any partially connected leaf of a symmetric assembly that conforms to neither Case 1 nor Case 2, must itself be a piece congruent to the leaf. We will refer to this property later as *partial leaf congruence*.

Here we note that none of the arguments so far have required P to be a leaf having at most one branch sibling; we will use that fact in the argument to follow. Let ℓ be the line collinear with segment $f(e_{QP})$, and let e_ℓ be the subset of Q that maps to the largest connected subset of $\ell \cap f(Q)$ containing $f(e_{QP})$. Consider the two disconnected sections of the boundary of Q between an endpoint of e_{PQ} and an endpoint of e_ℓ, which must each be more than an isolated point or exclusion from Case 1 would be violated. Piece P has at most one branch sibling, so at most one of these sections can be connected to a branch. Let q be an endpoint of e_ℓ in a section not connected to a branch.

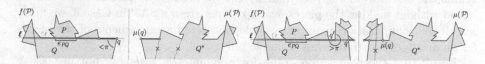

Fig. 7. Considering if $\mu(P)$ is a strict subset of Q^* and the boundary between e_{PQ} and q is a [Left] straight line or [Right] not a straight line.

Consider the boundary of Q between e_{QP} and q. Suppose this boundary were a line segment subset of e_Q, implying the internal angle of Q at q is less than π; see Fig. 7. Then $\mu(q)$ is in $f(Q^*)$ or else Q^* would connect to another piece somewhere on the segment between e_{QP} and q and construct an edge of the same length as a leaf connected to e_Q, contradicting exclusion from Case 2. If $\mu(q)$ is in $f(Q^*)$ and Q does not connect with any other piece at q, then $\mu(q)$ must be a vertex of $f(Q^*)$. Alternatively, q partially connects to a leaf through e_Q. By partial leaf congruence, the reflection of this leaf must itself be a congruent piece, so $\mu(q)$ is a vertex of $f(Q^*)$. In either case, the edge of Q^* adjacent to $\mu(q)$ contained in $\mu(e_Q)$ will have the same length as the subset of e_Q between q and a vertex of a leaf, contradicting exclusion from Case 2.

Thus, the boundary of Q between e_{QP} and q is not a line segment, so $f(Q)$ must cross ℓ, and the endpoint q' of e_Q in this section is a vertex of Q with internal angle greater than π; see Fig. 7. By the same argument as in the preceding paragraph, $\mu(q')$ must be in $f(Q^*)$, and if it were a vertex, we would have the same contradiction as before. However this time $\mu(q')$ need not be a vertex of $f(Q^*)$ because $f(Q^*)$ may extend past $\mu(q')$, with Q^* connecting to another piece on the other side of e_ℓ. However, the connection between these pieces will construct an edge that is the same length as an edge in either Q or a leaf connected to Q, and we have arrived at our final contradiction. So if P is not line symmetric, $\mu(P)$ is itself a piece of the SAP congruent to P.

Thus, our SAP has a leaf that is either a line symmetric piece, symmetric about the line of symmetry, and/or exists in a pair of two leaf pieces that are congruent and symmetric about the line of symmetry. If we remove such an identified leaf piece or pair from the SAP, what remains is a SAP with fewer pieces also admitting a symmetric assembly. Further, removing pieces cannot make the new SAP belong to one of the cases that the original SAP did not before. Repeatedly removing pieces using this process identifies every piece as either symmetric, or uniquely paired with a piece congruent to it, contradicting exclusion from Case 3.   □

Since every symmetric assembly can be classified as one of these cases, we can check for each case to decide if the SAP has a symmetric assembly. Given a SAP that does not satisfy Case 1 or Case 2, by Lemma 2 it must satisfy Case 3 if it has a symmetric assembly. Satisfying Case 3 is not sufficient to ensure a symmetric assembly. For example, two congruent regular polygons with many sides and a single regular star with many spikes cannot by themselves form a symmetric assembly though they satisfy Case 3 because no pair of edges can be joined without making the pieces overlap. Thus given a SAP in Case 3, we must search the configuration space of possible connected arrangements of the pieces for an arrangement that forms a simple polygon.

Recall that the connection graph for a symmetric assembly not in Case 1 must be a tree. For a SAP with k pieces, Cayley's formula says the number of distinct connection trees is k^{k-2} [1]. However, even if two pieces are connected, they could be connected through $O(n^2)$ different pairs of edges, so the number of different *edge distinguishing connection trees*, connection trees distinguishing between which pairs of edges are connected, can be no more than $n^{2k}k^k = O(n^{2k})$ (k is constant). As an instance of Case 3, $\mathcal{P}$ consists of one or more symmetric pieces, with the rest being congruent pairs. Let $\mathcal{D}_\mathcal{P}$ and $\mathcal{D}'_\mathcal{P}$ be maximal disjoint subsets of $\mathcal{P}$ such that there exists a matching $\eta : \mathcal{D}'_\mathcal{P} \to \mathcal{D}_\mathcal{P}$ between pieces in $\mathcal{D}_\mathcal{P}$ and $\mathcal{D}'_\mathcal{P}$ such that matched pairs are congruent. Let $\mathcal{S}_\mathcal{P}$ be the set of symmetric pieces in $\mathcal{P}$ not in $\mathcal{D}_\mathcal{P}$ or $\mathcal{D}'_\mathcal{P}$. Let $\mathcal{S}_T$ denote some subset of the symmetric pieces contained in $\mathcal{D}_\mathcal{P}$, and define a *trunk* to be a subset of symmetric pieces $\mathcal{R}_T = \mathcal{S}_\mathcal{P} \cup \mathcal{S}_T \cup \eta(\mathcal{S}_T)$ that can be connected into a simple polygon without overlap while aligning each of their lines of symmetry to a common line L (see Fig. 8). Define a *half tree* T to be an edge distinguishing connection tree on $\mathcal{R}_T \cup \mathcal{D}_\mathcal{P}$ such that every piece in $\mathcal{D}_\mathcal{P}$ connected to a piece R in $\mathcal{R}_T$ connects through an edge of R intersecting the same half-plane bounded by L. We call this half-plane the *connecting half-plane*, with the other half-plane the *free half-plane*. The reason we define half trees is if we can find a point in their configuration space for which pieces do not intersect and for which pieces in $\mathcal{D}_\mathcal{P}$ not in the trunk do not intersect the free half-plane, we can place the remaining congruent pieces in $\mathcal{D}_\mathcal{P} \setminus \mathcal{S}_T$ at the mirror image of their respective matched pairs to complete a symmetric assembly.

Let $\mathcal{T}_\mathcal{P}$ be the set of possible half trees. Let $\mathcal{L}_T$ be the set of undirected edges $\{P, Q\}$ where piece P is connected to piece Q in tree $T \in \mathcal{T}_\mathcal{P}$, and let $m = |\mathcal{L}_T| < k$. For a fixed edge distinguishing connection tree, the orientation of each piece is fixed as pieces may only translate along their specified connection. We want to define a set of intervals $\mathcal{I}_T\{P, Q\}$ where we could join e_P to e_Q while together forming a simple polygon, without overlap between P and Q. For each $\{P, Q\} \in \mathcal{L}_T$ with e_P and e_Q the respective connecting edges of P and Q with $\lambda(e_P) \geq \lambda(e_Q)$, let $\mathcal{I}_T\{P, Q\}$ be defined as follows. If P and Q are both in $\mathcal{R}_T$, let $\mathcal{I}_T\{P, Q\}$ be the empty set if $\texttt{join}(e_P, e_Q, d_{PQ})$ is the empty set and $\{d_{PQ}\}$ otherwise, where we use d_{PQ} to denote $|\lambda(e_P) - \lambda(e_Q)|/2$, the distance d would need to be in order to align the midpoints of e_P and e_Q. Alternatively if P or Q are not in $\mathcal{R}_T$, let $\mathcal{I}_T\{P, Q\}$ be the closure of the set of distances d for which $\texttt{join}(e_P, e_Q, d)$ is nonempty. The number of distinct intervals in $\mathcal{I}_T\{P, Q\}$ is at most linear in the number of vertices, $O(n)$. Any fixed arrangement of the pieces consistent with edge distinguishing connection tree T joins each pair of pieces by fixing one point in every $\mathcal{I}_T\{P, Q\}$, so the set of configurations is a subset of $\mathbb{R}^m$. Ignoring overlap between pieces that are not connected, the configuration space $\mathcal{C}_T$ of possible arrangements is equal to the cartesian product of $\mathcal{I}_T\{P, Q\}$ for every $\{P, Q\} \in \mathcal{L}_T$. Thus $\mathcal{C}_T$ is a set of $O(n^m)$ disjoint m-dimensional hyperrectangles in $\mathbb{R}^m$.

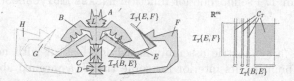

Fig. 8. An example showing a SAP $\mathcal{P}$ satisfying Case 3, with $\mathcal{S}_\mathcal{P} = \{A, B\}$, $\mathcal{D}_\mathcal{P} = \{C, E, F\}$, $\mathcal{D}'_\mathcal{P} = \{D, G, H\}$, $\mathcal{S}_T = \{C\}$, $\eta(\mathcal{S}_T) = \{D\}$, and trunk $\mathcal{R}_T = \{A, B, C, D\}$. $\mathcal{I}_T$ for two connected pieces in the trunk is just a single point as shown by the midpoint of their connection. Pieces not in the trunk have a degree of freedom sliding along their connection. $\mathcal{I}_T\{E, F\}$ is a single interval where F can attach to E, while $\mathcal{I}_T\{B, E\}$ is a four intervals. The right diagram shows $\mathcal{C}_T$ the cartesian product of each $\mathcal{I}_T$.

We now describe the subset of $\mathbb{R}^m$ where intersection occurs between two pieces that are not connected in T. If two pieces in a configuration overlap, by continuity there exist two edges e_P and e_Q from two distinct pieces P and Q that also intersect. The positions of e_P and e_Q are translations parameterized by a point in $\mathcal{C}_T$ and the region in which the two edges intersect is a convex region $\mathcal{X}_T\{e_P, e_Q\} \subset \mathbb{R}^m$ bounded by four hyperplanes forming the m-dimensional parallelogram representing the intersection of the two edges. For each $O(n^2)$ pair of edges from distinct pieces that are not connected, we can subtract each $\mathcal{X}_T\{e_P, e_Q\}$ from $\mathcal{C}_T$ to form $\mathcal{C}'_T$. If $\mathcal{C}'_T$ contains any point in its interior, then there exists a symmetric assembly since it will be a point in the configuration space avoiding overlap between pieces. However, the boundary of $\mathcal{C}'_T$ may contain configurations that are weakly simple as the boundaries of each $\mathcal{I}_T$ not between two pieces in $\mathcal{R}_T$ and the boundaries of each $\mathcal{X}_T$ all correspond to configuration containing non-simple touching between pieces. Thus we require $\mathcal{C}'_T$ to have a point on its interior unless all pieces exist in R_T, where $\mathcal{C}'_T$ may be a single point corresponding to a symmetric assembly.

Consider the function `hasAssemblyCase3` described in Algorithm 1.

Lemma 3. *Given symmetric assembly puzzle $\mathcal{P}$ that satisfies Case 3, function `hasAssemblyCase3`$\mathcal{P}$ returns TRUE if and only if $\mathcal{P}$ has a symmetric assembly, and terminates in $O(n^{5k})$ time.*

```
1  Function hasAssemblyCase3(P)
2  |  input  : Symmetric assembly puzzle P that satisfies Case 3.
3  |  output : TRUE if P has a symmetric assembly, FALSE otherwise.
4  |  for T ∈ T_P do
5  |  |   C'_T ← C_T
6  |  |   for {P, Q} ∈ L_T do
7  |  |   |   C'_T ← C'_T \ X_T{e_P, e_Q}
8  |  |   if int(C'_T) ≠ ∅ then
9  |  |   |   return TRUE
10 |  |   else if C'_T ≠ ∅ and D_P = ∅ then
11 |  |   |   return TRUE
12 |  return FALSE
```

Algorithm 1. Pseudocode for function hasAssemblyCase3($\mathcal{P}$)

Proof. If $\mathcal{P}$ has a symmetric assembly satisfying Case 3 with nonempty $\mathcal{D}_\mathcal{P}$, $\mathcal{C}'_T$ will have a point on its interior for some tree T as argued above; or if $\mathcal{D}_\mathcal{P}$ is empty, $\mathcal{C}'_T$ will be nonempty. There are $O(n^{2k})$ elements of $\mathcal{T}_\mathcal{P}$. There are $m = O(k)$ interval sets $\mathcal{I}_T\{P, Q\}$ each having computational complexity $O(n)$, so we can construct $\mathcal{C}_T$ naively in $O(n^k)$ time. The union X_T of the $O(n^2)$ regions $\mathcal{X}_T\{e_P, e_Q\}$, which are m-dimensional convex regions, has computational complexity at most $O(n^{2m})$, so the final computational complexity of $\mathcal{C}'_T = \mathcal{C}_T \setminus X_T$ is at most $O(n^{3m})$ and can be computed in as much time. Thus, the running time of hasAssemblyCase3 is bounded by $O(n^{5k})$. □

```
1  Function hasAssembly(P)
2  |  input  : Symmetric assembly puzzle P.
3  |  output : TRUE if P satisfies Case 1 or Case 2 or Case 3, FALSE otherwise.
4  |  for e_P ∈ E_P, e_Q ∈ E_Q, {P, Q} ⊂ P do
5  |  |   S ← join(e_P, e_Q, 0)
6  |  |   P' ← (P \ {P, Q}) ∪ {S}
7  |  |   if S ≠ ∅ and hasAssembly(P') then
8  |  |   |   return TRUE                              // Case 1
9  |  |   for e_R ∈ E_R, R ∈ P do
10 |  |   |   if λ(e_R) < λ(e_P) then
11 |  |   |   |   S ← join(e_P, e_Q, λ(e_R))
12 |  |   |   |   P' ← (P \ {P, Q}) ∪ {S}
13 |  |   |   |   if S ≠ ∅ and hasAssembly(P') then
14 |  |   |   |   |   return TRUE                       // Case 2
15 |  return hasAssemblyCase3(P)                       // Case 3
```

Algorithm 2. Pseudocode for function hasAssembly($\mathcal{P}$)

Our brute force algorithm hasAssembly $\mathcal{P}$ is described in Algorithm 2.

Lemma 4. *Function* hasAssembly($\mathcal{P}$) *returns* TRUE *if and only if* $\mathcal{P}$ *has a symmetric assembly that satisfies either Case 1, Case 2, or Case 3, and terminates in* $O(n^{5k})$ *time.*

Proof. We prove by induction. For the base case, $\mathcal{P}$ consists of only a single piece satisfying Case 3, which will drop directly to the last line of the algorithm checking Case 3 which, by Lemma 3 will evaluate correctly. Now suppose hasAssembly returns a correct evaluation for SAPs containing $k-1$ pieces. Then we show hasAssembly returns a correct evaluation for SAPs containing k pieces.

The outer **for** loop of hasAssembly cycles through every pair of directed edges $e_P = (p_1, p_2)$ and $e_Q = (q_1, q_2)$ taken from different pieces P and Q. For each pair, hasAssembly first checks to see if there exists a symmetric assembly for which e_P is connected to e_Q with p_1 coincident to q_1, which would satisfy Case 1. If one exists, then joining P and Q into one piece as described would produce a SAP $\mathcal{P}'$ with one fewer piece that also has a symmetric assembly. Then evaluating hasAssembly on the smaller instance will return correctly by induction. Since the outer **for** loop checks every possible pair of edges that could be joined in a symmetric assembly satisfying Case 1, hasAssembly will return TRUE if $\mathcal{P}$ satisfies Case 1.

Next hasAssembly checks to see if there exists a symmetric assembly for which e_P is connected to e_Q with p_1 and q_1 separated by a distance equal to the length of some other edge e_R in $\mathcal{P}$, which would satisfy Case 2. In the same way as with Case 1, both **for** loops check every possible pair of edges and that could be joined at every possible length that could produce a symmetric assembly satisfying Case 2, so hasAssembly will return TRUE if $\mathcal{P}$ satisfies Case 2.

Otherwise, no symmetric assembly exists satisfying Case 1 or Case 2. By Lemma 3, hasAssemblyCase3 correctly evaluates if $\mathcal{P}$ is in Case 3, so hasAssembly returns a correct evaluation for SAPs containing k pieces. Let $T(k)$ be the running time of hasAssembly on an instance with k pieces. Then the recurrence relation for hasAssembly is $T(k) = O(n^3)T(k-1) + O(n^{5k})$, where $O(n^{5k})$ is the running time given by Lemma 3. Running time for Case 3 dominates the recurrence relation so hasAssembly terminates in $O(n^{5k})$. $\qquad\square$

Now we can determining whether a symmetric assembly puzzle with a constant number of pieces has a symmetric assembly in polynomial time.

Proof (of Theorem 2). By Lemma 2, if the SAP has a symmetric assembly, it satisfies either Case 1, Case 2, or Case 3, and by Lemma 4 hasAssembly($\mathcal{P}$) can correctly determine if it has a symmetric assembly satisfying one of the cases in polynomial time, proving the claim. $\qquad\square$

Open questions include whether SAPs: are hard for simpler shapes (we conjecture SAPs containing only right triangles are still hard), are hard for non-simple target shapes with constant pieces, or are fixed-parameter tractable with respect to the number of pieces (we conjecture W[1]-hardness).

Acknowledgements. Many of the authors were introduced to symmetric assembly puzzles during the 30th Winter Workshop on Computational Geometry at the Bellairs Research Institute of McGill University, March 2015. Korman is supported in part by the ELC project (MEXT KAKENHI No. 24106008). Mitchell is supported in part by the National Science Foundation (CCF-1526406). Uno is supported in part by the ELC project (MEXT KAKENHI No. 15H00853).

References

1. Cayley, A.: A theorem on trees. Q. J. Math **23**, 376–378 (1889)
2. Demaine, E.D., Demaine, M.L.: Jigsaw puzzles, edge matching, and polyomino packing: connections and complexity. Graphs Comb. **23**(Suppl.), 195–208 (2007)
3. Fox-Epstein, E., Uehara, R.: The convex configurations of "Sei Shonagon Chie no Ita" and other dissection puzzles. In: 26th Canadian Conference on Computational Geometry (CCCG), pp. 386–389 (2014)
4. Iwase, N.: Symmetrix. In: 24th International Puzzle Party (IPP 24), p. 54. IPP24 Committee (2005, unpublished)
5. Slocum, J.: The Tangram Book: The Story of the Chinese Puzzle with over Puzzle to Solve. Sterling Publishing, New York (2000)
6. Wolter, J.D., Woo, T.C., Volz, R.A.: Optimal algorithms for symmetry detection in two and three dimensions. Vis. Comput. **1**(1), 37–48 (1985)
7. Yamamoto, H.: Personal communication (2014)

Simultaneous Approximation of Polynomials

Andrei Kupavskii[1] and János Pach[2($\boxtimes$)]

[1] EPFL, Lausanne and MIPT, Moscow, Russia
kupavskii@ya.ru
[2] EPFL, Lausanne and Rényi Institute, Budapest, Hungary
pach@cims.nyu.edu

Abstract. Let $\mathcal{P}_d$ denote the family of all polynomials of degree at most d in one variable x, with real coefficients. A sequence of positive numbers $x_1 \leq x_2 \leq \ldots$ is called $\mathcal{P}_d$-*controlling* if there exist $y_1, y_2, \ldots \in \mathbb{R}$ such that for every polynomial $p \in \mathcal{P}_d$ there exists an index i with $|p(x_i) - y_i| \leq 1$. We settle a problem of Makai and Pach (1983) by showing that $x_1 \leq x_2 \leq \ldots$ is $\mathcal{P}_d$-controlling if and only if $\sum_{i=1}^{\infty} \frac{1}{x_i^d}$ is divergent. The proof is based on a statement about covering the Euclidean space with translates of slabs, which is related to Tarski's plank problem.

1 Introduction

Let $\mathcal{F}$ be a class of real functions $\mathbb{R} \to \mathbb{R}$. We say that a sequence of positive numbers $x_1, x_2, x_3, \ldots$ is $\mathcal{F}$-*controlling* if there exist reals $y_1, y_2, \ldots$ with the property that for every $f \in \mathcal{F}$, one can find an i with

$$|f(x_i) - y_i| \leq 1.$$

In other words, a sequence $x_1, x_2, \ldots$ is $\mathcal{F}$-controlling if we can find $y_1, y_2, \ldots \in \mathbb{R}$ such that the points $p_1 = (x_1, y_1), p_2 = (x_2, y_2), \ldots \in \mathbb{R}^2$ simultaneously approximate all functions in $\mathcal{F}$, in the sense that the graph of every member $f \in \mathcal{F}$ gets (vertically) not farther than 1 to at least one point p_i. In this paper, we address the following question raised in [11]. Given a class of functions $\mathcal{F}$, how *sparse* an $\mathcal{F}$-controlling sequence can be? A similar question, motivated by a problem of Fejes Tóth [5], was studied in [4].

Let $\mathcal{P}_d$ denote the class of polynomials $\mathbb{R} \to \mathbb{R}$ of degree at most d. It was shown by Makai and Pach [11] that if a sequence of positive numbers $x_1 \leq x_2 \leq \ldots$ is $\mathcal{P}_d$-controlling, then the infinite series $\frac{1}{x_1^d} + \frac{1}{x_2^d} + \ldots$ is *divergent*. They conjectured that this condition is also sufficient for a sequence $x_1 \leq x_2 \leq \ldots$ to be $\mathcal{P}_d$-controlling (see Conjecture 3.2.B in [11]). The aim of this note is to prove this statement.

Theorem 1. *Let d be a positive integer and $x_1 \leq x_2 \leq \ldots$ be a monotone increasing infinite sequence of positive numbers. The sequence $x_1, x_2, \ldots$ is $\mathcal{P}_d$-controlling if and only if $\frac{1}{x_1^d} + \frac{1}{x_2^d} + \frac{1}{x_3^d} \ldots = \infty$.*

© Springer International Publishing AG 2016
J. Akiyama et al. (Eds.): JCDCGG 2015, LNCS 9943, pp. 193–203, 2016.
DOI: 10.1007/978-3-319-48532-4_17

We also generalize this result to other finitely generated function classes. Given $d + 1$ real functions, $f_0, f_1, \ldots, f_d : \mathbb{R}_+ \to \mathbb{R}_+$, let $\mathcal{L} = \mathcal{L}(f_0, \ldots, f_d)$ denote the set of all functions that can be obtained as *linear combinations* of them with real coefficients. That is,

$$\mathcal{L} = \{a_0 f_0 + \ldots + a_d f_d : a_0, \ldots, a_d \in \mathbb{R}\}.$$

Here $\mathbb{R}_+$ stands for the set of positive reals.

Theorem 2. *Let $d \geq 1$ be an integer, $x_0 > 0$, $\epsilon > 0$, and let $f_0, f_1, \ldots, f_d :$ $\mathbb{R}_+ \to \mathbb{R}_+$ be real functions that are monotone increasing for $x \geq x_0$ and bounded over every bounded subinterval of $\mathbb{R}_+$. Assume that the functions $F_j(x) = f_j(x)/(f_d(x))^{1-\epsilon}$ $(j = 0, \ldots, d - 1)$ are monotone decreasing for $x \geq x_0$ and tend to 0 as $x \to \infty$.*

An increasing sequence of positive numbers $x_1 \leq x_2 \leq \ldots$ is $\mathcal{L}(f_0, \ldots, f_d)$-controlling if and only if $\sum_{i=1}^{\infty} \frac{1}{f_d(x_i)} = \infty$.

Obviously, the functions $f_i(x) = x^i$ $(i = 0, 1, \ldots, d))$ meet the above requirements, so that Theorem 2 implies Theorem 1.

For the proof of Theorem 1, we will rephrase the question as a covering problem for slabs. A *slab* (sometimes called *plank* or *strip*) is the set of points S lying between two parallel hyperplanes in $\mathbb{R}^d$. The distance w between these two hyperplanes is called the *width* of the slab. We can write S as

$$S = \{\mathbf{x} \in \mathbb{R}^d : b - \frac{w}{2} \leq \langle \mathbf{v}, \mathbf{x} \rangle \leq b + \frac{w}{2}\},$$

for some unit vector $\mathbf{v}$ and real number b. We say that a sequence of slabs $S_1, S_2, \ldots$ permits a *translative covering* of a subset $\mathbb{R}^d$ if there are suitable translates S_i' of S_i $(i = 1, 2, \ldots)$ such that $\cup_{i=1}^{\infty} S_i' = \mathbb{R}^d$.

As it was shown in [11], Theorem 1 (and, in fact, Theorem 2, too) would easily follow from

Conjecture 1 ([3,11]). *Let d be a positive integer. A sequence of slabs in $\mathbb{R}^d$ with widths $w_1, w_2, \ldots$ permits a translative covering of $\mathbb{R}^d$ if and only if $\sum_{i=1}^{\infty} w_i = \infty$.*

The fact that this condition is *necessary* follows, for example, from Tarski's result [12] which states that the total width of any system of slabs the union of which covers a disk of unit diameter is at least 1. Tarski's "plank problem," whether this statement remains true in higher dimensions, remained open for almost twenty years. In 1950, Bang [1,2] answered this question in the affirmative. For $d = 2$, Conjecture 1 was proved by Makai and Pach [11] and, according to [6], independently, by Erdős and Straus (unpublished). (See [7,8] for some refinements.) For $d \geq 3$, the problem is open. Groemer [6] proved that any sequence of slabs in $\mathbb{R}^d$ with widths $w_1, w_2, \ldots$ satisfying

$$\sum_{i=1}^{\infty} w_i^{\frac{d+1}{2}} = \infty$$

permits a translative covering of $\mathbb{R}^d$. Recently, the authors of the present note [9] have come close to settling Conjecture 1 by replacing Groemer's *sufficient* condition with the weaker assumption

$$\limsup_{n \to \infty} \frac{w_1 + w_2 + \ldots + w_n}{\log(1/w_n)} > 0.$$

In particular, any sequence of slabs of widths $1, \frac{1}{2}, \frac{1}{3}, \ldots$ permits a translative covering of space.

To establish Theorem 1, it is enough to verify Conjecture 1 for special sequences of slabs, whose normal vectors lie on a moment curve. We will do precisely this in Sect. 2, by exploring the natural ordering of these vectors. In Sect. 3, we generalize our arguments to establish Theorem 2. The last section contains a few concluding remarks.

2   Proof of Theorem 1

We only have to prove the "if" part of the theorem.

Let $x_1 \le x_2 \le \ldots$ be a monotone increasing sequence of positive numbers with $\sum_i \frac{1}{x_i^d} = \infty$. We have to find a sequence of reals $y_1, y_2, \ldots$ such that for any polynomial $p(x) = \sum_{j=0}^{d} a_j x^j$ with real coefficients a_j, there exists a positive integer i with $|p(x_i) - y_i| \le 1$. Write $p(x)$ in the form $p(x) = \langle \mathbf{x}, \mathbf{a} \rangle$, where $\mathbf{x} = (1, x, \ldots, x^d)$, $\mathbf{a} = (a_0, a_1, \ldots, a_d) \in \mathbb{R}^{d+1}$, and $\langle . \rangle$ stands for the scalar product. Using this notation, we have $\mathbf{x}_i = (1, x_i, \ldots, x_i^d)$ and the inequality $|p(x_i) - y_i| \le 1$ can be rewritten as

$$y_i - 1 \le \langle \mathbf{x}_i, \mathbf{a} \rangle \le y_i + 1.$$

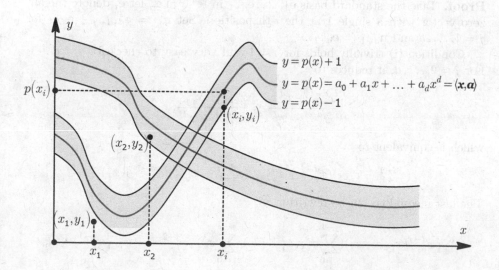

Fig. 1. Controlling polynomials of degree at most d.

For a fixed i, the locus of points $\mathbf{a} \in \mathbb{R}^{d+1}$ satisfying this double inequality is a slab $S_i \subset \mathbb{R}^{d+1}$ of width $w_i = \frac{2}{\|\mathbf{x}_i\|} = \frac{2}{(\sum_{j=0}^{d} x_i^{2j})^{1/2}}$, with normal vector $\mathbf{x}_i$. See Fig. 1. Since the condition $\sum_{i=1}^{\infty} \frac{1}{x_i^d} = \infty$ is equivalent to $\sum_{i=0}^{\infty} \frac{2}{\|\mathbf{x}_i\|} = \infty$, the sequence $x_1, x_2, \ldots$ is $\mathcal{P}_d$-controlling if and only if the sequence of slabs $S_1, S_2, \ldots$ permits a translative covering of $\mathbb{R}^{d+1}$.

If $x_i \leq 3$ for infinitely many (and, hence, for all) positive integers i, then for the widths of the corresponding slabs S_i we have $w_i > \frac{1}{3^d}$. Thus, these slabs permit a translative covering of $\mathbb{R}^{d+1}$, because each of them can be translated to cover any ball of diameter $\frac{1}{3^d}$.

Therefore, we can assume that $x_i > 3$ for all $i \geq m$. In fact, we can assume without loss of generality that $x_i > 3$ for all $i \geq 1$, otherwise we simply discard the first $m-1$ members of the sequence, and prove the theorem for the resulting sequence $x_m \leq x_{m+1} \leq \cdots$.

We are going to exploit the fact that the normal vectors $\mathbf{x}_i = (1, x_i, \ldots, x_i^d)$ of the slabs S_i lie on the moment curve $(1, x, x^2, \ldots, x^d)$. First, we need an auxiliary lemma.

Lemma 1. *Let d be a positive integer, let $3 \leq x_1 \leq x_2 \leq \ldots$ be a finite or infinite sequence of reals, and let $\mathbf{x}_i = (1, x_i, x_i^2, \ldots, x_i^d)$ for every i. Then there exist $d+1$ linearly independent vectors $\mathbf{u}_1, \ldots, \mathbf{u}_{d+1} \in \mathbb{R}^{d+1}$ such that for every i $(i = 1, 2, \ldots)$ and j $(j = 1, 2, \ldots, d+1)$, we have*

$$(i) \qquad \frac{\langle \mathbf{x}_{i+1}, \mathbf{u}_1 \rangle}{\langle \mathbf{x}_i, \mathbf{u}_1 \rangle} \leq \frac{\langle \mathbf{x}_{i+1}, \mathbf{u}_j \rangle}{\langle \mathbf{x}_i, \mathbf{u}_j \rangle},$$

$$(ii) \qquad \langle \mathbf{x}_i, \mathbf{u}_j \rangle \geq \frac{1}{3} \|\mathbf{x}_i\| \|\mathbf{u}_j\|.$$

Proof. Take the standard basis $\mathbf{e}_1, \ldots, \mathbf{e}_{d+1}$ in $\mathbb{R}^{d+1}$, i.e., let $\mathbf{e}_i$ denote the all-zero vector with a single 1 at the i-th position. Set $\mathbf{u}_j := \mathbf{e}_{d+1-j} + \mathbf{e}_{d+1}$ for $j = 1, \ldots, d$ and $\mathbf{u}_{d+1} := \mathbf{e}_{d+1}$.

Condition (i) trivially holds for $j = 1$ and very easy to check for $j = d + 1$. For $j = 2, \ldots, d$, it reduces to

$$\frac{x_{i+1}^{d-1} + x_{i+1}^d}{x_i^{d-1} + x_i^d} \leq \frac{x_{i+1}^{d-j} + x_{i+1}^d}{x_i^{d-j} + x_i^d},$$

which is equivalent to

$$(x_{i+1}^{d-1} + x_{i+1}^d)(x_i^{d-j} + x_i^d) \leq (x_{i+1}^{d-j} + x_{i+1}^d)(x_i^{d-1} + x_i^d).$$

The last inequality can be rewritten as

$$x_{i+1}^{d-j} x_i^{d-j}(x_{i+1} - x_i)\left(\sum_{k=0}^{j-1} x_{i+1}^k x_i^{j-1-k} + \sum_{k=0}^{j-2} x_{i+1}^k x_i^{j-2-k} - x_{i+1}^{j-1} x_i^{j-1}\right) \leq 0,$$

or, dividing both sides by $x_{i+1}^{d-j} x_i^{d-j}(x_{i+1} - x_i)$, as

$$\sum_{k=0}^{j-1} x_{i+1}^k x_i^{j-1-k} + \sum_{k=0}^{j-2} x_{i+1}^k x_i^{j-2-k} - x_{i+1}^{j-1} x_i^{j-1} \leq 0.$$

Using the fact $x_{i+1} \geq x_i$, and bounding from above each sum by its largest term multiplied by the number of terms, we obtain that the left-hand side of the last inequality is at most

$$j x_{i+1}^{j-1} + (j-1) x_{i+1}^{j-2} - x_{i+1}^{j-1} x_i^{j-1} < x_{i+1}^{j-1} (2j - 1 - x_i^{j-1}).$$

As $x_i \geq 3$, the right-hand side of this inequality is always negative and (i) holds.

It remains to verify condition (ii). Taking into account that $x_i \geq 3$, we have

$$\langle \mathbf{x}_i, \mathbf{u}_{d+1} \rangle = x_i^d \geq \frac{1}{2} \|\mathbf{x}_i\| = \frac{1}{2} \|\mathbf{x}_i\| \|\mathbf{u}_{d+1}\|.$$

On the other hand, for $j = 1, \ldots, d$, we obtain

$$\langle \mathbf{x}_i, \mathbf{u}_j \rangle = x_i^{d-j} + x_i^d \geq \frac{1}{2} \|\mathbf{x}_i\| \geq \frac{1}{3} \|\mathbf{x}_i\| \|\mathbf{u}_j\|.$$

This completes the proof of Lemma 1.   □

In order to establish Theorem 1, it is enough to prove that there is a constant c depending on d such that any system of slabs S_i $(i = 1, \ldots, n)$ in $\mathbb{R}^{d+1}$ whose normal vectors are $(1, x_i, \ldots, x_i^d)$ for some $3 \leq x_1 \leq x_2 \leq \ldots \leq x_n$ and whose total width is at least c, permits a translative covering of a ball of unit diameter. This is an immediate corollary of Lemma 1 and the following assertion.

Lemma 2. *For every positive integer d, for any system of $d + 1$ linearly independent vectors $\mathbf{u}_1, \ldots, \mathbf{u}_{d+1}$ in $\mathbb{R}^{d+1}$, and for any $\gamma > 0$, there is a constant c with the following property.*

Given any system of slabs S_i $(i = 1, \ldots, n)$ in $\mathbb{R}^{d+1}$, whose normal vectors $\mathbf{x}_i$ satisfy the conditions

$$(i) \qquad \frac{\langle \mathbf{x}_{i+1}, \mathbf{u}_1 \rangle}{\langle \mathbf{x}_i, \mathbf{u}_1 \rangle} \leq \frac{\langle \mathbf{x}_{i+1}, \mathbf{u}_j \rangle}{\langle \mathbf{x}_i, \mathbf{u}_j \rangle},$$

$$(ii) \qquad \langle \mathbf{x}_i, \mathbf{u}_j \rangle \geq \gamma \|\mathbf{x}_i\| \|\mathbf{u}_j\|$$

for every i and j, and whose total width $\sum_{i=1}^n w_i$ is at least c, the slabs S_i permit a translative covering of a $(d + 1)$-dimensional ball of unit diameter.

Proof. Instead of covering a ball of unit diameter, it will be more convenient to cover the simplex Δ with one vertex in the origin $\mathbf{0}$ and the others at the points (vectors) $\mathbf{u}_j$ $(j = 1, \ldots, d + 1)$. By properly scaling these vectors, if necessary, we can assume that Δ contains a ball of unit diameter.

We place the slabs one by one. See Fig. 2. We place S_1', a translate of S_1, so that one of its boundary hyperplanes passes through $\mathbf{0}$ and the other one cuts a simplex Δ_1 out of the cone Γ of all linear combinations of the vectors

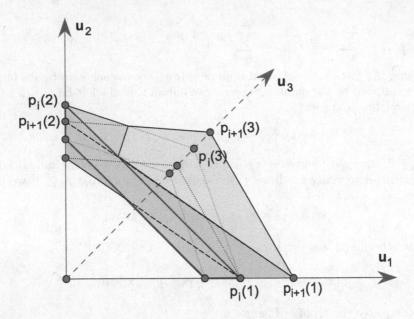

Fig. 2. We place the slabs one by one.

$\mathbf{u}_1, \ldots, \mathbf{u}_{d+1}$ with positive coefficients. According to assumption (ii), we have $\langle \mathbf{x}_1, \mathbf{u}_j \rangle > 0$ for every j. Therefore, S_1' does not separate Γ into two cones: $S_1' \cap \Gamma$ is indeed a simplex Δ_1.

Suppose that we have already placed $S_1', \ldots, S_i'$, the translates of $S_1, \ldots, S_i$, so that their union covers a simplex Δ_i with one vertex at the origin, and the others along the $d + 1$ half-lines that span the cone Γ. We also assume that the facet of Δ_i opposite to the origin is a boundary hyperplane of S_i'. Let $\mathbf{p}_i(j)$ denote the vertex of Δ_i that belongs to the open half-line parallel to $\mathbf{u}_j$ emanating from 0 $(j = 1, \ldots, d + 1)$.

Next, we place a translate S_{i+1}' of S_{i+1} so that one of its boundary hyperplanes, denoted by π, passes through $\mathbf{p}_i(1)$, and the other one, π', cuts the half-line parallel to $\mathbf{u}_1$ at a point $\mathbf{p}_{i+1}(1)$ with $\|\mathbf{p}_{i+1}(1)\| > \|\mathbf{p}_i(1)\|$. That is, $\mathbf{p}_{i+1}(1)$ is further away from the origin than $\mathbf{p}_i(1)$ is. Let $\mathbf{p}_{i+1}(2), \ldots, \mathbf{p}_{i+1}(d+1)$ denote the intersection points of π' with the half-lines parallel to $\mathbf{u}_2, \ldots, \mathbf{u}_{d+1}$, respectively, and let Δ_{i+1} be the simplex induced by the vertices $\mathbf{0}, \mathbf{p}_{i+1}(1), \ldots, \mathbf{p}_{i+1}(d+1)$.

We have to verify that Δ_{i+1} is entirely covered by the slabs $S_1', \ldots, S_{i+1}'$. By the induction hypothesis, Δ_i was covered by the slabs $S_1', \ldots, S_i'$. Thus, it is sufficient to check that the hyperplane π intersects every edge $\mathbf{0}\mathbf{p}_i(j)$ of Δ_i, for $j = 1, \ldots, d + 1$. Let $\alpha_j \mathbf{u}_j$ be the intersection point of π with the half-line parallel to $\mathbf{u}_j$, and let $\mathbf{p}_i(j) = \beta_j \mathbf{u}_j$. We have to prove that $\alpha_j \leq \beta_j$.

By definition, we have $\langle \mathbf{x}_{i+1}, \mathbf{p}_i(1) - \alpha_j \mathbf{u}_j \rangle = 0$ and $\langle \mathbf{x}_i, \mathbf{p}_i(1) - \beta_j \mathbf{u}_j \rangle = 0$. From here, we get

$$\frac{\alpha_j}{\beta_j} = \frac{\langle \mathbf{x}_{i+1}, \mathbf{p}_i(1) \rangle}{\langle \mathbf{x}_{i+1}, \mathbf{u}_j \rangle} \bigg/ \frac{\langle \mathbf{x}_i, \mathbf{p}_i(1) \rangle}{\langle \mathbf{x}_i, \mathbf{u}_j \rangle} = \frac{\langle \mathbf{x}_{i+1}, \mathbf{p}_i(1) \rangle}{\langle \mathbf{x}_i, \mathbf{p}_i(1) \rangle} \bigg/ \frac{\langle \mathbf{x}_{i+1}, \mathbf{u}_j \rangle}{\langle \mathbf{x}_i, \mathbf{u}_j \rangle} = \frac{\langle \mathbf{x}_{i+1}, \mathbf{u}_1 \rangle}{\langle \mathbf{x}_i, \mathbf{u}_1 \rangle} \bigg/ \frac{\langle \mathbf{x}_{i+1}, \mathbf{u}_j \rangle}{\langle \mathbf{x}_i, \mathbf{u}_j \rangle}.$$

In view of assumption (i) of the lemma, the right-hand side of the above chain of equations is at most 1, as required.

Observe that during the whole procedure the uncovered part of the cone Γ always remains convex and, hence, connected. In the nth step, $\cup_{i=1}^n S_i' \supset \Delta_n$. By the construction, $\mathbf{p}_i(1)$ lies at least w_i farther away from the origin along the half-line parallel to $\mathbf{u}_1$ than $\mathbf{p}_{i-1}(1)$ does. Thus, we have

$$\|\mathbf{p_n}(1)\| \geq \sum_{i=1}^n w_i \geq c.$$

Using the fact that $\langle \mathbf{x}_n, \mathbf{p}_n(j) - \mathbf{p}_n(1) \rangle = 0$ for every $j \geq 2$, and taking into account assumption (ii), we obtain

$$\|\mathbf{p}_n(j)\| \geq \frac{\langle \mathbf{x}_n, \mathbf{p}_n(j) \rangle}{\|\mathbf{x}_n\|} = \frac{\langle \mathbf{x}_n, \mathbf{p}_n(1) \rangle}{\|\mathbf{x}_n\|} \geq \gamma \|\mathbf{p_n}(1)\| \geq \gamma c.$$

Thus, if c is sufficiently large, we have $\|\mathbf{p}_n(j)\| \geq \|\mathbf{u}_j\|$. This means that Δ_n contains the simplex Δ defined in the first paragraph of this proof. Hence, it also contains a ball of unit diameter, as required.   □

3   Proof of Theorem 2

In this section, we extend the technique used in the proof of Theorem 1 to establish Theorem 2.

As in the proof Theorem 1, we can write any function $l = \sum_{k=0}^d a_k f_k \in \mathcal{L}(f_0, \ldots, f_d)$ as $l(x) = \langle \mathbf{x}, \mathbf{a} \rangle$, where $\mathbf{x} = (f_0(x), f_1(x), \ldots, f_d(x))$ and $\mathbf{a} = (a_0, a_1, \ldots, a_d) \in \mathbb{R}^{d+1}$. As before, we only have to prove the "if" part of the theorem, which is equivalent to the fact that the slabs $S_i \subset \mathbb{R}^{d+1}$ with normal vector $\mathbf{x}_i = (f_0(x_i), \ldots, f_d(x_i))$ and width

$$w_i = \frac{2}{\|\mathbf{x}_i\|} = \frac{2}{(\sum_{k=0}^d f_k^2(x_i))^{1/2}} \geq \frac{2}{\sqrt{d} f_d(x_i)},$$

for $i = 1, 2, \ldots$, permit a translative covering of $\mathbb{R}^{d+1}$. Again, it is enough to consider the case when $\lim_{i \to \infty} x_i = \infty$, otherwise each slab S_i contains a ball of diameter at least

$$\frac{2}{\sqrt{d} f_d(\lim_{i \to \infty} x_i)} > 0.$$

We follow the scheme of the proof of Theorem 1. According to Lemma 2, it is enough to show that there exist $d + 1$ linearly independent vectors $\mathbf{u}_1, \ldots, \mathbf{u}_{d+1}$ that satisfy conditions (i) and (ii) with $\mathbf{x}_i = (f_0(x_i), \ldots, f_d(x_i))$ and with a suitable constant $\gamma > 0$. We can assume without loss of generality that x_1, and hence all x_is, are so large that they satisfy $x_1 \geq x_0$ and the inequalities

$$\frac{f_j(x)}{f_d(x)} \leq \frac{f_j(x_1)}{f_d(x_1)} \leq \frac{1}{\sqrt{d}}, \tag{1}$$

for every $x \geq x_1$ and $j = 0, \ldots, d-1$. To see this, observe that $f_j(x)/f_d(x) = F_j(x)/f_d^e(x)$ is monotone decreasing in x, because F_j is monotone decreasing, while f_d is monotone increasing.

Let $\mathbf{e}_1, \ldots, \mathbf{e}_{d+1}$ be the standard basis in $\mathbb{R}^{d+1}$. For $1 \leq j \leq d+1$, set

$$\mathbf{u}_j := \sum_{k=1}^{d+1} \mathbf{e}_k - \frac{1}{2}\mathbf{e}_{d+2-j}.$$

In other words, all coordinates of $\mathbf{u}_j$ are 1, with the exception of the $(d+2-j)$-th coordinate, which is $\frac{1}{2}$.

By definition, we have $\langle \mathbf{x}_i, \mathbf{u}_j \rangle \geq \frac{1}{2}f_d(x_i)$ and $\|\mathbf{u}_j\| < \sqrt{d+1}$. It follows from (1) that $\frac{f_j(x_i)}{f_d(x_i)} \leq \frac{1}{\sqrt{d}}$ for $j \neq d$, so that

$$\|\mathbf{x}_i\| \leq \left(\sum_{k=0}^{d} f_k^2(x_i) \right)^{1/2} \leq \sqrt{2}f_d(x_i).$$

Hence, for every i and j,

$$\langle \mathbf{x}_i, \mathbf{u}_j \rangle \geq \frac{1}{2}f_d(x_i) \geq \frac{1}{2\sqrt{2}}\|\mathbf{x}_i\| \geq \frac{1}{2\sqrt{2(d+1)}}\|\mathbf{x}_i\|\|\mathbf{u}_j\|.$$

Therefore, condition (ii) in Lemma 2 is satisfied with $\gamma = \frac{1}{2\sqrt{2(d+1)}}$.

It remains to verify condition (i). For the rest of the argument, fix j ($1 \leq j \leq d+1$). We have to show that for every i ($i = 1, 2, \ldots$), the inequality

$$\frac{\langle \mathbf{x}_{i+1}, \mathbf{u}_1 \rangle}{\langle \mathbf{x}_i, \mathbf{u}_1 \rangle} \leq \frac{\langle \mathbf{x}_{i+1}, \mathbf{u}_j \rangle}{\langle \mathbf{x}_i, \mathbf{u}_j \rangle}$$

holds. For $j = 1$, the statement is trivial. Therefore, we may suppose that $j > 1$. Next, we want to get rid of $f_d(x)$ in the left hand side, keeping both the numerator and denominator positive. The above inequality is equivalent to the following:

$$\frac{\langle \mathbf{x}_{i+1}, \mathbf{u}_1 \rangle - \frac{1}{2}\langle \mathbf{x}_{i+1}, \mathbf{u}_j \rangle}{\langle \mathbf{x}_i, \mathbf{u}_1 \rangle - \frac{1}{2}\langle \mathbf{x}_i, \mathbf{u}_j \rangle} \leq \frac{\langle \mathbf{x}_{i+1}, \mathbf{u}_j \rangle}{\langle \mathbf{x}_i, \mathbf{u}_j \rangle}.$$

Using the notation

$$\phi(x) = \frac{1}{2}\sum_{k=0}^{d-1} f_k(x) + \frac{1}{4}f_{d+1-j}(x), \qquad \psi(x) = \sum_{k=0}^{d-1} f_k(x) - \frac{1}{2}f_{d+1-j}(x),$$

the above inequality may be rewritten as

$$\frac{\phi(x_{i+1})}{\phi(x_i)} \le \frac{f_d(x_{i+1}) + \psi(x_{i+1})}{f_d(x_i) + \psi(x_i)}. \tag{2}$$

Before checking that (2) is true, let us summarize the properties of the functions ϕ and ψ we need:

1. $\phi(x_{i+1})/\phi(x_i) \le f_d^{1-\epsilon}(x_{i+1})/f_d^{1-\epsilon}(x_i)$ for the constant $\epsilon > 0$ from Theorem 2,
2. $\psi(x_{i+1}) \le c f_d^{1-\epsilon}(x_{i+1})$ for a constant $c > 0$, and
3. $\psi(x_{i+1}) \ge \psi(x_i)$.

By the monotonicity of F_k, we have $f_k(x_{i+1})/f_k(x_i) \le f_d^{1-\epsilon}(x_{i+1})/f_d^{1-\epsilon}(x_i)$, for $k = 0, \ldots, d-1$. Now property 1 follows from the fact that, if $a_0, \ldots, a_{d-1}, b_0, \ldots, b_{d-1}, t$ are positive numbers satisfying $a_0/b_0 \le t, \ldots, a_{d-1}/b_{d-1} \le t$, then $(a_0 + \ldots + a_{d-1})/(b_0 + \ldots + b_{d-1}) \le t$. Using that $\lim_{x\to\infty} F_k(x) = 0$ for $k = 0, \ldots, d-1$, we get property 2. Property 3 is a direct consequence of our assumption that each f_k $(k = 0, 1, \ldots)$ is monotone increasing for $x \ge x_0$.

We have to verify (2). In view of property 1, it is sufficient to show

$$\frac{f_d^{1-\epsilon}(x_{i+1})}{f_d^{1-\epsilon}(x_i)} \le \frac{f_d(x_{i+1}) + \psi(x_{i+1})}{f_d(x_i) + \psi(x_i)},$$

which is equivalent to

$$\psi(x_i) f_d^{1-\epsilon}(x_{i+1}) - \psi(x_{i+1}) f_d^{1-\epsilon}(x_i) \le f_d(x_i) f_d^{1-\epsilon}(x_{i+1}) \left(\left(\frac{f_d(x_{i+1})}{f_d(x_i)} \right)^\epsilon - 1 \right),$$

or, in a slightly different form,

$$\psi(x_i) f_d^{1-\epsilon}(x_{i+1}) - \psi(x_{i+1}) f_d^{1-\epsilon}(x_i) \le f_d(x_i) f_d^{1-\epsilon}(x_{i+1}) \left(\left(1 + \frac{f_d^{1-\epsilon}(x_{i+1}) - f_d^{1-\epsilon}(x_i)}{f_d^{1-\epsilon}(x_i)} \right)^{\frac{\epsilon}{1-\epsilon}} - 1 \right).$$

Replacing the left-hand side by a larger quantity (taking property 3 into account) and the right-hand side by a smaller one (applying the inequality $(1 + x)^\alpha \ge 1 + \alpha x$, valid for all $\alpha, x \ge 0$), we obtain the stronger inequality

$$\psi(x_{i+1})(f_d^{1-\epsilon}(x_{i+1}) - f_d^{1-\epsilon}(x_i)) \le f_d(x_i) f_d^{1-\epsilon}(x_{i+1}) \left(\frac{\epsilon}{1-\epsilon} \frac{f_d^{1-\epsilon}(x_{i+1}) - f_d^{1-\epsilon}(x_i)}{f_d^{1-\epsilon}(x_i)} \right). \tag{3}$$

Thus, it is sufficient to prove (3). Rearranging the terms, we obtain

$$\psi(x_{i+1}) \le \frac{\epsilon}{1-\epsilon} f_d^\epsilon(x_i) f_d^{1-\epsilon}(x_{i+1}).$$

By property 2, we have $\psi(x_{i+1}) \le c f_d^{1-\epsilon}(x_{i+1})$, so that it is enough to check that

$$cf_d^{1-\epsilon}(x_{i+1}) \leq \frac{\epsilon}{1-\epsilon} f_d^\epsilon(x_i) f_d^{1-\epsilon}(x_{i+1}),$$

that is, $c \leq \frac{\epsilon}{1-\epsilon} f_d^\epsilon(x_i)$. As $f_d(x)$ is an increasing function for $x \geq x_0$, the last inequality is satisfied if we choose x_1 (and, hence, all other x_i) sufficiently large.

This completes the proof of (3) and (2), and so the proof of Theorem 2. □

4   Concluding Remarks

1. As was mentioned in the Introduction, Conjecture 1 is known to be true in the plane. Moreover, in [11] a stronger statement was proved: there exists a constant c such that every collection of strips with total width at least c permits a translative covering of a disk of diameter 1. In view of this, one can make the following even bolder conjecture.

Conjecture 2. *For any positive integer d, there exists a constant c depending on d such that every collection of slabs in $\mathbb{R}^d$ of total width at least c permits a translative covering of a unit diameter d-dimensional ball.*

Suppose Conjecture 1 is true for a positive integer d. Answering a question in [11], Ruzsa [10] proved that then, for the same value of d, Conjecture 2 also holds. Thus, the two conjectures are equivalent.

2. Given a class $\mathcal{F}$ of functions $\mathbb{R} \to \mathbb{R}$, we say that a sequence of positive numbers $x_1 \leq x_2 \leq \ldots$ is *strongly $\mathcal{F}$-controlling* if there exist reals $y_1, y_2, \ldots$ with the property that, for every $\varepsilon > 0$ and every $f \in \mathcal{F}$, one can find an i with

$$|f(x_i) - y_i| \leq \varepsilon.$$

It is easy to see that the condition in Theorem 1 is sufficient to guarantee that the sequence $x_1, x_2, \ldots$ is strongly $\mathcal{P}_d$-controlling. Theorem 2 can also be strengthened analogously.

3. The aim of this paper was to find necessary and sufficient conditions for a sequence of numbers to be $\mathcal{L}$-controlling, where $\mathcal{L} = \mathcal{L}(f_1, \ldots, f_d)$ is the class of functions that can be obtained as linear combinations of $f_1, \ldots, f_d$. We reduced this problem to a question about covering $\mathbb{R}^d$ with translates of certain slabs. However, the two problems are not necessarily equivalent. For example, we have noticed that the slabs obtained at this reduction had some special properties: apart from their widths, their normal vectors were also prescribed. This enabled us to cover $\mathbb{R}^d$ with their translates, even if we do not know whether such a covering exists for *every* system of slabs with the same widths.

Nevertheless, in a more complicated sense, the two problems are equivalent.

Theorem 3. *Given a positive integer d, and a sequence of positive numbers $x_1, x_2, \ldots$, define a family $\mathcal{F} = \mathcal{F}(d, x_1, x_2, \ldots)$ of d-tuples of functions $f_1, \ldots, f_d : \mathbb{R} \to \mathbb{R}$ as*

$$\mathcal{F} = \{(f_1, \ldots, f_d) : \sum_{j=1}^d f_j^2(x_i) = x_i^2 \text{ for all } i\}.$$

Then a sequence of slabs with widths $x_1, x_2, \ldots$ *permits a translative covering of* $\mathbb{R}^d$ *if and only if* $x_1, x_2, \ldots$ *is* $\mathcal{L}(f_1, \ldots, f_d)$*-controlling for every d-tuple* $(f_1, \ldots, f_d) \in \mathcal{F}$*, where*

$$\mathcal{L}(f_1, \ldots, f_d) = \{a_1 f_1 + \ldots + a_d f_d : a_1, \ldots, a_d \in \mathbb{R}\}.$$

Acknowledgements. Research of the first author is supported in part by the grant N 15-01-03530 of the Russian Foundation for Basic Research. The research of the second author is partially supported by Swiss National Science Foundation Grants 200020-144531 and 200020-162884.

References

1. Bang, T.: On covering by parallel-strips. Mat. Tidsskr. B. **1950**, 49–53 (1950)
2. Bang, T.: A solution of the "plank problem,". Proc. Am. Math. Soc. **2**, 990–993 (1951)
3. Brass, P., Moser, W., Pach, J.: Research Problems in Discrete Geometry. Springer, Heidelberg (2005)
4. Erdős, P., Pach, J.: On a problem of L. Fejes Tóth. Discrete Math. **30**(2), 103–109 (1980)
5. Fejes Tóth, L.: Remarks on the dual of Tarski's plank problem. Matematikai Lapok **25**, 13–20 (1974). (in Hungarian)
6. Groemer, H.: On coverings of convex sets by translates of slabs. Proc. Am. Math. Soc. **82**(2), 261–266 (1981)
7. Groemer, H.: Covering and packing properties of bounded sequences of convex sets. Mathematika **29**, 18–31 (1982)
8. Groemer, H.: Some remarks on translative coverings of convex domains by strips. Canad. Math. Bull. **27**(2), 233–237 (1984)
9. Kupavskii, A., Pach, J.: Translative covering of the space with slabs, manuscript
10. Ruzsa, I.Z.: Personal communication
11. Makai, E., Pach, J.: Controlling function classes and covering Euclidean space. Stud. Scient. Math. Hungarica **18**, 435–459 (1983)
12. Tarski, A.: Uwagi o stopniu równoważności wielokątów. Parametr **2**, 310–314 (1932). (in Polish)

Distance Geometry on the Sphere

Leo Liberti[1](✉), Grzegorz Swirszcz[2], and Carlile Lavor[3]

[1] CNRS LIX, École Polytechnique, 91128 Palaiseau, France
liberti@lix.polytechnique.fr
[2] IBM Research, Yorktown Heights, USA
swirszcz@us.ibm.com
[3] IMECC, University of Campinas, 13081-970 Campinas-SP, Brazil
clavor@ime.unicamp.br

Abstract. The Distance Geometry Problem asks whether a given weighted graph has a realization in a target Euclidean space $\mathbb{R}^K$ which ensures that the Euclidean distance between two realized vertices incident to a same edge is equal to the given edge weight. In this paper we look at the setting where the target space is the surface of the sphere $\mathbb{S}^{K-1}$. We show that the Distance Geometry Problem is almost the same in this setting, as long as the distances are Euclidean. We then generalize a theorem of Gödel about the case where the distances are spherical geodesics, and discuss a method for realizing cliques geodesically on a K-dimensional sphere.

1 Introduction

The Distance Geometry Problem (DGP), discussed at length in the surveys [10,14,16], is as follows. Given a positive integer K and a simple undirected graph $G = (V, E)$ weighted by an edge weight function $d : E \to \mathbb{R}_+$, determine whether there is a realization $x : V \to \mathbb{R}^K$ such that:

$$\forall \{u, v\} \in E \quad \|x_u - x_v\|_2 = d_{uv}. \tag{1}$$

The DGP is relevant to many applications: determining the shape of proteins from nuclear magnetic resonance data, localizing mobile sensors in wireless networks, designing efficient time synchronization protocols, controlling fleets of unmanned underwater vehicles, and more. It is auxiliary to other problems, such as the control of a multi-joint robotic arm, the rigidity of a bar-and-joint architecture structure, the completion of a matrix so that it is positive semidefinite, the visualization of high-dimensional data points [22].

The aim of this paper is to discuss the DGP on the sphere $\mathbb{S}^{K-1}$. Specifically, we emphasize two relatively straightforward observations which have a very high impact in realizing graphs on spheres using both Euclidean and geodesic

L. Liberti—Partly supported by the French national research agency ANR under the "Bip:Bip" project under contract ANR-10-BINF-0003.
C. Lavor—The support of FAPESP and CNPq is gratefully acknowledged.

J. Akiyama et al. (Eds.): JCDCGG 2015, LNCS 9943, pp. 204–215, 2016.
DOI: 10.1007/978-3-319-48532-4_18

distances, and use them to derive a method for realizing cliques geodesically on a sphere.

The DGP problem was shown to be **NP**-hard [24] in $\mathbb{R}^K$ where $K = 1$, by reduction from PARTITION, and even for any fixed K with only a handful of edge weight values, by reduction from 3-SATISFIABILITY. A similar reduction from PARTITION was also used to show **NP**-hardness of the subclass consisting of certain Henneberg type 1 graphs, namely graphs with a vertex order ensuring that:

- the first K vertices form a clique;
- every vertex v of rank greater than K in the order is adjacent to the vertices of ranks $v - 1, \ldots, v - K$.

This class, also called ^{K}DMDGP, is relevant in the study of protein conformation [9].

The ^{K}DMDGP is usually solved using a worst-case exponential time Branch-and-Prune (BP) [12] algorithm, which is precise, reliable and efficient notwithstanding the **NP**-hardness of the problem. It was shown in [17] that the structure of the symmetry group of the partial reflections in the realizations of a given problem instance can be found efficiently. In [11] it was shown that this group can be used to count the number of incongruent realizations of a given ^{K}DMDGP graph. In [15] the latter result was used to show that the BP is actually Fixed-Parameter Tractable (FPT) on protein graphs, and that the parameter could be fixed at a very low value for all tested proteins. This essentially yields a polynomial time behaviour of the BP when used to realize protein graphs in $\mathbb{R}^3$, and explains the efficiency of the BP on these graphs.

Instances from other applications have different structures which can also be exploited. Mobile sensor networks usually have at least two or three "anchors", i.e. sensors which are actually fixed, and whose position is known; most often, anchor locations are likely to be evenly distributed among the mobile sensors, in order to control load peaks. This appears to have a good numerical effect on Semidefinite Programming (SDP) algorithms when solving an SDP formulation of the DGP [7].

Flexible graphs can be realized using a plethora of heuristic and approximate approaches, some of which are based on local Nonlinear Programming (NLP) solution algorithms [13,20], and some others on different paradigms, see e.g. [1,26].

Given the wealth of knowledge on solving the DGP in a Euclidean space $\mathbb{R}^K$, it would be desirable to be able to extend some of this knowledge to other spaces or manifolds. One specific application-related motivation for looking at the sphere $\mathbb{S}^{K-1}$ is that it is a natural setting for the problem of completing partial correlation or covariance matrices, which arises in the financial sector [25].

2   Realizing Cliques in $\mathbb{R}^K$

The fundamental "building block" for realizing graphs in $\mathbb{R}^K$ are cliques on $K + 1$ vertices. In general, a 1-clique is a vertex, which can be realized in zero dimensions; a 2-clique is an edge, which can be realized in one dimension; a 3-clique is a triangle, which can be realized in two dimensions, as long as the edge weights satisfy the triangular inequality; a 4-clique is a tetrahedron, which can be realized in three dimensions, as long as the triangular and simplex inequalities are satisfied; larger cliques can be realized as long as the corresponding Cayley-Menger determinant [2,18], which is proportional to the square of the signed volume, is appropriately signed.

2.1 Recursive Realization Process

We can obtain a $(K + 1)$-clique from a K clique by adding a new vertex v to V, and edges of the form $\{u, v\}$ (for $u \in V$) to E. This recursive construction of cliques can be exploited to define a realization algorithm for $(K + 1)$-cliques in $\mathbb{R}^K$: number the vertices so that $V = \{v_1, \ldots, v_{K+1}\}$, assume (inductively) that the positions for $v_1, \ldots, v_K$ in $\mathbb{R}^K$ are known to be $x_1, \ldots, x_K$, and find the position y for v_{K+1} using K-lateration; the induction starts by setting x_1 at the origin.

2.2 K-Lateration

The fundamental building block for the algorithm in Sect. 2.1 is the (well known) process of K-lateration — a generalization of trilateration — i.e. the process of computing one of the vertices of a triangle from the two other vertices and the side lengths. Whereas K-lateration is usually applied to realizations in $\mathbb{R}^{K-1}$ [3,4], we apply it here to $\mathbb{R}^K$, which requires a further step [8]. We start with the squared distance system:

$$\forall i \leq K \quad \|y - x_i\|_2^2 = d_{i,K+1}^2, \tag{2}$$

where $x_i \in \mathbb{R}^K$ and $d_{i,K+1}$ are known. Equation (2) is trivially obtained by squaring Eq. (1). We re-write Eq. (2) as follows:

$$\|y\|_2^2 - 2x_1\,y = d_{1,K+1}^2 - \|x_1\|_2^2 \qquad ([1])$$

$$\vdots$$

$$\|y\|_2^2 - 2x_i\,y = d_{i,K+1}^2 - \|x_i\|_2^2 \qquad ([i])$$

$$\vdots$$

$$\|y\|_2^2 - 2x_K\,y = d_{K,K+1}^2 - \|x_K\|_2^2 \qquad ([K]),$$

where we denote the i-th equation of the system by $[i]$. We can now eliminate the square terms $\|y\|_2^2$ by forming the surrogate system $[i] - [j]$, where j is any given number in $\{1, \ldots, K\}$. If we fix $j = K$ without loss of generality, we obtain:

$$2(x_K - x_1)\,y = d_{1,K+1}^2 - d_{K,K+1}^2 - \|x_1\|_2^2 + \|x_K\|_2^2 \qquad ([1] - [K])$$

$$\vdots$$

$$2(x_K - x_{K-1})\,y = d_{K-1,K+1}^2 - d_{K,K+1}^2 - \|x_{K-1}\|_2^2 + \|x_K\|_2^2 \qquad ([K-1] - [K]),$$

which is a linear system which can written as $Ay = b$ for appropriate A, b, where A is a $(K - 1) \times K$ matrix, and $b \in \mathbb{R}^K$.

The locus of points for y can be obtained by intersecting the affine space $Ay = b$ and one of the K spheres described by the equations in Eq. (2). Without loss of generality, we again take the K-th sphere:

$$\left. \begin{array}{r} Ay = b \\ \|y - x_K\|^2 = d_{K,K+1}^2. \end{array} \right\} \tag{3}$$

2.3   Assumptions on the Rank of A

If A has full rank, then $\mathrm{rk}(A) = K - 1$. Since A has K columns, $Ay = b$ describes a line in $\mathbb{R}^K$. Hence, the intersection Eq. (3) can either be empty (if the line is disconnected from the sphere), consist of exactly one point (if the line is tangent to the sphere), or consist of two points otherwise. If the application warrants the assumption that solutions do exist (as in the case of proteins), then Eq. (3) has either one or two solutions. If we have no further knowledge of the data at hand, then we can reasonably assume that Eq. (3) has two solutions almost surely.

If $\mathrm{rk}(A) = K - 2$ or less, then $Ay = b$ describes a plane or hyperplane in $\mathbb{R}^K$. The intersection of a hyperplane with a sphere could be empty, or consist of only one point, or consist of uncountably many points. Since we are realizing a clique, and cliques are not flexible graphs, we can discount the latter possibility. If it consists of only one point, then the realization can be shown to be rigid, but infinitesimally flexible (think e.g. of a "flat triangle" realized in the plane as part of a line, which happens whenever the triangular inequality is satisfied at equality). Since the set of rank deficient $(K - 1) \times K$ matrices has Lebesgue measure zero in the set of all $(K - 1) \times K$ matrices, if we have no further knowledge of the data at hand, we can again reasonably assume that A has full rank almost surely [14].

2.4   Finding the Intersection of a Line and a Sphere

We now assume that A has full rank. We use $Ay = b$ as a dictionary: we identify a $(K-1) \times (K-1)$ nonsingular submatrix B of A, and partition the columns of A as $(B|N)$, where N is a single column. For simplicity of notation we identify the columns with their indices, and thus correspondingly partition y into (y_B, y_N), where y_N, called a *nonbasic variable*, is a single scalar. The linear system $Ay = b$ can therefore be written as $By_B + Ny_N = b$, which allows us to write the *basic variables* y_B in function of the nonbasic y_N:

$$y_B = B^{-1}b - B^{-1}Ny_N. \tag{4}$$

Now we use Eq. (4) to replace y_B in the sphere equation $\|y - x_K\|^2 = d_{K,K+1}^2$ in Eq. (3), and obtain a quadratic equation in the single variable y_N. The discriminant of this equation could be either negative, or zero, or positive. The first case corresponds to Eq. (3) having an empty intersection; the second to

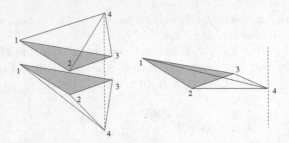

Fig. 1. Two reflected realizations of a 4-clique in $\mathbb{R}^3$ (left), which may coincide for certain values of the edge weights (right).

a single intersection point (the line is tangent to the sphere) and the third to two intersection points. In terms of realizing the $(K + 1)$-st clique vertex, the realization does not exist in the first case, corresponds to a "flat simplex" in the second case (i.e. a simplex which is realized in a lower dimensional space, see Fig. 1, right), or to two possible positions y^+, y^- for x_{K+1}, leading to two possible clique realizations x^+, x^-, which turn out to be reflection of each other w.r.t. the hyperplane spanned by $x_1, \ldots, x_K$ (see Fig. 1, left).

2.5 An Efficient Algorithm

The algorithm for realizing $(K - 1)$-cliques in $\mathbb{R}^K$ should now be clear: when $x_1, \ldots, x_K$ are known, we compute y: if it does not exist, the clique cannot be realized; if there are two distinct points, any of them can be chosen; if they coincide, the realization occurs in an affine lower dimensional subspace. Now $x_1, \ldots, x_K$ can be computed recursively, and we set $x_1 = \mathbf{0}$. To make this algorithm deterministic, we can give any rule to choose between the two points for x_{K+1} (for example, we can always choose y_N^+). This is a polynomial time algorithm in K. Note that, in most applications, K is fixed, so we can treat this as an $O(1)$ algorithm.

3   The Branch-and-Prune Algorithm

The BP algorithm applies a similar idea to ^{K}DMDGP graphs, defined on page 2. The initial K-clique is realized in $\mathbb{R}^K$ in $O(1)$ as per Sect. 2.5. Thereafter, the order ensures that each later vertex v is adjacent to at least its K immediate predecessors. Therefore v can be realized according to Sect. 2.4 in two points x_v^+, x_v^- which are reflections of each other w.r.t. $x_{v-1}, \ldots, x_{v-K}$. We check whether the points x_v^+, x_v^- are feasible with respect to any further edge distance to vertices $u < v - K$, and remove the infeasible ones. We then recurse the process on $v + 1$ on the set of feasible points: we do not recurse at all if neither x^+, x^- are feasible; we recurse once if only one point is feasible, and we recurse twice if both are feasible. The algorithm terminates when $v = |V|$ [9,12].

In practice, BP is currently the only algorithm which can find all incongruent solutions to a given ^{K}DMDGP graph. Moreover, it is the fastest, and is also very reliable. It scales up to realize protein backbone graphs tens of thousands of vertices, which it can realize in a few seconds of a common last generation laptop [23].

3.1   Complexity

The BP defines a binary search tree. At level v, this tree contains all possible positions for vertex v. Every path from a leaf to the root defines a possible realization for the input graph. The complexity of the BP algorithm has the following extrema: if the number of calls which yield two feasible points is bounded by a polynomial in the instance size, then the search tree has a bounded tree width, and the BP is a polynomial time algorithm. Otherwise, is it exponential.

Typically, protein graphs have a combinatorial explosion at the beginning of the sequence, say up to the vertex having rank r. Then the folds of the protein ensure the that there are sufficiently many edges in the graph to guarantee that only polynomially many calls determine the feasibility of both points x_v^+, x_v^- at level v. This yields a complexity $O(2^r p(||(G, d)||))$, where p is a polynomial in the size of the input (G, d), which causes BP to be FPT on a class of graphs which includes all proteins we tested.

3.2   Number of Solutions

Since cliques are rigid graphs and ^{K}DMDGP instances consist of sequences of rigidly connected cliques defined by the vertex order, ^{K}DMDGP graphs are rigid. However, in view of the fact that there may be up to two positions for each vertex v in any branch of the BP tree, most ^{K}DMDGP instances do not have unique realizations, but rather a finite set X of possible realizations modulo translations and rotations. We were able to explicitly describe the invariant group of X [17, 21], which is isomorphic to a certain cartesian product of copies of the cyclic group C_2. We then used it to determine $|X|$ efficiently from the edge set E [11]. It turns out that $|X|$ is always a power of two, as long as the full rank assumptions given in Sect. 2.3 hold.

4   The DGP on the Sphere

We now turn to the DGP on the sphere $\mathbb{S}^{K-1}$, meaning that we constrain any realization x to belong to the surface of the sphere. We first discuss the case where the edge weights are realized as Euclidean distances in $\mathbb{S}^{K-1}$ embedded in $\mathbb{R}^K$, meaning that each edge is realized as a segment which crosses the interior of the sphere. We then discuss the case where the edge weights are realized as geodesic distances.

4.1 Euclidean Distances

In this section we tackle the DGP where x is constrained to belong to the surface of the sphere $\mathbb{S}^{K-1}$, i.e.:

$$\forall v \in V \quad \|x_v\| = 1. \tag{5}$$

Since realize edges $\{u, v\}$ as segments of Euclidean length d_{uv}, the system in Eq. (1) holds. In particular, K-lateration can be simplified using Eq. (5):

$$\forall i \leq K \quad \|y - x_i\|_2^2 = d_{i,K+1}^2$$
$$\Rightarrow \quad \|y\|^2 - 2x_i\, y + \|x_i\|^2 = d_{i,K+1}^2$$
$$\text{(by Eq. 5 applied to } y, x_i) \quad \Rightarrow \quad 2 - 2x_i\, y = d_{i,K+1}^2$$
$$\Rightarrow \quad x_i\, y = 1 - \frac{1}{2} d_{i,K+1}^2,$$

which is a linear system $Ay = b$, where A is a square $K \times K$ matrix and $y = x_{K+1}$.

As in Sect. 2.3, we can make assumptions on the rank of A being full, which brings us immediately to a spherical K-lateration process yielding $y = A^{-1}b$, which has a unique solution. Note that the algebraic derivation above holds even if the original system is infeasible, whereas $Ay = b$ always has a unique solution as long as A has full rank. This occurs because the derivation above is necessary but not sufficient, i.e. the linear system $Ay = b$ is implied by Eqs. (1) and (5), but does not imply them univocally. For sufficiency, y needs to be verified feasible with respect to Eqs. (1) and (5). If so, then y is a possible valid realization of the $(K + 1)$-st vertex of the clique; otherwise, the input graph is a NO instance of the corresponding DGP.

With the full rank assumption, the difference between K-lateration in $\mathbb{R}^K$ and $\mathbb{S}^{K-1}$ is exactly the same as that between $\mathbb{R}^K$ and $\mathbb{R}^{K-1}$: in the former case the linear system is undetermined, and describes a line in $\mathbb{R}^K$, whereas in the latter it only describes a point in $\mathbb{R}^{K-1}$. Accordingly, in $\mathbb{R}^K$ we need to intersect the line with a sphere of Eq. (1) in order to obtain at most two points, whereas in $\mathbb{R}^{K-1}$ and $\mathbb{S}^{K-1}$ we do not.

In view of Sect. 3, this difference translates to ^{K}DMDGP graphs realized in a Euclidean space as follows: if vertices are adjacent to K immediate predecessors but not necessarily $K + 1$, then we have to realize the graph using the BP algorithm, which has a worst-case exponential behaviour, and finds an exponential number of incongruent realizations. If vertices can be guaranteed to be adjacent to at least $K + 1$ immediate predecessors, the BP can be shown to work in polynomial time (in fact, linear in the number of recursion calls, each of which has polynomial complexity in K).

The procedure on the sphere which is analogous to K-lateration in $\mathbb{R}^K$ (yielding exponential behaviour in the BP), is K-lateration in $\mathbb{S}^K$, embedded in $\mathbb{R}^{K+1}$. In this setting the system $Ay = b$ derived above is $K \times (K + 1)$, and therefore again describes a line in $\mathbb{R}^{K+1}$, which must be intersected with one of the spheres in either Eq. (1) or (5) (the latter giving rise to easier algebraic derivations), in order to obtain at most two points in $\mathbb{R}^{K+1}$.

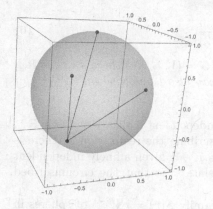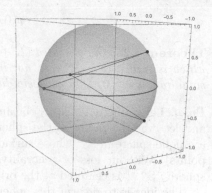

Fig. 2. A tetrahedron in a sphere (left) and two reflected triangles in a sphere (right).

Example 4.1. *Realizing a tetrahedron on $\mathbb{S}^2$ with Euclidean distances by K-lateration yields a unique point, whereas realizing a triangle on $\mathbb{S}^2$ yields at most two points (see Fig. 2). Comparing with $\mathbb{R}^3$, it would take the distances to four known points to determine the solution for the last point uniquely, whereas the distances to three known points only suffice to determine at most two positions, each of which is a reflection of the other.* □

Summarizing, in order to realize x_{K+1} from $x_1, \ldots, x_K$ on $\mathbb{S}^{K-1}$ or $\mathbb{S}^K$ using Euclidean distances, it suffices to remark that the norm constraints $\|x_K\|^2 = 1$ are quadratic constraints which can be used in conjunction with the original DGP system in Eq. (1).

4.2   Geodesic Distances

Not many people know that Kurt Gödel performed research in Distance Geometry (DG) in his youth. Two of the talks he gave at Karl Menger's seminar [19] are about DG, and also appear in [5]. Specifically, we are interested in [6], titled *Über die metrische Einbettbarkeit der Quadrupel des R_3 in Kugelflächen*, translated as *On the isometric embeddability of quadruples of points of R_3 in the surface of a sphere*. Apparently, Gödel had been working to solve a question posed by Laura Klanfer in a previous *colloquium*, i.e. whether an affinely independent quadruplet of points in $\mathbb{R}^3$ can be realized on the surface of a scaling of $\mathbb{S}^2$ so that the geodesic distances between the realized points have the same length as the Euclidean distances between the given points. Gödel managed to reply in the positive by means of a clever fixed point argument in $\mathbb{R}^3$ and $\mathbb{S}^2$.

In the following, we present a (rather trivial) generalization of Gödel's DG theorem to an arbitrary dimension K. We first remark that, for any $K > 1$, there is a unique shortest curve, called *geodesic*, between any two points on the surface of $\mathbb{S}^K$. Moreover, by elementary trigonometry the length c of the chord subtending a geodesic of length α on a sphere of radius $\frac{1}{\rho}$ (for some $\rho > 0$) is given by

$$c_\rho(\alpha) = \frac{2}{\rho} \sin \frac{\alpha\rho}{2}. \tag{6}$$

Theorem 4.2. *Any weighted $(K+1)$-clique $G = (V, E, d)$, where $d : E \to \mathbb{R}_+$, which is realizable in $\mathbb{R}^K$ but not in $\mathbb{R}^{K-1}$, can also be realized on $r\mathbb{S}^{K-1}$ (for some radius $r > 0$) with geodesic distances.*

Proof. Let $x = (x_1, \ldots, x_{K+1})$ be an affinely independent realization of G in $\mathbb{R}^K$, and let $\bar{r}$ be the radius of the sphere circumscribing the realization x (there is a unique sphere in $\mathbb{R}^K$ whose surface contains $K+1$ given affinely independent points). Without loss of generality, we translate x so that the circumscribed sphere $\bar{r}\mathbb{S}^{K-1}$ is centered at the origin.

The idea is to deform this sphere into a family $S(r) = r\mathbb{S}^{K-1}$ of spheres in function of r ranging in a certain open interval specified below. This also deforms the realization x to a continuous map of realizations on $S(r)$. We then find a value r^* which makes the lengths of the geodesics on $S(r^*)$ equal to the lengths of the chords in $S(\bar{r})$. The nontrivial part of the argument shows that such an r^* exists. Its existence will be implied by a fixed point argument on an appropriate function of the inverse ρ of the radius r.

Let $\tau(\rho)$ be the realization on $S(r)$ mapped from x as r decreases. More precisely, we let $\tau(\rho)$ be the realization of G with edges weighted by the function $c_\rho(d)$, meaning that the weight of the edge $\{u, v\} \in E$ is $c_\rho(d_{uv})$. We now define $\phi : \mathbb{R}_+ \to \mathbb{R}_+$ so that $\frac{1}{\phi(\rho)}$ is the radius of the sphere circumscribed about $\tau(\rho)$.

The parameter ρ is a measure of "how close the sphere is to being flat": it is easy to see that, as ρ tends to zero, r tends to infinity (yielding a sphere with zero curvature, where the chord and the geodesic lengths are equal), which means that $c_\rho(d_{uv})$ tends to d_{uv} for all edges $\{u, v\} \in E$. This implies that $\tau(\rho)$ tends to the realization x of G in $\mathbb{R}^K$. Since x exists, we can define $\tau(0) = x$ and $\phi(0) = 1/\bar{r}$.

We now claim that ϕ has a fixed point in the open interval $I = (0, \pi/\alpha)$, where $\alpha = \max_{\{u,v\} \in E} d_{uv}$ (see Lemma 4.3 for the proof). So let ρ^* be the fixed point of ϕ, namely $\phi(\rho^*) = \rho^*$. What this means is that $r^* = \frac{1}{\rho^*}$ is the radius of the sphere circumscribed about $\tau(\rho^*)$. In turn, $\tau(\rho^*)$ is a realization of G where the edges are weighted by the length of the chords subtending geodesics of length d_{uv} (for all $\{u, v\} \in E$) with respect to a radius r^*. A moment's reflection on this long sentence should convince the reader that this is the same as saying that $\tau(\rho^*)$ is a realization of G on the surface of a sphere $r^*\mathbb{S}^{K-1}$ where the edges are realized as geodesics of length d_{uv}. $\qquad\square$

Lemma 4.3. *The function ϕ defined in the proof of Theorem 4.2 has a fixed point in the open interval $I = (0, \pi/\alpha)$, where α is the maximum edge weight of the given clique graph G.*

Proof. First notice that $\tau(\rho)$ is defined in terms of c_ρ, and c_ρ is continuous over α for each $\rho > 0$ by definition (see Eq. (6)). Note that $\tau(0)$ exists since it is equal to x by definition. Since $\tau(\rho)$ is defined as the realization of G weighted by $c_\rho(d)$ over a sphere of radius $1/\rho$, τ is a continuous map in some open interval $J = (0, \varepsilon)$ for some $\varepsilon > 0$, since 0 is in the closure of J. Therefore $\bar{\rho} = \max\{\rho \in I \mid \tau(\rho)$ is defined$\}$ exists by continuity of τ. We look at two mutually exclusive cases: $\bar{\rho} = \pi/\alpha$ and $\bar{\rho} < \pi/\alpha$.

(i) If $\bar{\rho} = \pi/\alpha$, then $\tau(\bar{\rho})$ is defined and its longest edge has length $c_{\bar{\rho}}(\alpha) = \frac{2\alpha}{\pi}$. Hence the radius of the sphere circumscribed around $\tau(\bar{\rho})$ is greater than $c_{\bar{\rho}}(\alpha)/2$, i.e. greater than $\alpha/\pi = 1/\bar{\rho}$, which implies $\phi(\bar{\rho}) < \bar{\rho}$. On the other hand, we have $\phi(0) = 1/\bar{r} > 0$, so the intermediate value theorem ensures that $\exists \rho^* \in (0, \bar{\rho})$ such that $\phi(\rho^*) = \rho^*$.

(ii) Assume now $\bar{\rho} < \pi/\alpha$, and suppose that $\tau(\bar{\rho})$ is an affinely independent realization in $\mathbb{R}^K$. Then, for each $\tilde{\rho}$ in an arbitrary small neighbourhood around $\bar{\rho}$, $\tau(\tilde{\rho})$ must exists by continuity; then there must be some $\tilde{\rho} > \bar{\rho}$ where $\tau(\tilde{\rho})$ is defined, which contradicts the maximality of $\bar{\rho}$. So the realization $\tau(\bar{\rho})$ is affinely dependent, which means that its circumscribed sphere is flat, i.e. that $\phi(\bar{\rho}) = 0 < \bar{\rho}$. Together with $\phi(0) = 1/\bar{r} > 0$, this shows that there is $\rho^* < \pi/\alpha$ such that $\phi(\rho^*) = \rho^*$, which concludes the lemma.   □

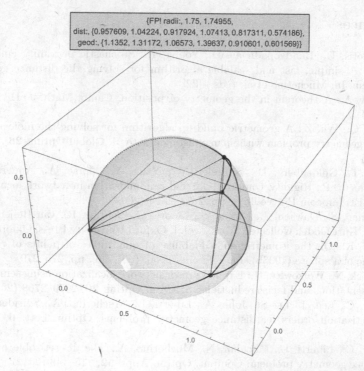

Fig. 3. Gödel's theorem yields a method for computing geodesic realizations. This picture shows the fixed point.

4.3   Putting It All Together

The results of Sects. 4.1 and 4.2 yield a method for realizing cliques in $\mathbb{S}^{K-1}$ with geodesic distances: first, numerically solve the fixed point equation $\phi(\rho) = \rho$ to obtain a value of ρ^*; then use Eq. (6) to compute new Euclidean distances corresponding to the the given geodesic distances, and finally realize the clique on a sphere of radius ρ^* using the new Euclidean distances. This will yield the required realization $\tau(\rho^*)$ (see Fig. 3).

5   Conclusion

This paper emphasizes two relatively easy observations about extending the considerable theoretical developments of the DGP to the setting of a spherical surface. The first observation applies to Euclidean distances, and amounts to noticing that the unit norm constraint can be exploited together with the DGP constraints. The second observation concerns geodesic distances, and yields an extension to $\mathbb{S}^{K-1}$ of a result of Gödel's in $\mathbb{S}^2$. The two observation yield a method for realizing cliques on a sphere with geodesic distances.

References

1. Agrafiotis, D., Bandyopadhyay, D., Young, E.: Stochastic proximity embedding (SPE): a simple, fast and scalable algorithm for solving the distance geometry problem. In: Mucherino et al. (eds.) [22]
2. Cayley, A.: A theorem in the geometry of position. Camb. Math. J. **II**, 267–271 (1841)
3. Dong, Q., Wu, Z.: A geometric build-up algorithm for solving the molecular distance geometry problem with sparse distance data. J. Global Optim. **26**, 321–333 (2003)
4. Eren, T., Goldenberg, D., Whiteley, W., Yang, Y., Morse, A., Anderson, B., Belhumeur, P.: Rigidity, computation, and randomization in network localization. In: IEEE Infocom Proceedings, pp. 2673–2684 (2004)
5. Feferman, S., Dawson, J., Kleene, S., Moore, G., Solovay, R., van Heijenoort, J. (eds.): Kurt Gödel: Collected Works, vol. I. Oxford University Press, Oxford (1986)
6. Gödel, K.: On the isometric embeddability of quadruples of points of r_3 in the surface of a sphere (1933b). In: Feferman et al. (eds.) [6], pp. 276–279
7. Krislock, N., Wolkowicz, H.: Explicit sensor network localization using semidefinite representations and facial reductions. SIAM J. Optim. **20**, 2679–2708 (2010)
8. Lavor, C., Lee, J., Lee-St. John, A., Liberti, L., Mucherino, A., Sviridenko, M.: Discretization orders for distance geometry problems. Optim. Lett. **6**, 783–796 (2012)
9. Lavor, C., Liberti, L., Maculan, N., Mucherino, A.: The discretizable molecular distance geometry problem. Comput. Optim. Appl. **52**, 115–146 (2012)
10. Lavor, C., Liberti, L., Maculan, N., Mucherino, A.: Recent advances on the discretizable molecular distance geometry problem. Eur. J. Oper. Res. **219**, 698–706 (2012)

11. Liberti, L., Lavor, C., Alencar, J., Abud, G.: Counting the number of solutions of ^{k}DMDGP instances. In: Nielsen, F., Barbaresco, F. (eds.) GSI 2013. LNCS, vol. 8085, pp. 224–230. Springer, Heidelberg (2013)

12. Liberti, L., Lavor, C., Maculan, N.: A branch-and-prune algorithm for the molecular distance geometry problem. Int. Trans. Oper. Res. **15**, 1–17 (2008)

13. Liberti, L., Lavor, C., Maculan, N., Marinelli, F.: Double variable neighbourhood search with smoothing for the molecular distance geometry problem. J. Global Optim. **43**, 207–218 (2009)

14. Liberti, L., Lavor, C., Maculan, N., Mucherino, A.: Euclidean distance geometry and applications. SIAM Rev. **56**(1), 3–69 (2014)

15. Liberti, L., Lavor, C., Mucherino, A.: The discretizable molecular distance geometry problem seems easier on proteins. In: Mucherino et al. [22]

16. Liberti, L., Lavor, C., Mucherino, A., Maculan, N.: Molecular distance geometry methods: from continuous to discrete. Int. Trans. Oper. Res. **18**, 33–51 (2010)

17. Liberti, L., Masson, B., Lavor, C., Lee, J., Mucherino, A.: On the number of realizations of certain Henneberg graphs arising in protein conformation. Discrete Appl. Math. **165**, 213–232 (2014)

18. Menger, K.: New foundation of Euclidean geometry. Am. J. Math. **53**(4), 721–745 (1931)

19. Menger, K. (ed.): Ergebnisse eines Mathematischen Kolloquiums. Springer, Wien (1998)

20. Moré, J., Wu, Z.: Global continuation for distance geometry problems. SIAM J. Optim. **7**(3), 814–846 (1997)

21. Mucherino, A., Lavor, C., Liberti, L.: Exploiting symmetry properties of the discretizable molecular distance geometry problem. J. Bioinf. Comput. Biol. **10**, 1242009(1–15) (2012)

22. Mucherino, A., Lavor, C., Liberti, L., Maculan, N. (eds.): Distance Geometry: Theory, Methods, and Applications. Springer, New York (2013)

23. Mucherino, A., Lavor, C., Liberti, L., Talbi, E.G.: A parallel version of the Branch & Prune algorithm for the molecular distance geometry problem. In: ACS/IEEE International Conference on Computer Systems and Applications (AICCSA10), pp. 1–6. IEEE, Hammamet (2010)

24. Saxe, J.: Embeddability of weighted graphs in k-space is strongly NP-hard. In: Proceedings of 17th Allerton Conference in Communications, Control and Computing, pp. 480–489 (1979)

25. van der Schans, M., Boer, A.: A heuristic for completing covariance and correlation matrices. Technical report 2013–01, ORTEC Finance (2013)

26. Tenenbaum, J., de Silva, V., Langford, J.: A global geometric framework for nonlinear dimensionality reduction. Science **290**, 2319–2322 (2000)

The Sigma Chromatic Number of the Circulant Graphs $C_n(1,2)$, $C_n(1,3)$, and $C_{2n}(1,n)$

Paul Adrian D. Luzon, Mari-Jo P. Ruiz, and Mark Anthony C. Tolentino[✉]

Ateneo de Manila University, Katipunan Ave., Loyola Heights,
Quezon City 1108, Philippines
paulluzon48@yahoo.com, {mruiz,mtolentino}@ateneo.edu

Abstract. For a non-trivial connected graph G, let $c : V(G) \to \mathbb{N}$ be a vertex coloring of G. For each $v \in V(G)$, the *color sum* of v, denoted by $\sigma(v)$, is defined to be the sum of the colors of the vertices adjacent to v. If $\sigma(u) \neq \sigma(v)$ for every two adjacent $u, v \in V(G)$, then c is called a *sigma coloring* of G. The minimum number of colors required in a sigma coloring of G is called its *sigma chromatic number* and is denoted by $\sigma(G)$. In this paper, we determine the sigma chromatic numbers of three families of circulant graphs: $C_n(1,2)$, $C_n(1,3)$, and $C_{2n}(1,n)$.

Keywords: Neighbor-distinguishing coloring · Sigma coloring · Circulant graphs

1 Introduction

A vertex or edge coloring c of a graph G is said to be *neighbor-distinguishing* if c produces a vertex labelling under which every pair of adjacent vertices in G are assigned distinct labels. While the most studied example of such a coloring is proper vertex coloring, various neighbor-distinguishing colorings have also been introduced and discussed in the literature. In particular, examples of neighbor-distinguishing vertex colorings are presented in [2,5].

In [3], Chartrand et al. introduced and explored the following neighbor-distingushing vertex coloring that makes use of sums of colors.

Definition 1 (Chartrand et al. [3]). *An example is shown in Fig. 1. For a non-trivial connected graph G, let $c : V(G) \to \mathbb{N}$ be a vertex coloring of G. For each $v \in V(G)$, the **color sum** of v, denoted by $\sigma(v)$, is defined to be the sum of the colors of the vertices adjacent to v. If $\sigma(u) \neq \sigma(v)$ for every two adjacent $u, v \in V(G)$, then c is called a **sigma coloring** of G. The minimum number of colors required in a sigma coloring of G is called its **sigma chromatic number** and is denoted by $\sigma(G)$.*

Along with important properties of sigma coloring, Chartrand et al. also determined in [3] the sigma chromatic numbers of paths, cycles, bipartite, and complete multipartite graphs. In [6], Dehghan et al. proved complexity results for

© Springer International Publishing AG 2016
J. Akiyama et al. (Eds.): JCDCGG 2015, LNCS 9943, pp. 216–227, 2016.
DOI: 10.1007/978-3-319-48532-4_19

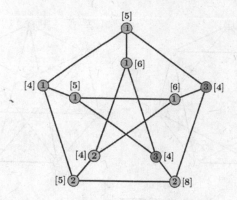

Fig. 1. A sigma coloring of a graph

the sigma coloring problem; most notably, they showed that it is **NP**-complete to decide whether $\sigma(G) = 2$ for a given 3-regular graph G.

In this paper, we focus on the sigma chromatic numbers of circulant graphs defined below.

Definition 2 [1]. *Given a subset D of the nonzero elements of the cyclic group $\mathbb{Z}_n$ of the integers modulo n, the **circulant graph** $C_n(D) = G(\mathbb{Z}_n, D)$ has $\mathbb{Z}_n$ as a vertex set, and ij as an edge if and only if $i - j \in D \cup (-D)$.*

In particular, we determine the sigma chromatic numbers of three families of circulant graphs: $C_n(1,2)$, $C_n(1,3)$, and $C_{2n}(1,n)$. Note that while most circulant graphs are 4-regular, $C_{2n}(1,n)$ is 3-regular for any $n \geq 2$. Our main results are as follows.

Theorem 3. *Let $n \geq 6$ be an integer. Then*

$$\sigma\left(C_n(1,2)\right) = \begin{cases} 2, & n = 6k, k \in \mathbb{N}, \\ 3, & otherwise. \end{cases}$$

Remark 4. When $n = 3, 4, 5$, $C_n(1,2)$ is isomorphic to the complete graph K_n; hence, $\sigma\left(C_n(1,2)\right) = n$ in these cases.

Theorem 5. *Let $n \geq 3$ be an integer. Then*

$$\sigma\left(C_n(1,3)\right) = \chi(C_n(1,3)) = \begin{cases} 2, & n \text{ is even}, \\ 3, & n \text{ is odd}, n \neq 5, \\ 5, & n = 5. \end{cases}$$

Theorem 6. *Let $n \geq 3$ be a positive integer. Then*

$$\sigma\left(C_{2n}(1,n)\right) = \chi\left(C_{2n}(1,n)\right) = \begin{cases} 2, & n \text{ is odd}, \\ 3, & n \text{ is even}. \end{cases}$$

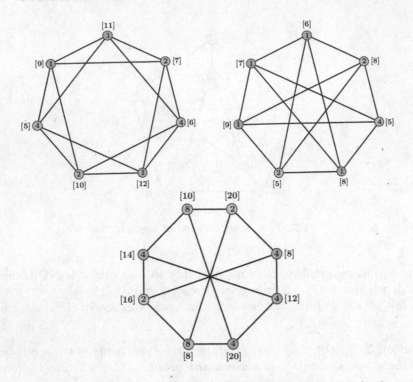

Fig. 2. Optimal sigma colorings of $C_7(1,2)$, $C_7(1,3)$, and $C_8(1,4)$.

Examples are shown in Fig. 2. The remaining sections of this paper are devoted to the proofs of Theorem 3 (Sect. 2), Theorem 5 (Sect. 3), and Theorem 6 (Sect. 4). Our proofs make use of the following relationship between the sigma chromatic number and the chromatic number of a graph.

Theorem 7 (Chartrand, Okamoto, Zhang [3]). *For every graph* G,

$$\sigma(G) \leq \chi(G).$$

The above theorem gives an upper bound on the sigma chromatic number of a graph provided that its chromatic number is known. However, the determination of the chromatic number is not easy even when restricted to the class of circulant graphs ([4]). Results on the chromatic numbers of certain families of circulant graphs can be found in [1,7]; among those results, the following will be valuable in our proofs.

Theorem 8 (Heuberger [7]). *Let* $D = \{a, b\}$ *be a generating subset of* $\mathbb{Z}_n = \{0, 1, ..., n-1\}$ *not containing* 0. *If* $\chi(n, D)$ *is the chromatic number of the circulant graph* $C_n(D)$, *then*

$$\chi(n,D) = \begin{cases} 2, & a \text{ and } b \text{ are both odd and } n \text{ is even,} \\ 4, & b = 2a \text{ or } a = 2b, \text{ and } n \not\equiv 0 \pmod 3 \text{ or} \\ & n = 13 \text{ and } D \text{ is equivalent to } \{1,5\}, \\ 5, & n = 5, \\ 3, & \text{otherwise.} \end{cases}$$

Finally, we present here the definition of a color block that was also defined in [3]. Our proofs rely on observing the behavior of these blocks in a sigma coloring.

Definition 9. *For a cyclic sequence* $s : c_1, c_2, ..., c_k, c_1$ *of* k *colors (not necessarily distinct from each other), we define a* **block** *of* s *as a maximal subsequence of* s *of the same color. Moreover, for a positive integer* $j \leq k$, *we define a* j-*block of* s *as a block of* s *of length* j *and we write* $(aa \cdots a)$ $(j \ a's)$ *to denote a* j-*block of color* a.

2 The Circulant Graphs $C_n(1,2)$

For $n \geq 6$, denote set of vertices of $C_n(1,2)$ by V_n and label them as in Fig. 3.

Fig. 3. Labelling of the vertices of $C_n(1,2)$

Let a, b be distinct positive integers and $c : V_n \to \{a, b\}$ be a vertex coloring of $C_n(1,2)$; define the cyclic sequence $C : c(v_1), c(v_2), ..., c(v_n), c(v_1)$.

Proposition 10. *Any block of* C *must be of length 1 or 2.*

Proof. It is easy to check that a block of C must be of length less than 6. Moreover, Fig. 4 shows that a block of C cannot have length 5 or 4.

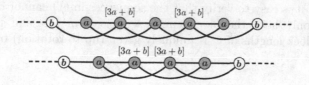

Fig. 4. C does not have a block of length 4 or 5.

Without loss of generality, suppose $(c(v_2) \ c(v_3) \ c(v_4)) = (aaa)$ is a 3-block of C. Then $c(v_1) = c(v_5) = b$. Moreover, since c is a sigma coloring, $c(v_n)$ and $c(v_6)$ must be equal to a. This implies that $\sigma(v_2) = \sigma(v_4)$ (see Fig. 5), which is a contradiction. Hence, C must not have a 3-block.

Therefore, the blocks in C can only be of length 1 or 2. $\square$

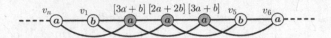

Fig. 5. C does not have a block of length 3.

Lemma 11. $\sigma\left(C_n(1,2)\right) = 2$ *if and only if* $n = 6k$, *where* $k \in \mathbb{N}$.

Proof. Let c and C be as defined above. If B_1, B_2, B_3, and B_4 are four consecutive blocks of C and $\ell(B_i)$ denotes the length of each block, then the quadruple $(\ell(B_1), \ell(B_2), \ell(B_3), \ell(B_4))$ is said to be a *block-length sequence* of C. By Proposition 10, C can only have blocks of length 1 or 2; hence, C has only 16 possible block-length sequences.

Consider a 2-block $(c(v_2)c(v_3))$ of C. Then $c(v_1) = c(v_4) = b$ and since c is a sigma coloring, $(c(v_n), c(v_5))$ is equal to (a, b) or (b, a). This implies that any block-length sequence of C involving a 2-block must be of the form $(1, 1, 2, 2)$ or $(2, 2, 1, 1)$ (Fig. 6).

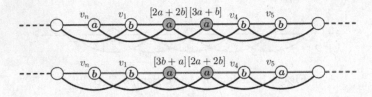

Fig. 6. A 2-block of C induces the block-length sequence $(1, 1, 2, 2)$ or $(2, 2, 1, 1)$.

Consequently, the only possible block-length sequences of C are the following:

$$(1, 1, 1, 1), \quad (1, 1, 1, 2), \quad (2, 1, 1, 1), \tag{1}$$

$$(1, 1, 2, 2), \quad (1, 2, 2, 1), \quad (2, 1, 1, 2), \quad (2, 2, 1, 1). \tag{2}$$

Among these, it is easy to verify that the sequences in (1) cannot occur. Therefore, C can only have the block-length sequences in (2); and the complete sequence of block lengths of C is uniquely given (up to rotation) by

$$1, 1, 2, 2, \dots, 1, 1, 2, 2. \tag{3}$$

It follows that, in this case, the number of vertices in $C_n(1,2)$ must be divisible by 6; that is, $n = 6k, k \in \mathbb{N}$.

Conversely, if $n = 6k, k \in \mathbb{N}$, then any coloring c that induces a sequence of block lengths equivalent to (3) is a sigma coloring of $C_n(1,2)$. This concludes the proof. $\qquad\square$

Lemma 12. *The circulant graph* $C_n(1,3)$ *has a sigma 3-coloring for all* $n \geq 6$.

Proof. We divide the proof into three cases.

Case 1: $n \equiv 0 \pmod{3}$

Define the coloring c of the vertices of $C_n(1,2)$ by

$$c(v_i) = \begin{cases} 1, & i \equiv 1 \pmod{3}, \\ 2, & i \equiv 2 \pmod{3}, \\ 4, & i \equiv 0 \pmod{3}. \end{cases} \tag{4}$$

Then the color sums of the vertices are given by

$$\sigma(v_i) = \begin{cases} 12, & i \equiv 1 \pmod{3}, \\ 10, & i \equiv 2 \pmod{3}, \\ 6, & i \equiv 0 \pmod{3}. \end{cases} \tag{5}$$

Consequently, $\sigma(v_i) = \sigma(v_j)$ if and only if $i \equiv j \pmod{3}$; since v_i is not adjacent to v_j in this case, c is a sigma coloring of $C_n(1,2)$.

Case 2: $n \equiv 1 \pmod{3}$

It is easy to verify that the same coloring c in (4) is also a sigma 3-coloring of $C_n(1,2)$ for this case.

Case 3: $n \equiv 2 \pmod{3}$

Define the coloring c of the vertices of $C_n(1,2)$ as follows: if $i = 1, 2, ..., n-4$, $c(v_i)$ is given by (4) while for the remaining vertices, we have $c(n-3) = 1$, $c(n-2) = 2$, and $c(n-1) = c(n) = 4$. Then $\sigma(v_i)$ is also given by (5) if $i = 2, 3, ..., n-6$ while

$$\sigma(v_i) = \begin{cases} 14, & i = 1, \\ 5, & i = n-5, \\ 9, & i = n-4 \text{ or } n, \\ 11, & i = n-3, \\ 10, & i = n-2, \\ 8, & i = n-1. \end{cases}$$

Consequently, c is a sigma coloring of $C_n(1,2)$. □

Lemmas 11 and 12 imply that $\sigma\left(C_n(1,2)\right) = 2$ if and only if $n \equiv 0 \pmod{6}$; otherwise, $\sigma\left(C_n(1,2)\right) = 3$. This proves Theorem 3.

3   The Circulant Graphs $C_n(1,3)$

For $n \geq 3$, denote the vertex set of $C_n(1,3)$ by V_n and label its vertices as in Fig. 7. Let a, b be distinct positive integers and $c : V_n \rightarrow \{a, b\}$ be a vertex coloring of $C_n(1,3)$; define the cyclic sequence C by

$$C : c(v_1), c(v_2), ..., c(v_n), c(v_1)$$

and denote the collection of all 2-blocks in C by $\mathcal{B}_2$.

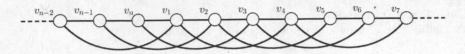

Fig. 7. Labelling of the vertices of $C_n(1,3)$

Definition 13. *Let $B \in \mathcal{B}_2$. Then B is said to be of*

1. *Type 1 if there is a unique 2-block that is adjacent to B,*
2. *Type 2 if there is a unique 2-block that is one 1-block away from B, and*
3. *Type 3 if there is a unique 4-block that is adjacent to B.*

The following lemma shows that $\mathcal{B}_2$ is the disjoint union of the 2-block types.

Lemma 14. *Suppose $n \geq 10$. Then any 2-block of C belongs to exactly one of the three 2-block types.*

Proof. Without loss of generality, we assume that $c(v_2) = c(v_3) = a$ and that $B := (c(v_2)\, c(v_3)) = (aa)$ is a 2-block of C. It follows that $c(v_1) = c(v_4) = b$.

We consider cases depending on the value of $(c(v_{n-1}), c(v_n), c(v_5), c(v_6))$. Note that since c is a sigma-coloring of $C_n(1,3)$, this quadruple cannot be equal to any of the following: (a,a,a,a), (a,a,b,b), (b,a,a,b), (b,b,a,a), (a,b,b,a), and (b,b,b,b).

Type 1: Suppose $(c(v_{n-1}), c(v_n), c(v_5), c(v_6)) = (a,b,a,a)$. Since c is a sigma-coloring of $C_n(1,3)$, it follows that $c(v_7) = a$. It is then clear that B is a 2-block of Type 1 and not of Type 2 nor 3 (Fig. 8).

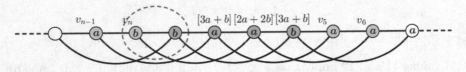

Fig. 8. The block $(c(v_n)\, c(v_1))$ is the only 2-block adjacent to the 2-block B.

Similarly, it can be shown that B becomes a 2-block of Type 1 if the quadruple $(c(v_{n-1}), c(v_n), c(v_5), c(v_6))$ is equal to one the following: (a,a,b,a), (a,b,a,b), (b,a,b,a), (a,b,b,b), and (b,b,b,a).

Type 2: Suppose $(c(v_{n-1}), c(v_n), c(v_5), c(v_6)) = (a,a,a,b)$. Since c is a sigma-coloring of $C_n(1,3)$, it follows that $c(v_7) = a$ and $c(v_{n-2}) = b$. It is then clear that B is a 2-block of Type 2 and not of Type 1 nor 3 (Fig. 9).

Similarly, it can be shown that B becomes a 2-block of Type 2 if the quadruple $(c(v_{n-1}), c(v_n), c(v_5), c(v_6))$ is equal to (b,a,a,a).

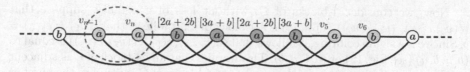

Fig. 9. The block $(c(v_{n-1})\ c(v_n))$ is the only 2-block that is a 1-block away from B.

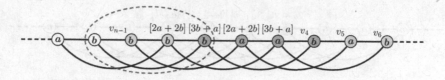

Fig. 10. The block $(c(v_{n-3})\ c(v_{n-2})\ c(v_{n-1})\ c(v_n))$ is the only 4-block that is adjacent to the 2-block B.

Type 3: Suppose $(c(v_{n-1}), c(v_n), c(v_5), c(v_6)) = (b, b, a, b)$. Since c is a sigma-coloring of $C_n(1,3)$, it follows that $c(v_{n-2}) = b$ and $c(v_{n-3}) = a$. It is then clear from Fig. 10 that B is a 2-block of Type 3 and neither of Types 1 nor 2.

Similarly, it can be shown that B becomes a 2-block of Type 3 if the quadruple $(c(v_{n-1}), c(v_n), c(v_5), c(v_6))$ is equal to (b, a, b, b).

Therefore, the 2-block B must belong to exactly one of the three 2-block types. □

Proof of Theorem 5. For the case $n = 5$, note that $C_5(1,3) \equiv K_5$ and $\sigma(C_5(1,3)) = \chi(C_5(1,3)) = 5$. Now, suppose $n \geq 3$ is even. Then by Theorem 8, $C_n(1,3)$ is bipartite; hence, $\sigma(C_n(1,3)) = 2$.

Now, since the cases $n = 3, 5, 7, 9$ can easily be checked, we can simply assume that $n \geq 11$ is odd. As before, let $c : V_n \to \{a, b\}$ be a vertex coloring of $C_n(1,3)$ and define the cylic sequence $C : c(v_1), c(v_2), ..., c(v_n), c(v_1)$. Let $\mathcal{B}_2$ and $\mathcal{B}_4$ denote the collection of all 2-blocks and 4-blocks of C, respectively. First, note that any block of C must be of length at most seven; moreover, C does not have any 6-block (Fig. 11). Hence, the only possible block lengths of C are 1, 2, 3, 4, 5, and 7.

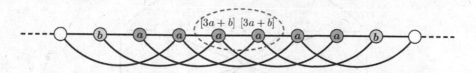

Fig. 11. The cyclic sequence C cannot have a 6-block.

Since $n \geq 10$, Lemma 14 holds and by definition, it follows that there is an even number of Type 1 or Type 2 2-blocks in $\mathcal{B}_2$.

Now, we consider 4-blocks of C; without loss of generality, suppose that $(c(v_2) \, c(v_3) \, c(v_4) \, c(v_5)) = (aaaa)$ is a 4-block of C. It follows that $c(v_1) = c(v_6) = b$. Since c is a sigma coloring, it follows that $(c(v_{n-1}), c(v_n), c(v_7), c(v_8))$ is equal to (b, a, b, a) (see Fig. 12) or (a, b, a, b). This implies that any 4-block of C is adjacent to exactly one 2-block; that is, the number of 4-blocks of C is equal to the number of Type 3 2-blocks of C.

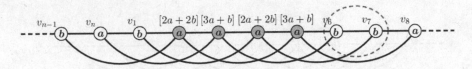

Fig. 12. Any 4-block of C is adjacent to exactly one 2-block.

The above observations imply that $|\mathcal{B}_2 \cup \mathcal{B}_4|$ is even. Since C must have an even number of blocks, it must also have an even number of blocks that are of odd length (i.e. of length 1, 3, 5, 7). But this implies that the number n of vertices in $C_n(1,3)$ must be even, which contradicts the assumption that n is odd.

Therefore, when $n \geq 3$ ($n \neq 5$) is odd, $C_n(1,3)$ cannot have a sigma coloring using only two colors. Since $\chi(C_n(1,3)) = 3$ in this case, we must have $\sigma(C_n(1,3)) = 3$ for $n \geq 3$ ($n \neq 5$).   □

4   The Circulant Graphs $C_{2n}(1, n)$

For $n \geq 3$, we draw and label the vertices of $C_{2n}(1, n)$ as in Fig. 13. Let a, b be distinct positive integers and $c : V_n \rightarrow \{a, b\}$ be a vertex coloring of $C_n(1, 2)$; we define the cyclic sequence $C : c(v_0), c(v_1), ..., c(v_n), c(v_0)$. We will need the following lemmas in the proof of Theorem 6.

Lemma 15. *If n is even, then C has an even number of 2-blocks.*

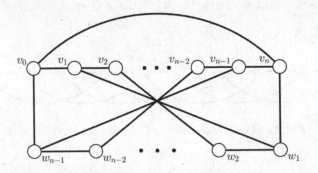

Fig. 13. Labelling of the vertices of $C_{2n}(1, n)$

Proof. Our proof is by contradiction. Suppose C has an odd number of 2-blocks. Since n is even, C must have a subsequence of the form $B_1 \, S \, B_2$ where S is a sequence of an odd number of 2-blocks and B_1, B_2 are blocks that are not of length 2. Without loss of generality, we assume that S begins at the vertex $v_{\alpha+1}$ and ends at $v_{\beta-1}$ for $\alpha, \beta = 0, 1, ..., n$; moreover, suppose $c(v_{\alpha+1}) = c(v_{\beta-1}) = a$. Hence, $(c(v_\alpha), c(v_{\alpha+1}), ..., c(v_{\beta-1}), c(v_\beta)) = (b, a, a, b, b, a, a, ..., a, a, b)$.

The proof involves checking the following cases depending on the value of α and β:

I. $\alpha < \beta$
  I.A. $0 < \alpha < \beta < n$
  I.B. $0 = \alpha < \beta < n$
  I.C. $0 < \alpha < \beta = n$
  I.D. $0 = \alpha < \beta = n$
II. $\beta < \alpha$
  II.A. $0 < \beta < \alpha < n$

II.B. $0 = \beta < \alpha < n$
II.C. $0 < \beta < \alpha = n$
II.D. $0 = \beta < \alpha = n$
III. $\beta = \alpha$
III.A. $0 < \alpha = \beta < n - 2$
III.B. $\alpha = \beta \in \{0, n-2, n-1, n\}$

The main idea of the proof relies on the observation that the sequence S induces an alternating coloring of a sequence of vertices in $\{w_1, w_2, ..., w_{n-1}\}$. We present here the proof only for Case I.A; the other cases can be proven similarly.

Assume $0 < \alpha < \beta < n$. For $i = \alpha + 1, \alpha + 2, ..., \beta - 1$, we observe that $\sigma(v_i) = a + b + c(w_i)$. Since c is a sigma coloring of $C_{2n}(1, n)$, we must have $c(w_i) \neq c(w_{i+1})$ for $i = \alpha + 1, ..., \beta - 2$; that is, the vertices $w_{\alpha+1}, ..., w_{\beta-1}$ are colored alternately. By symmetry, we can assume that $c(w_{\alpha+1}) = a$; then we obtain the partial coloring in Fig. 14.

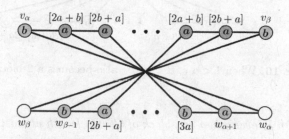

Fig. 14. A partial coloring of $C_{2n}(1, n)$ when $0 < \alpha < \beta < n$

Since $\sigma(w_{\alpha+1}) = a + b + c(w_\alpha) \neq 2a + b = \sigma(v_\alpha)$, we must have $c(w_\alpha) = b$. Consequently, $c(v_{\alpha-1}) = b$ and $\sigma(v_\alpha) = 2b + a$ since $\sigma(v_\alpha) \neq \sigma(v_{\alpha+1})$. We now consider two subcases:

1. Assume $\alpha = 1$. Then $\sigma(v_\alpha) = \sigma(v_1) = 2b + a$ and $\sigma(w_\alpha) = \sigma(w_1) = c(v_1) + c(v_n) + c(w_2) = b + c(v_n) + a$. Since $\sigma(v_1) \neq \sigma(w_1)$, we must have $c(v_n) = a$ (see Fig. 15). It follows that $B_1 = (c(v_0) \, c(v_1)) = (bb)$ is also a 2-block, a contradiction.

2. Assume $\alpha > 1$. Then $c(w_{\alpha-1}) = a$ since $\sigma(w_\alpha) \neq \sigma(v_\alpha)$. Moreover, $c(v_{\alpha-2}) = a$ since $\sigma(v_\alpha - 1) \neq \sigma(v_\alpha)$. It follows that $B_1 = (c(v_{\alpha-1}) \, c(v_\alpha)) = (bb)$ is also a 2-block, a contradiction (Fig. 16).   □

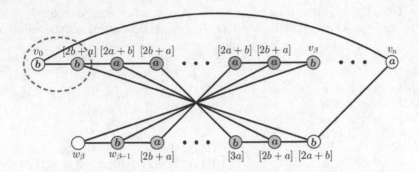

Fig. 15. When $1 = \alpha < \beta < n$, B_1 also becomes a 2-block.

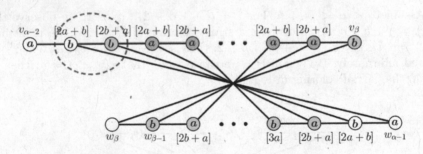

Fig. 16. When $1 < \alpha < \beta < n$, B_1 also becomes a 2-block.

Lemma 16. *If n is even, then any block of C with length at least three is of odd length.*

Proof. The proof is also by contradiction. Suppose B is a block of C with length that is even and at least four. Similar to the situation in Lemma 15, such a block induces an alternate coloring of a sequence of vertices in $\{w_1, w_2, ..., w_{n-1}\}$. Then the rest of the proof can be done in the same way as in Lemma 15.   □

Proof of Theorem 6. Let $n \geq 3$ be an integer and consider the circulant graph $C_{2n}(1, n)$. When n is even, it follows from Theorem 8 that $C_{2n}(1, n)$ is bipartite; hence, in this case, $\sigma(C_{2n}(1, n)) = 2$.

Now, suppose $n \geq 4$ is even. We will show that $C_{2n}(1, n)$ has no sigma 2-coloring via contradiction. Suppose such a sigma 2-coloring c exists; define the cyclic sequence C : $c(v_0), c(v_1), ..., c(v_n), c(v_1)$. Then Lemma 16 implies that C can only have blocks of length $1, 2, 3, 5, 7, ...$ while Lemma 15 implies that C has an even number of 2-blocks. Since C must have an even number of blocks, it must also have an even number of blocks with odd length. But this implies that the number of vertices in C must be even, which is a contradiction because C has an odd number $(n + 1)$ of vertices. Hence, $C_{2n}(1, n)$ cannot have a sigma 2-coloring; and since $\chi(C_{2n}(1, n)) = 3$, it follows that $\sigma(C_{2n}(1, n)) = 3$ as well. $\square$

5   Future Work

Let $n \geq 6$ and $k \geq 2$ be integers. Theorems 7 and 8 imply that the circulant graph $C_n(1, k)$ is sigma 3-colorable except possibly when (a) $k = 2$ and $n \not\equiv 0 \ (mod \ 3)$, or (b) when $C_n(1, k) = C_{13}(1, 5)$. These exceptions are, in fact, also sigma 3-colorable; this follows from Theorem 3 for (a) and from an easy verification for (b). Therefore, any circulant graph $C_n(1, k)$, $n \geq 5, k \geq 2$, is sigma 3-colorable. The authors believe that this property also holds for more general graphs.

Conjecture 17. Let $n \geq 6$ and $D = \{a, b\}$ be a generating subset of $\{0, 1, ..., n - 1\}$ not containing 0. Then any connected circulant graph $C_n(D)$ is sigma 3-colorable.

The following is a more general claim; here, we note that the graphs concerned are 4-colorable (Brook's theorem).

Conjecture 18. Any connected 4-regular graph of order at least 6 is sigma 3-colorable.

References

1. Barajas, J., Serra, O.: On the chromatic number of circulant graphs. Discrete Math. **309**, 5687–5696 (2009)
2. Chartrand, G., Okamoto, F., Rasmussen, C.W., Zhang, P.: The set chromatic number of a graph. Discuss. Math. **29**, 545–561 (2009)
3. Chartrand, G., Okamoto, F., Zhang, P.: The sigma chromatic number of a graph. Graphs Comb. **26**, 755–773 (2010)
4. Codenotti, B., Gerace, I., Vigna, S.: Hardness results and spectral techniques for combinatorial problems on circulant graphs. Linear Algebra Appl. **285**, 123–142 (1998)
5. Czerwiński, S., Grytczuk, J., Zelazny, W.: Lucky labelings of graph. Inf. Proc. Lett. **109**, 1078–1081 (2009)
6. Dehghan, A., Sadeghi, M.R., Ahadi, A.: The complexity of the sigma chromatic number of cubic graphs, March 2014. arXiv:1403.6288v1 [math.CO]
7. Heuberger, C.: On planarity and colorability of circulant graphs. Discrete Math. **268**, 153–169 (2003)

A Polynomial-Space Exact Algorithm
for TSP in Degree-6 Graphs

Norhazwani Md Yunos[1,2(✉)], Aleksandar Shurbevski[1], and Hiroshi Nagamochi[1]

[1] Department of Applied Mathematics and Physics, Kyoto University, Kyoto, Japan
{wanie,shurbevski,nag}@amp.i.kyoto-u.ac.jp
[2] Technical University of Malaysia Malacca, Malacca, Malaysia

Abstract. This paper presents the first polynomial-space exact algorithm specialized for the TSP in graphs with degree at most 6. We develop a set of branching rules to aid the analysis of the branching algorithm. Using the measure-and-conquer method, we show that when applied to an n-vertex graph with degree at most 6, the algorithm has a running time of $O^*(3.0335^n)$, which is still advantageous over other known polynomial-space algorithms for the TSP in general graphs.

Keywords: Traveling salesman problem · Exact exponential algorithm · Branch-and-Reduce · Measure-and-Conquer

1 Introduction

The Traveling Salesman Problem is one of the most extensively studied problems in combinatorial optimization. Besides being a well-known NP-hard combinatorial optimization problem, it also has a great practical importance. Present-day computers have only limited memory and algorithms which use exponential execution space will run out of memory well before they run out of time. For this reason, we limit this exposition to algorithms which require polynomially bounded execution space.

Gurevich and Shelah [6] gave the first polynomial-space exact algorithm for the TSP, whose running time in a general n-vertex graph is bounded by $O^*(4^n n^{\log n})$, where the O^* notation suppresses polynomial factors. This time bound has only recently been improved, but only for graphs of limited degree. From this viewpoint, let degree-i graph stand for a graph in which vertices have at most i incident edges. Note, for any graph with maximum degree at most d, the TSP can be solved in $O(n(d-1)^n)$ time and $O(dn)$ space by generating paths from a vertex.

A number of studies have been done focusing on the TSP in degree-bounded graphs. Currently the fastest polynomial-space algorithms for the TSP in degree-3 and degree-4 graphs were given by Xiao and Nagamochi [12,14], running in time $O^*(1.2312^n)$ and $O^*(1.692^n)$, respectively. Other previous studies of the TSP in degree-3 and degree-4 graphs have been done by Eppstein [3], Iwama and Nakashima [7], and Liskiewicz and Schuster [8].

© Springer International Publishing AG 2016
J. Akiyama et al. (Eds.): JCDCGG 2015, LNCS 9943, pp. 228–240, 2016.
DOI: 10.1007/978-3-319-48532-4_20

To the best of our knowledge, presently the only investigation on the TSP in graphs of degree up to 5 has been done by Md Yunos et al. [9], giving an $O^*(2.4723^n)$-time algorithm. Furthermore, there exist no reports in the literature of exact algorithms specialized to the TSP in degree-6 graphs. Therefore this paper presents the first algorithm for the TSP in degree-6 graphs, and shows that the algorithm runs in $O^*(3.0335^n)$ time. Due to space limitation, we omit the details of technical proofs of the result, which can be found in the technical reports by Md Yunos et al. [10,11]. Note the time bound claimed in the former technical report [10] is incorrect, and the present correct bound, $O^*(3.0335^n)$ is given in the latter technical report [11].

The remainder of this paper is organized as follows; Sect. 2 overviews the basic notation used in this paper. Section 3 describes the details of our polynomial-space branching algorithm. Section 4 establishes a framework for analyzing the proposed algorithm, and we proceed with the analysis. Finally, Sect. 5 makes some concluding remarks.

2   Preliminaries

For a graph G, let $V(G)$ (resp., $E(G)$) denote the set of vertices (resp., the set of edges) in G. A vertex u is a neighbor of a vertex v if v and u are adjacent by an edge uv in $E(G)$. We denote the set of neighbors of a vertex v by $N(v)$, and denote by $d(v)$ the cardinality $|N(v)|$ of $N(v)$, also called the *degree* of v. For a subset of vertices $W \subseteq V(G)$, let $N(v; W) = N(v) \cap W$. For a subset of edges $E' \subseteq E(G)$, let $N_{E'}(v) = N(v) \cap \{u \mid uv \in E'\}$, and let $d_{E'}(v) = |N_{E'}(v)|$. Analogously, let $N_{E'}(v; W) = N_{E'}(v) \cap W$, and $d_{E'}(v, W) = |N_{E'}(v, W)|$. Also, for a subset E' of $E(G)$, we denote by $G - E'$ the graph $(V, E \setminus E')$ obtained from G by removing the edges in E'.

We employ a known generalization of the TSP introduced by Eppstein [3], named the *forced* TSP. We define an instance $I = (G, F)$ that consists of an ordered pair of a simple, edge weighted, undirected graph G, and a subset F of edges in G, called *forced*. For brevity, throughout this paper let U denote $E(G) \setminus F$. A vertex is called *forced* if exactly one of its incident edges is forced. Similarly, it is called *unforced* if no forced edge is incident to it. Vertices incident with 2 or more forced edges are special cases and treated separately. A Hamiltonian cycle in G is called a *tour* if it passes through all the forced edges in F. Under these conditions, the forced TSP requests to find a minimum cost of a tour of an instance (G, F).

In this paper, we assume that the maximum degree of a vertex in G is at most 6. We refer to a forced (resp., unforced) vertex of degree i as type fi (resp., ui). Vertices of degree 1 and 2 are treated as special cases, and we examine eight types of vertices in an instance (G, F), namely u6, f6, u5, f5, u4, f4, u3 and f3-vertices. For each $i \in \{3, 4, 5, 6\}$, let V_{fi} (resp., V_{ui}) denote the set of fi-vertices (resp., ui-vertices) in (G, F).

3   A Polynomial-Space Branching Algorithm

Our algorithm consists of two major steps which are repeated iteratively. In the first step, the algorithm applies reduction rules until no further reduction is possible. In the second step, the algorithm applies branching rules in a reduced instance to search for a solution.

3.1   Reduction Rules

Reduction is a process of transforming an instance to a smaller instance. It takes polynomial time to construct a solution of an original instance from a solution of a smaller instance obtained through reduction.

If an instance has no tour, we call it *infeasible*. Observation 1 below gives two sufficient conditions for an instance to be infeasible. These two conditions will be checked when executing the reduction rules.

Observation 1. *If one of the following conditions holds, then the instance* (G, F) *is infeasible.*

(i) $d(v) \leq 1$ *for some vertex* $v \in V(G)$.
(ii) $d_F(v) \geq 3$ *for some vertex* $v \in V(G)$.

In this paper, we apply two reduction rules as stated in Md Yunos et al. [9, Lemma 2]. The reduction rules as stated in Observation 2 preserve the minimum cost of a tour in an instance, and they are applied in each of the branching operations.

Observation 2. *Each of the following reductions preserves the feasibility and a minimum cost tour of an instance* (G, F).

(i) *If* $d(v) = 2$ *for a vertex* v, *then add to* F *any unforced edge incident to the vertex* v; *and*
(ii) *If* $d(v) > 2$ *and* $d_F(v) = 2$ *for a vertex* v, *then remove from* G *any unforced edge incident to the vertex* v.

Our reduction algorithm is described in Fig. 1. An instance (G, F) is called *reduced* if it does not satisfy any of the conditions in Observation 1 and 2.

3.2   Branching Rules

Our branching algorithm is based on a set of branching rules. Without loss of generality, let v be a vertex of degree 6 and t its neighbor via an unforced edge. The choice of an edge to branch on plays a key role in the analysis of our branching algorithm. To this effect, in an instance (G, F), we assign the following priority in choosing an unforced edge $e = vt$ to branch on. Forced vertices take precedence over unforced ones, and for the choice of t, vertices of lower degree take precedence over vertices of higher degree. A pair of neighbors vt with no

---

Input: An instance (G, F).
Output: A reduced instance (G', F') of (G, F); or a message for the infeasibility of (G, F), which evaluates to ∞.

Initialize $(G', F') := (G, F)$;
while (G', F') is not a reduced instance **do**
    If there is a vertex v in (G', F') such that $d(v) \leq 1$ or $d_{F'}(v) \geq 3$ **then**
        Output message "Infeasible"
    Elseif there is a vertex v in (G', F') such that $2 = d(v) > d_{F'}(v)$ **then**
        Let $E^\dagger$ be the set of unforced edges incident to all such vertices;
        $F' := F' \cup E^\dagger$
    Elseif there is a vertex v in (G', F') such that $d(v) > d_{F'}(v) = 2$ **then**
        Let $E^\dagger$ be the set of unforced edges incident to all such vertices;
        $G' := G' - E^\dagger$
    Endif
End while;
Output (G', F').

Fig. 1. Algorithm Red(G, F).

neighbor in common has highest priority, and the priority decreases as the size of the common neighborhood increases. If the graph has a degree-6 vertex, then at least one unforced edge $e = vt$ of highest priority exists, and it is called *optimal*. Otherwise, the degrees of all vertices in the given instance are at most 5, and we can make use of an algorithm specialized to TSP instances of maximum degree up to 5, e.g., the polynomial-space algorithm of Md Yunos et al. [9]. We refer to this priority in choosing an edge $e = vt$ to branch on as the *branching rules*. A list giving the above priorities is given in Fig. 2, where the condition (c-i) with minimum index i is optimal, over all unforced edges vt in (G, F).

The collective set of branching rules for conditions c-1 to c-19 are illustrated in Fig. 3. Details of our branching algorithm are described in Fig. 4.

4 Analysis

4.1 Analysis Framework

To effectively analyze the branching algorithm of Fig. 4, we use the measure-and-conquer method as introduced by Fomin et al. [4]. Given an instance $I = (G, F)$ of the forced TSP, we assign a nonnegative weight $\omega(v)$ to each vertex $v \in V(G)$ according to its type, and use the sum of weights of all vertices in the graph G as a measure $\mu(I)$ of the instance $I = (G, F)$, that is,

$$\mu(I) = \sum_{v \in V(G)} \omega(v). \tag{1}$$

It is important for the analysis to find a measure which satisfies the following properties

(c-1) $v \in V_{f6}$ and $t \in N_U(v; V_{f3})$ such that $N_U(v) \cap N_U(t) = \emptyset$;
(c-2) $v \in V_{f6}$ and $t \in N_U(v; V_{f3})$ such that $N_U(v) \cap N_U(t) \neq \emptyset$;
(c-3) $v \in V_{f6}$ and $t \in N_U(v; V_{u3})$;
(c-4) $v \in V_{f6}$ and $t \in N_U(v; V_{f4})$ such that $N_U(v) \cap N_U(t) = \emptyset$;
(c-5) $v \in V_{f6}$ and $t \in N_U(v; V_{f4})$ such that $N_U(v) \cap N_U(t) \neq \emptyset$;
    (I) $|N_U(v) \cap N_U(t)| = 1$; and
    (II) $|N_U(v) \cap N_U(t)| = 2$;
(c-6) $v \in V_{f6}$ and $t \in N_U(v; V_{u4})$;
(c-7) $v \in V_{f6}$ and $t \in N_U(v; V_{f5})$ such that $N_U(v) \cap N_U(t) = \emptyset$;
(c-8) $v \in V_{f6}$ and $t \in N_U(v; V_{f5})$ such that $N_U(v) \cap N_U(t) \neq \emptyset$;
    (I) $|N_U(v) \cap N_U(t)| = 1$;
    (II) $|N_U(v) \cap N_U(t)| = 2$; and
    (III) $|N_U(v) \cap N_U(t)| = 3$;
(c-9) $v \in V_{f6}$ and $t \in N_U(v; V_{u5})$;
(c-10) $v \in V_{f6}$ and $t \in N_U(v; V_{f6})$ such that $N_U(v) \cap N_U(t) = \emptyset$;
(c-11) $v \in V_{f6}$ and $t \in N_U(v; V_{f6})$ such that $N_U(v) \cap N_U(t) \neq \emptyset$;
    (I) $|N_U(v) \cap N_U(t)| = 1$;
    (II) $|N_U(v) \cap N_U(t)| = 2$;
    (III) $|N_U(v) \cap N_U(t)| = 3$; and
    (IV) $|N_U(v) \cap N_U(t)| = 4$;
(c-12) $v \in V_{f6}$ and $t \in N_U(v; V_{u6})$;
(c-13) $v \in V_{u6}$ and $t \in N_U(v; V_{f3})$;
(c-14) $v \in V_{u6}$ and $t \in N_U(v; V_{u3})$;
(c-15) $v \in V_{u6}$ and $t \in N_U(v; V_{f4})$;
(c-16) $v \in V_{u6}$ and $t \in N_U(v; V_{u4})$;
(c-17) $v \in V_{u6}$ and $t \in N_U(v; V_{f5})$;
(c-18) $v \in V_{u6}$ and $t \in N_U(v; V_{u5})$; and
(c-19) $v \in V_{u6}$ and $t \in N_U(v; V_{u6})$.

Fig. 2. Preference conditions for choosing a branching edge, whose illustrations are given in Fig. 3.

(i) $\mu(I) = 0$ if and only if I can be solved in polynomial time; and
(ii) If I' is a sub-instance of I obtained through a reduction or a branching operation, then $\mu(I') \leq \mu(I)$.

We call a measure μ satisfying conditions (i) and (ii) above a *proper measure*.

We perform the time analysis of the branching algorithm via appropriately constructed recurrences related to the measure $\mu = \mu(I)$ of an instance $I = (G, F)$, for each branching rule of the algorithm. Let $T(\mu)$ denote the number of nodes in the search tree generated by our algorithm when invoked on the instance I with measure μ. Let I' and I'' be instances obtained from I by a branching operation, and let $a \leq \mu(I) - \mu(I')$ and $b \leq \mu(I) - \mu(I'')$ be lower bounds on the amounts of decrease in measure. The values of a and b will be nonnegative for a proper measure μ. We call (a, b) the *branching vector* of the branching operation, and this implies the linear recurrence:

$$T(\mu) \leq T(\mu - a) + T(\mu - b). \tag{2}$$

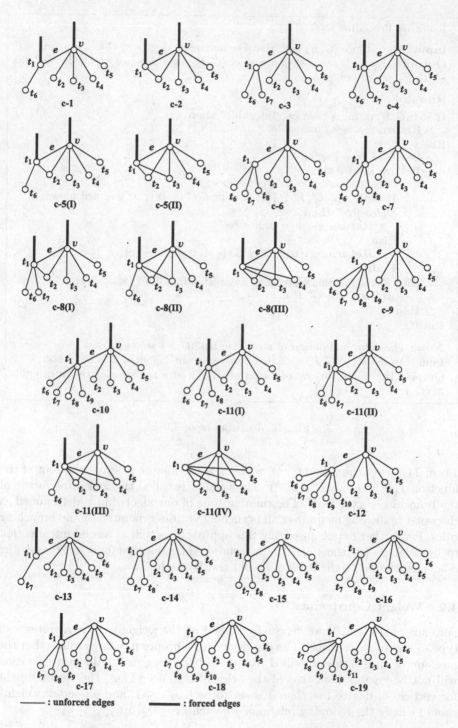

Fig. 3. Illustration of the branching rules in Fig. 2.

Recursive Procedure tsp6(G, F).

Input: An instance (G, F) such that the maximum degree of G is at most 6.
Output: The minimum cost of a tour of (G, F); or a message for the infeasibility of (G, F), which evaluates to ∞.

Run Red(G, F);
If Red(G, F) outputs message "Infeasible" **then**
    Return message "Infeasible"
Else
    Let $(G', F') := $ Red(G, F);
    If $V_{u6} \cup V_{f6} \neq \emptyset$ **then**
        Choose an optimal unforced edge e;
        If both tsp6$(G', F' \cup \{e\})$ and tsp6$(G' - \{e\},\ F')$ return message "Infeasible" **then**
            Return message "Infeasible"
        Else
            Return min{tsp6$(G', F' \cup \{e\})$, tsp6$(G' - \{e\},\ F')$}
        Endif
    Else /* the maximum degree of any vertex in (G', F') is at most 5 */
        Return tsp5(G', F')
    Endif
Endif.

Note: The input and output of algorithm tsp5(G, F) are as follows
Input: An instance (G, F) such that the maximum degree of G is at most 5.
Output: The minimum cost of a tour of (G, F); or a message for the infeasibility of (G, F).

Fig. 4. Algorithm tsp6(G, F).

Then, $T(\mu)$ is of the form $O(\tau^\mu)$, where τ is the unique positive real root of the function $f(x) = 1 - \left(x^{-a} + x^{-b}\right)$. The value τ is called the *branching factor* of the branching vector (a, b). The running time of our algorithm is determined as the worst branching factor over all branching vectors generated by the branching rules. For further details justifying this approach, as well as a solid introduction to branching algorithms and the measure-and-conquer method in general, the reader is referred to the book of Fomin and Kratsch [5].

4.2 Weight Constraints

For each $i \in \{3, 4, 5, 6\}$ we denote by w_i and w_i' the weight $\omega(v)$ of a vertex v of type ui and fi, respectively. The conditions for a proper measure require that the measure of an instance obtained through a branching or a reduction operation will not be greater than that of the original instance. Thus, the vertex weight for vertices of degree less than 3 is set to be 0, $w_6 \leq 1$, and all vertex weights should satisfy the following relations; for each $i \in \{3, 4, 5, 6\}$,

$$w_i' \leq w_i, \tag{3}$$

and for each $i \in \{4, 5, 6\}$,

$$w_{i-1} \leq w_i, \tag{4}$$

$$w'_{i-1} \leq w'_i. \tag{5}$$

As a result of the reduction and branching operations of Figs. 1 and 4, the degree of some vertices will decrease, while the degree of other vertices will remain unchanged. A forced edge will never be eliminated, neither by the reduction nor by branching operations. Conversely, an unforced edge may be removed or become forced by a reduction or a branching operation. Thus, the measure of an instance obtained through a reduction or branching operation will not be greater than that of the original instance. Lemma 1 shows that with respect to the algorithms given in Figs. 1 and 4, setting vertex weights which satisfy the conditions of Eqs. (3) to (5) is sufficient to obtain a proper measure. We can prove this lemma similarly as in Md Yunos et al. [9, Lemma 3].

Lemma 1. *If the weights of vertices are chosen as in Eqs. (3) to (5), then the measure $\mu(I)$ never increases as a result of the reduction or the branching operations of Figs. 1 and 4.*

To simplify some arguments, we introduce the following notation. For each $i \in \{3, 4, 5, 6\}$, let Δ_i denote $w_i - w'_i$, let $\Delta_{i,i-1}$ denote $w_i - w_{i-1}$, and let $\Delta'_{i,i-1}$ denote $w'_i - w'_{i-1}$. As a result of Eqs. (3) to (5), the values of each Δ_i, $\Delta_{i,i-1}$ and $\Delta'_{i,i-1}$ cannot be less than 0.

In the remainder of the analysis, for an optimal edge $e = vt_1$, we refer to $N_U(v)$ by $\{t_1, t_2, \ldots, t_a\}$, $a = d_U(v)$, and to $N_U(t_1) \setminus \{v\}$ by $\{t_{a+1}, t_{a+2}, \ldots, t_{a+b}\}$, $b = d_U(t_1) - 1$. We assume without loss of generality that $t_{1+i} = t_{a+i}$ for $i = 1, 2, \ldots, c$, where $c = |N_U(v) \cap N_U(t_1)|$ is the number of common neighbors of v and t_1.

If there exists an f3-vertex t_{a+i} in $N_U(t_1) \setminus \{v\}$, let $x \in N_U(t_{a+i}) \setminus \{v, t_1\}$. We see that the choice of vertex x is unique, because t_{a+i} is of type f3 and $|N_U(t_{a+i}) \setminus \{v, t_1\}| = 1$. This vertex x will play a key role in our analysis, as shown in Fig. 5.

4.3   Main Result

We choose a vertex weight function $\omega(v)$ as follows:

$$\omega(v) = \begin{cases} w_6 = 1 & \text{for a u6-vertex } v \\ w'_6 = 0.502801 & \text{for an f6-vertex } v \\ w_5 = 0.815641 & \text{for a u5-vertex } v \\ w'_5 = 0.421871 & \text{for an f5-vertex } v \\ w_4 = 0.580698 & \text{for a u4-vertex } v \\ w'_4 = 0.311647 & \text{for an f4-vertex } v \\ w_3 = 0.262796 & \text{for a u3-vertex } v \\ w'_3 = 0.149646 & \text{for an f3-vertex } v \\ 0 & \text{otherwise.} \end{cases} \tag{6}$$

The vertex weight function $\omega(v)$ given in Eq. (6) is obtained as a solution to a quasiconvex program, according to the method introduced by Eppstein [2]. All the branching vectors are in fact constraints in the quasiconvex program.

Lemma 2. *If the vertex weight function $\omega(v)$ is set as in Eq. (6), then each branching operation in Fig. 4 has a branching factor not greater than 3.033466.*

A proof of Lemma 2 can be derived analytically by analyzing the branching vectors which result by applying the branching and reduction operations. For the sake of space, in Sect. 4.4 we demonstrate the analysis of a single branching rule, case c-13. The reader is referred to the technical report by Md Yunos et al. [10] for a complete case analysis.

From Lemma 2, we get our main result as stated in Theorem 1.

Theorem 1. *The TSP in an n-vertex graph G with degree at most 6 can be solved in $O^*(3.0335^n)$ time and polynomial space.*

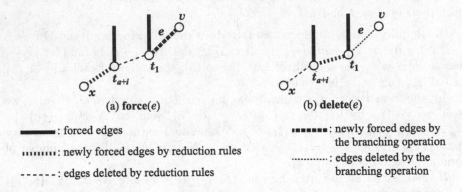

$$\text{(a) } \mathbf{force}(e) \qquad\qquad\qquad \text{(b) } \mathbf{delete}(e)$$

▬▬ : forced edges

▪▪▪▪▪▪▪ : newly forced edges by the branching operation

⸳⸳⸳⸳⸳⸳⸳⸳ : newly forced edges by reduction rules

⸳⸳⸳⸳⸳⸳⸳ : edges deleted by the branching operation

‒ ‒ ‒ ‒ ‒ : edges deleted by reduction rules

Fig. 5. Illustration of (a) newly forced and (b) deleted edge by a branching operation and reduction rules for an f3 vertex t_{a+i}.

4.4   Case Analysis of the Branching Operation for Case c-13

In interest of space, we will only illustrate the process of deriving the branching vectors for only one case, case c-13.

Case c-13. None of the previous branching rules c-1 to c-12 can be applied, and there exist vertices $v \in V_{u6}$ and $t_1 \in N_U(v; V_{f3})$ (see Fig. 6): We branch on the edge vt_1. Note that $N_U(t_1) \setminus \{v\} = \{t_7\}$.

In the branch of **force**(vt_1), the edge vt_1 will be added to F' by the branching operation, and the edge t_1t_7 will be deleted from G' by the reduction rules. Hence the weight of vertex v decreases by Δ_6, and the weight of vertex t_1 decreases by w'_3.

In the branch of **delete**(vt_1), the edge vt_1 will be deleted from G' by the branching operation, and the edge t_1t_7 will be added to F' by the reduction

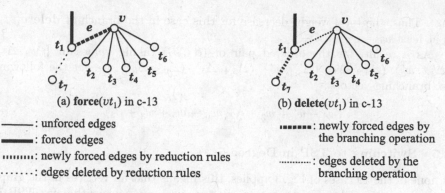

(a) **force**(vt_1) in c-13   (b) **delete**(vt_1) in c-13

———— : unforced edges

━━━━ : forced edges

••••••••• : newly forced edges by reduction rules

••••••• : edges deleted by reduction rules

■■■■■■■ : newly forced edges by
the branching operation

•••••••••• : edges deleted by the
branching operation

Fig. 6. Illustration of branching rule c-13, where vertex $v \in V_{u6}$ and $t_1 \in N_U(v; V_{f3})$.

rules. Hence the weight of vertex v decreases by $\Delta_{6,5}$, and the weight of vertex t_1 decreases by w_3'.

There are two cases for the vertex type of vertex t_7; 1) vertex t_7 is of type f3, 2) otherwise. We analyze the branches **force** and **delete** for these two cases separately.

First we analyze the case where the vertex t_7 is an f3-vertex (see Fig. 5). Recall that in this case, we denote by x the unique good neighbor of t_7 different from t_1. In the branch of **force**(vt_1), the edge xt_7 will be added to F' by the reduction rules. Hence the weight of vertex t_7 decreases by w_3'. Note that the vertex x cannot be an f6 vertex, otherwise one of the branching rules c-1 to c-12 would be applicable. If x is an f3-vertex (resp., u3, f4, u4, f5, u5, or a u6-vertex), then the weight decrease α_1 of x will be w_3' (resp., Δ_3, w_4', Δ_4, w_5', Δ_5, and Δ_6). Thus the total weight decrease for this case in the branch of **force**(vt_1) is at least $w_6 - w_6' + w_3' + w_3' + \alpha_1$.

In the branch of **delete**(vt_1), the edge xt_7 will be deleted from G' by the reduction rules. Hence the weight of vertex t_7 decreases by w_3'. If x is an f3-vertex (resp., u3, f4, u4, f5, u5, or a u6-vertex), then the weight decrease β_1 of x will be w_3' (resp., w_3, $\Delta_{4,3}'$, $\Delta_{4,3}$, $\Delta_{5,4}'$, $\Delta_{5,4}$, and $\Delta_{6,5}$). Thus the total weight decrease for this case in the branch of **delete**(vt_1) is at least $w_6 - w_5 + w_3' + w_3' + \beta_1$.

As a result, for the ordered pair (α_1, β_1) taking values in $\{(w_3', w_3'), (\Delta_3, w_3), (w_4', \Delta_{4,3}'), (\Delta_4, \Delta_{4,3}), (w_5', \Delta_{5,4}'), (\Delta_5, \Delta_{5,4}), (\Delta_6, \Delta_{6,5})\}$, we get the following seven branching vectors:

$$(w_6 - w_6' + 2w_3' + \alpha_1, \; w_6 - w_5 + 2w_3' + \beta_1). \tag{7}$$

Next we examine the case where the vertex t_7 is not an f3-vertex. In the branch of **force**(vt_1), if t_7 is a u3-vertex (resp., f4, u4, f5, u5, or a u6-vertex), then the weight decrease α_2 of t_7 will be w_3 (resp., $\Delta_{4,3}'$, $\Delta_{4,3}$, $\Delta_{5,4}'$, $\Delta_{5,4}$, and $\Delta_{6,5}$). Thus, the total weight decrease for this case in the branch of **force**(vt_1) is at least $w_6 - w_6' + w_3' + \alpha_2$.

In the branch of **delete**(vt_1), if t_7 is a u3-vertex (resp., f4, u4, f5, u5, or a u6-vertex), then the weight decrease β_2 of t_7 will be Δ_3 (resp., w_4', Δ_4, w_5', Δ_5, and

Δ_6). Thus, the total weight decrease for this case in the branch of **delete**(vt_1) is at least $w_6 - w_5 + w'_3 + \beta_2$.

As a result, for the ordered pair of (α_2, β_2) taking values in $\{(w_3, \Delta_3), (\Delta'_{4,3}, w'_4), (\Delta_{4,3}, \Delta_4), (\Delta'_{5,4}, w'_5), (\Delta_{5,4}, \Delta_5), (\Delta_{6,5}, \Delta_6)\}$, we get the following six branching vectors:

$$(w_6 - w'_6 + w'_3 + \alpha_2, \ w_6 - w_5 + w'_3 + \beta_2). \tag{8}$$

4.5 Switching to TSP in Degree 5

If none of the 19 cases of Fig. 3 applies, this means that all vertices in the graph have degree 5 or less. In that case, we can use a fast algorithm for TSP in degree-5 graphs, called tsp5(G, F) to solve the remaining instances. Xiao and Nagamochi [13, Lemma 3] have shown how to leverage results obtained by a measure-and-conquer analysis, and that an algorithm can be used as a sub-procedure. We can get a non-trivial time bound on this sub-procedure if we know the respective weight setting mechanism. We calculate the maximum ratio of the vertex weights for the TSP in degree-5 graphs and the TSP in degree-6 graphs, and this will become a constraint in the quasiconvex program whose solution gives us the respective vertex weights.

Here we use the $O^*(2.4723^n)$-time algorithm by Md Yunos et al. [9], where the weights of vertices in degree-5 graphs are set as follows: for an f3-vertex, $\hat{w}'_3 = 0.183471$, for a u3-vertex, $\hat{w}_3 = 0.322196$, for an f4-vertex, $\hat{w}'_4 = 0.347458$, for a u4-vertex, $\hat{w}_4 = 0.700651$, for an f5-vertex, $\hat{w}'_5 = 0.491764$, and for a u5-vertex, $\hat{w}_5 = 1$. Let $\kappa = \max\left\{\frac{0.183471}{w'_3}, \frac{0.322196}{w_3}, \frac{0.347458}{w'_4}, \frac{0.700651}{w_4}, \frac{0.491764}{w'_5}, \frac{1}{w_5}\right\}$. For this step, the running time bound is

$$T(\mu(I)) \leq O\left(2.472232^\kappa\right). \tag{9}$$

4.6 Overall Analysis

As a result, the branching factor of each of the branching vectors does not exceed 3.033466, and the tight constraints are in conditions c-4, c-7, c-15, c-16, c-18 and the switching constraint of Eq. (9). This completes a proof of Theorem 1.

5 Conclusion

In this paper, we presented an exact algorithm for the TSP in degree-6 graphs. We use a similar technique as in the algorithm of the TSP in degree-5 graphs by Md Yunos et al. [9]. The greatest challenge in obtaining a non-trivial time-bound for the algorithm is to derive a proper case analysis. Namely we have to choose a good set of branching rules so that the rule giving the largest reduction will be executed first. But we can only calculate the reduction of each branching rule after a complete case analysis, and this interdependence makes the problem ever more difficult with graphs of higher degree, making it a considerable step from the authors' result on the TSP in degree-5 graphs [9], and even more challenging to extend this approach to the TSP in degree-7 graphs.

On the other hand, the above results offer a purely theoretically improved time-bound on the running time. Following recent results by Akiba and Iwata [1] indicating that algorithms with theoretically improved running time can indeed be also superior in practice, it would be of much interest to implement the proposed algorithms, and evaluate their empirical performance. This line of research would also open the question of devising bounding rules which might not have impact on the theoretical bound on the running time, might contribute meaningfully to the algorithm's performance in practice.

References

1. Akiba, T., Iwata, Y.: Branch-and-reduce exponential/FPT algorithms in practice: a case study of vertex cover. In: Proceedings of the 17th Workshop on Algorithm Engineering and Experiments (ALENEX), pp. 70–81. SIAM (2015)
2. Eppstein, D.: Quasiconvex analysis of backtracking algorithms. In: Proceedings of the 15th Annual ACM-SIAM Symposium On Discrete Algorithms (SODA 2004), pp. 781–790. ACM Press (2004)
3. Eppstein, D.: The traveling salesman problem for cubic graphs. J. Graph Algorithms Appl. **11**(1), 61–81 (2007)
4. Fomin, F.V., Grandoni, F., Kratsch, D.: A measure & conquer approach for the analysis of exact algorithms. J. ACM (JACM), **56**(6), Article no. 25 (2009)
5. Fomin, F.V., Kratsch, D.: Exact Exponential Algorithms. Springer, Heidelberg (2010)
6. Gurevich, Y., Shelah, S.: Expected computation time for Hamiltonian path problem. SIAM J. Comput. **16**(3), 486–502 (1987)
7. Iwama, K., Nakashima, T.: An improved exact algorithm for cubic graph TSP. In: Lin, G. (ed.) COCOON 2007. LNCS, vol. 4598, pp. 108–117. Springer, Heidelberg (2007)
8. Liskiewicz, M., Schuster, M.R.: A new upper bound for the traveling salesman problem in cubic graphs. Arxiv preprint (2012)
9. Md Yunos, N., Shurbevski, A., Nagamochi, H.: A polynomial-space exact algorithm for TSP in degree-5 graphs. In: The 12th International Symposium on Operations Research and Its Applications (ISORA 2015), pp. 45–58 (2015)
10. Md Yunos, N., Shurbevski, A., Nagamochi, H.: A polynomial-space exact algorithm for TSP in degree-6 graphs. Technical report 2015–003, Department of Applied Mathematics and Physics, Kyoto University (2015). http://www.amp.i.kyoto-u.ac.jp/tecrep/ps_file/2015/2015-003.pdf
11. Md Yunos, N., Shurbevski, A., Nagamochi, H.: Time bound on polynomial-space exact algorithms for TSP in degree-5 and degree-6 graphs. Technical report, 2015–004, Department of Applied Mathematics and Physics, Kyoto University (2015). http://www.amp.i.kyoto-u.ac.jp/tecrep/ps_file/2015/2015-004.pdf
12. Xiao, M., Nagamochi, H.: An exact algorithm for TSP in degree-3 graphs via circuit procedure and amortization on connectivity structure. Algorithmica **74**(2), 713–741 (2016)

13. Xiao, M., Nagamochi, H.: Exact algorithms for maximum independent set. In: Cai, L., Cheng, S.-W., Lam, T.-W. (eds.) Algorithms and Computation. LNCS, vol. 8283, pp. 328–338. Springer, Heidelberg (2013)
14. Xiao, M., Nagamochi, H.: An improved exact algorithm for TSP in graphs of maximum degree 4. Theory Comput. Syst. **58**(2), 241–272 (2016)

Topological Graph Layouts
into a Triangular Prism

Miki Miyauchi[✉]

NTT Communication Science Laboratories, NTT Corporation, Atsugi-shi, Japan
miyauchi.miki@lab.ntt.co.jp

Abstract. Prism layouts are special cases of track layouts of graphs. A *triangular prism layout for graphs* is a graph layout into a triangular prism that carries the vertices along the three crests between two triangles of the prism and the edges in the three rectangular surfaces such that no two edges cross in the interior of the surfaces. Also, a *topological* prism layout for graphs is defined so that edges are allowed to cross the crests. As for topological prism layouts, it is desirable to have good bounds on number of edge-crossings over crests for various classes of graphs. This paper constructs two-color-edge topological triangular prism layouts for complete bipartite graphs with fewer edge-crossings over the crests than previous results.

Keywords: Graph layout · Bipartite graph · Graph subdivision · Track layout · Prism layout

1 Introduction

A graph $G_{m,n}$ is a *bipartite graph* having *partite sets* A with m vertices and B with n vertices if $V(G) = A \cup B$, $A \cap B = \emptyset$ and each edge joins a vertex of A with a vertex of B. A bipartite graph $G_{m,n}$ is *complete* if $G_{m,n}$ contains all edges joining vertices in distinct sets. A complete bipartite graph is denoted by $K_{m,n}$.

An *ordering* of a set S is a total order $<_\sigma$ on S. It will be convenient to interchange "σ" and "$<_\sigma$" when there is no ambiguity. A *vertex ordering* of a graph G is an ordering σ of the vertex set $V(G)$.

A *vertex k-coloring* of a graph G is a partition $\{V_i : 1 \le i \le k\}$ of $V(G)$ such that for every edge $vw \in E(G)$, if $v \in V_i$ and $w \in V_j$ then $i \ne j$. Suppose that each color class V_i is ordered by $<_i$. Then the ordered set $(V_i, <_i)$ is called a *track*, and $\{(V_i, <_i) : 1 \le i \le k\}$ is a *k-track assignment* of G.

The *span* of an edge vw in a track assignment $\{(V_i, <_i) : 1 \le i \le k\}$ is $|i - j|$ where $v \in V_i$ and $w \in V_j$. An *X-crossing* in a track assignment consists of two edges vw and xy such that $v, x \in V_i$, $w, y \in V_j$, $v <_i x$ and $y <_j w$, for distinct colors i and j. An *edge 2-coloring* of G is simply a partition $\{E_0, E_1\}$ of $E(G)$. An edge $vw \in E_i$ is said to be *colored i*. A *$(2, k)$-track layout* of G consists of a k-track assignment of G and an edge 2-coloring of G with no monochromatic

© Springer International Publishing AG 2016
J. Akiyama et al. (Eds.): JCDCGG 2015, LNCS 9943, pp. 241–246, 2016.
DOI: 10.1007/978-3-319-48532-4_21

X-crossing. A graph admitting a $(2, k)$-track layout is called a $(2, k)$-*track graph*. Track layouts were introduced by V. Dujmović, P. Morin, and D. R. Wood [1,2].

If $e = uv$ is an edge of G, then e is said to be *subdivided* when it is replaced by the edges uw and wv. We call the new vertex w the *division vertex*. If every edge of G is subdivided, the resulting graph is called the *subdivision graph* of G. Note that a graph is also considered to be a subdivision of itself.

This paper studies $(2,3)$ track layouts of graph subdivisions. Dujmović and Wood [3] showed the following theorem:

Theorem 1 (Dujmović and Wood [3]). *Every graph G has a $(2,3)$-track subdivision of G with $2\lceil \log \mathrm{qn}(G)\rceil + 1$ division vertices per edge, where $\mathrm{qn}(G)$ is the queue number of G.*

The definition of the *queue number* is as follows. In a vertex ordering σ of a graph G, let $L(e)$ and $R(e)$ denote the endpoints of each edge $e \in E(G)$ such that $L(e) <_\sigma R(e)$. Consider two edges $e, f \in E(G)$ with no common endpoint such that $L(e) <_\sigma L(f)$. If $L(e) <_\sigma L(f) <_\sigma R(f) <_\sigma R(e)$ then e and f *nest*. A *queue* is a set of edges $E \subset E(G)$ such that no two edges in E nest. For an integer $d > 0$, a *d-queue layout* of G consists of a vertex ordering σ of G and a partition $\{E_i : 1 \le i \le d\}$ of $E(G)$ such that each E_i is a queue in σ. The *queue number* $\mathrm{qn}(G)$ of a graph G is the minimum d such that there is a d-queue layout of G. As for queue layout see [5,6] etc. There is a summary of bounds on the queue numbers for various kinds of graph families in [7].

As for Theorem 1, Dujmović and Wood [3] also showed that the order of the number of division vertices is optimal. Thus, to find a track layout with fewer division vertices for various kinds of graph families become an interesting problem.

This paper deals with the number of division vertices of bipartite graphs. For the queue number of a complete bipartite graph $K_{m,n}$, Heath and Rosenberg [6] showed the following theorem:

Theorem 2 (Heath and Rosenberg [6])

$$\mathrm{qn}(K_{m,n}) = \min\left(\lceil m/2\rceil, \lceil n/2\rceil\right).$$

Applying Theorem 2 to Theorem 1, we have the following corollary:

Corollary 1. *Every complete bipartite graph $K_{m,n}$ $(m \ge n)$ has a $(2,3)$-track subdivision with*

$$2\lceil\log\lceil n/2\rceil\rceil + 1 = \begin{cases} 2\lceil\log(n+1) - \log 2\rceil + 1 & (\textit{if } n \textit{ is odd}) \\ 2\lceil\log n - \log 2\rceil + 1 & (\textit{if } n \textit{ is even}) \end{cases}$$

division vertices per edge.

This paper improves this result and shows the following theorem:

Theorem 3. *Every bipartite graph $G_{m,n}$ has a $(2,3)$-track subdivision with $\lceil\log n\rceil - 1$ division vertices per edge, where $m \ge n$.*

Here, we define a new graph layout named prism layout for graphs. A *triangular prism layout for graphs* is a graph layout into a triangular prism that carries the vertices along the three crests between two triangles of the prism and the edges in the three rectangular surfaces such that no two edges with same color cross in the interior of the surfaces. Also, a *topological* prism layout for graphs is defined so that edges are allowed to cross the crests. As for topological prism layouts, it is desirable to have good bounds on number of edge-crossings over crests for various classes of graphs.

Then, a $(2,3)$-track layout of graph subdivisions can be regarded as a two-color-edge topological graph layout into a triangular prism. Therefore this paper constructs topological triangular prism layouts for complete bipartite graphs with fewer edge-crossings over the crests than previous results.

The proof of Theorem 3 is similar to that of V. Dujmović and D. R. Wood's result [3], however, it becomes simpler by capitalizing the character of bipartite graphs.

2   Proof of Theorem 3

In this section, we will construct a $(2,3)$-track layout of a subdivision of $G_{m,n}$ with $\lceil \log n \rceil - 1$ division vertices per edge.

Let $S = \{0,1\}$ be the binary alphabet and S^* the set of all strings over S of length at most h $(h > 0)$. If $s \in S^*$ has length k $(0 \le k \le h)$, then write

$$s = s_1 s_2 \ldots s_k$$

where s_i is the character of s in position i. Order the elements of S by $0 < 1$. Define a *breadth-first ordering* $<_*$ on S^* as follows: For $s = s_1 \ldots s_i, t = t_1 \ldots t_j \in S^*$, define $s <_* t$ when either of the following conditions holds.

1. $i < j$.
2. $i = j$, $s_1 \ldots s_{i-1} = t_1 \ldots t_{i-1}$, $s_i < t_i$.
3. $i = j (> 1)$, $s_1 \ldots s_{i-1} <_* t_1 \ldots t_{i-1}$.

As an example, the strings of length at most 2 are ordered as follows:

$$\epsilon <_* 0 <_* 1 <_* 00 <_* 01 <_* 10 <_* 11$$

where ϵ denotes the empty string.

Define $k = \lceil \log n \rceil$. A number s $(0 \le s < n)$ has a unique representation as a string in $S^k \subseteq S^*$ using binary representation, where S^k is the set of all elements of length k. For a number s, use the representation $s_1 \ldots s_k$ for its binary representation, where s_1 is the highest-order digit. For a string $s = s_1 \ldots s_k$ in S^k let $s(i)$ be the string consisting of the first i letters of s, that is,

$$s(i) = s_1 \ldots s_i$$

and $s(0)$ be the empty string ϵ.

Consider a subdivision $G^*_{m,n}$ of $G_{m,n}$ made by subdividing each edge

$$(a_s, b_t) \in E(G_{m,n})$$

$$(a_s \in A, b_t \in B, 0 \le s < m, 0 \le t < n)$$

by adding $k - 1$ vertices between a_s and b_t. We label these vertices in the order from a_s to b_t as follows;

$$a_s = (a_s, b_t; 0), (a_s, b_t; 1), \ldots, (a_s, b_t; k) = b_t.$$

where $(a_s, b_t; 0)$ is identified with a_s and $(a_s, b_t; k)$ is identified with b_t.

Since a $(2, 3)$-track layout of $G^*_{m,n}$ corresponds to a $(2, 3)$-track subdivision of $G_{m,n}$ by regarding vertices $V(G^*_{m,n}) - V(G_{m,n})$ as division vertices, we will construct a $(2, 3)$-track layout of $G^*_{m,n}$. First we define a vertex ordering σ of $V(G^*_{m,n})$ and then add 2 colors 0 and 1 to edges of $G^*_{m,n}$ so that there is no monochromatic X-crossing.

For two vertices $(a_s, b_t; i), (a_p, b_q; j) \in V(G^*_{m,n})$, we define $(a_s, b_t; i) <_\sigma (a_p, b_q; j)$ when one of the following three conditions holds:

1. $t(i) <_* q(j)$.
2. $t(i) = q(j)$ and $s < p$.
3. $t(i) = q(j)$ and $s = p$, $t < q$.

As for an edge coloring of $G^*_{m,n}$, let an edge $((a_s, b_t; i - 1), (a_s, b_t; i))$ be colored t_i.

Lemma 1. *Let $V_i = \{(a_s, b_t; i) : a_s \in A, b_t \in B, 0 \le s < m, 0 \le t < n\}$ for $0 \le i \le k$. For every bipartite graph $G_{m,n}$, the subdivision $G^*_{m,n}$ of $G_{m,n}$ has the $(2, k + 1)$-track layout defined by the family of the ordered sets $\{(V_i, \sigma \mid_{V_i}) : 0 \le i \le k\}$ and the edge 2-coloring we mentioned above. Moreover the maximum span is one, and the number of division veritices per edge is $\lceil \log n \rceil - 1$.*

Proof. Note that the ordering σ is a total order. Thus the family of the ordered sets $\{(V_i, \sigma) : 0 \le i \le k\}$ is a $k + 1$-track assignment of $G^*_{m,n}$.

Next, we show that this track layout is legal, i.e., no two edges in this track assignment $\{(V_i, \sigma)\}$ form a monochromatic X-crossing.

Let $((a_s, b_t; i - 1), (a_s, b_t; i))$ and $((a_p, b_q; j - 1), (a_p, b_q; j))$ be two edges in $E(G^*_{m,n})$ $(0 < i, j \le k)$ that form an X-crossing. Note that by the definition of the vertex ordering,

$$(a_s, b_t; i - 1) <_\sigma (a_s, b_t; i) \quad \text{and} \quad (a_p, b_q; j - 1) <_\sigma (a_p, b_q; j).$$

We may assume that the endpoints of the two edges are laid out from left to right in the order

$$(a_s, b_t; i - 1) <_\sigma (a_p, b_q; j - 1) \quad \text{and} \quad (a_p, b_q; j) <_\sigma (a_s, b_t; i). \qquad (*)$$

We want to show that the two division edges have different colors, that is, $t_i \neq q_j$. From the assumption $(*)$ and the definition of the vertex-ordering, we have

$$t(i-1) \leq_* q(j-1) \text{ and } q(j) \leq_* t(i).$$

These inequalities hold only when $i = j$. Suppose $t(i-1) <_* q(i-1)$. Then by the definition of the vertex ordering we have $t(i) <_* q(i)$ which contradicts the assumption. Thus we have $t(i-1) = q(i-1)$ and $q(i) \leq_* t(i)$. If $q_i = t_i$, then $q(i) = t(i)$. By the definition 2 or 3 of the vertex ordering σ and

$$(a_s, b_t; i-1) <_\sigma (a_p, b_q; i-1),$$

we have either $s < p$ or $s = p, t < q$, respectively. Thus, we have

$$(a_s, b_t; i) <_\sigma (a_p, b_q; i),$$

which contradicts the assumption $(*)$. Therefore $q_i \neq t_i$. Therefore we have proved that this track layout is legal.

Moreover by the definition of the adjacency relations we can easily find that the maximum span of the $(2, k+1)$-track layout of $G^*_{m,n}$ is one. Also, each edge (a_s, b_t) of $G_{m,n}$ is divided by adding $\lceil \log n \rceil - 1$ division points in the subdivision $G^*_{m,n}$. Thus, we have Lemma 1.   ∎

The following "wrapping" algorithm (Lemma 2) is implicitly proved by Felsner et al. [4] and generalized by Dujmović and Wood [3] to span s $(s \geq 1)$.

Lemma 2 (Dujmović and Wood [3]). *If a graph G has a $(2, k)$-track layout with maximum span one, then G has a $(2, 3)$-track layout.*

Applying Lemma 1 to Lemma 2, we find that $G^*_{m,n}$ has a $(2, 3)$-track layout. Moreover, applying wrapping algorithm used in the proof of Lemma 2 to the $(2, k+1)$-track layout which we construct in the proof of Lemma 1, we can prove Theorem 4.

Proof of Theorem 3. Construct a vertex three-coloring of $G^*_{m,n}$ by merging tracks $\{V_i : i \equiv j (\text{mod} 3)\}$ for each j, $(0 \leq j < 3)$. Then we have $(2, 3)$-track assignment $\{(V_i, \sigma) : i = 0, 1, 2\}$ of $G^*_{m,n}$. We show that this $(2, 3)$-track assignment and the edge-coloring we defined above form a $(2, 3)$-track layout.

Let $((a_s, b_t; i-1), (a_s, b_t; i))$ and $((a_p, b_q; j-1), (a_p, b_q; j))$ be two edges in $E(G^*_{m,n})$ $(0 < i, j \leq k)$ that form an X-crossing. We may assume that the endpoints of the two edges are laid out from left to right in the order

$$(a_s, b_t; i-1) <_\sigma (a_p, b_q; j-1) \text{ and } (a_p, b_q; j) <_\sigma (a_s, b_t; i).$$

From the above assumption and the definition of the vertex ordering, this inequality holds only when $i = j$. In this case, these two edges in the $(2, 3)$-track layout are laid out in the same way as in the original $(2, k)$-track layout. Therefore these two edges do not form a monochromatic X-crossing.

This wrapping algorithm does not change the number of division vertices for each edge, thus this $(2, 3)$-track layout also has $\lceil \log n \rceil - 1$ division vertices per edge.   ∎

3 Conclusion

This paper defines a new graph layout named prism layout for graphs. Also, we construct a two-color-edge topological triangular prism layout for complete bipartite graphs with fewer edge-crossings than previous result. We don't know whether this result is best possible or not. To find better topological prism layouts for complete bipartite graphs is still an interesting problem.

References

1. Dujmović, V., Morin, P., Wood, D.R.: Layout of graphs with bounded tree-width. SIAM J. Comput. **34**(3), 553–579 (2005)
2. Dujmović, V., Pór, A., Wood, D.R.: Track layouts of graphs. Discrete Math. Theor. Comput. Sci. **6**(2), 497–522 (2004)
3. Dujmović, V., Wood, D.R.: Stacks, queues and tracks: layouts of graph subdivisions. Discrete Math. Theor. Comput. Sci. **7**, 155–202 (2005)
4. Felsner, S., Liotta, G., Wismath, S.K.: Straight-line drawings on restricted integer grids in two and three dimensions. J. Graph Algorithms Appl. **7**(4), 363–398 (2003)
5. Heath, L.S., Leighton, F.T., Rosenberg, A.L.: Comparing queues and stacks as mechanisms for laying out graphs. SIAM J. Discrete Math. **5**(3), 398–412 (1992)
6. Heath, L.S., Rosenberg, A.L.: Laying out graphs using queues. SIAM J. Comput. **21**(5), 927–958 (1992)
7. Wood, D.R.: Queue layouts of graph products and powers. Discrete Math. Theor. Comput. Sci. **7**(1), 255–268 (2005)

On the Competition Numbers
of Diamond-Free Graphs

Yoshio Sano[✉]

Division of Information Engineering, Faculty of Engineering,
Information and Systems, University of Tsukuba, Ibaraki 305-8573, Japan
sano@cs.tsukuba.ac.jp

Abstract. In this note, we give a short proof for a theorem on the competition numbers of diamond-free graphs: If a graph G is diamond-free, then the competition number of G is bounded above by $2 + \frac{1}{2} \sum_{v \in V_{ns}(G)} (\theta_V(N_G(v)) - 2)$, where $V_{ns}(G)$ denotes the set of non-simplicial vertices in G and $\theta_V(N_G(v))$ denotes the minimum number of cliques that cover all the neighbors of a vertex v in G.

Keywords: Competition graph · Competition number · Diamond-free graph · Edge clique cover · Essential vertex

2010 Mathematics Subject Classification: 05C20 · 05C76

1 Introduction

The notion of a competition graph was introduced by Cohen [1] in connection with a problem in ecology. The *competition graph* $C(D)$ of a digraph D is the (simple undirected) graph which has the same vertex set as D and has an edge between two distinct vertices u and v if and only if there exists a vertex x in D such that (u, x) and (v, x) are arcs of D. Roberts [7] observed that any graph G together with sufficiently many isolated vertices is the competition graph of an acyclic digraph. The *competition number* $k(G)$ of a graph G is defined to be the smallest nonnegative integer k such that G together with k isolated vertices added is the competition graph of an acyclic digraph. It is not easy in general to compute the exact value of the competition number for an arbitrary graph G. Indeed, Opsut [5] showed that the computation of the competition number of a graph is an NP-hard problem. See [8] for the current best lower bound that holds for the competition numbers of arbitrary graphs, and see [2] for the current best upper bound that holds for the competition numbers of arbitrary graphs.

We use the following notation and terminology. For a digraph D and a vertex v of D, let $N_D^-(v) := \{u \in V(D) \mid (u, v) \in A(D)\}$. For a graph G and a vertex v of G, let $N_G(v) := \{u \in V(G) \mid uv \in E(G)\}$, and let $N_G[v] := N_G(v) \cup \{v\}$. We also denote by the same symbol $N_G(v)$ (resp. $N_G[v]$) the subgraph of G induced by $N_G(v)$ (resp. $N_G[v]$) if there is no confusion. A *clique* of a graph G

© Springer International Publishing AG 2016
J. Akiyama et al. (Eds.): JCDCGG 2015, LNCS 9943, pp. 247–252, 2016.
DOI: 10.1007/978-3-319-48532-4_22

is a complete subgraph of G. A *vertex clique cover* of a graph G is a family of cliques of G such that each vertex of G is contained in a clique in the family. The minimum size of a vertex clique cover of a graph G is called the *vertex clique cover number* of G and is denoted by $\theta_V(G)$. A vertex v in a graph G is said to be *isolated* if $\theta_V(N_G(v)) = 0$. A vertex v in a graph G is said to be *simplicial* if $\theta_V(N_G(v)) \leq 1$, and v is said to be *non-simplicial* if $\theta_V(N_G(v)) \geq 2$.

The *line graph* of a graph H is the simple graph $L(H)$ defined by $V(L(H)) = E(H)$ and $E(L(H)) = \{ee' \mid e \neq e', e \cap e' \neq \emptyset\}$. A graph G is called a *line graph* if there exists a graph H such that G is isomorphic to the line graph of H. In 1982, Opsut [5] showed the following:

Theorem 1.1 ([5]). *For a line graph G, $k(G) \leq 2$.*

There are several graph classes containing line graphs. In 2012, Park and Sano [6] showed that the competition numbers of *generalized line graphs* are also bounded from the above by two.

Theorem 1.2 ([6]). *For a generalized line graph G, $k(G) \leq 2$.*

A *quasi-line graph* is a graph G in which $\theta_V(N_G(v)) \leq 2$ holds for any vertex v of G. In 2014, McKay *et al.* [4] showed the following:

Theorem 1.3 ([4]). *For a quasi-line graph G, $k(G) \leq 2$.*

A *diamond* is the complete tripartite graph $K_{1,1,2}$. A graph G is said to be *diamond-free* if G does not contain a diamond as an induced subgraph. Recently, Kim *et al.* [3] gave an upper bound for the competition numbers of diamond-free graphs.

Theorem 1.4 ([3]). *Let G be a diamond-free graph. Then*

$$k(G) \leq 1 + \frac{1}{2}\varepsilon(G) + \frac{1}{2} \sum_{v \in V_{\mathrm{ns}}(G)} (\theta_V(N_G(v)) - 2),$$

where $V_{\mathrm{ns}}(G)$ denotes the set of non-simplicial vertices in G, and $\varepsilon(G)$ is defined by

$$\varepsilon(G) := \min\{2, \min\{\theta_V(N_G(v)) \mid v \in V(G)\}\}.$$

Note that the upper bound given in Theorem 1.4 is sharp in the sense that any nontrivial triangle-free connected graph attains the equality.

In this note, we give a short proof for the above theorem by using induction on $|V(G)| + |E(G)|$ instead of induction on $|V(G)|$ which is used in [3].

2   Preliminaries

Lemma 2.1. *Let G be a diamond-free graph and let K be a maximal clique of G. Then the graph obtained from G by deleting all the edges of K is diamond-free.*

Proof. Since G is diamond-free, K does not share an edge with any other maximal clique in G. Therefore, diamonds cannot be made by deleting all the edges of K from G. Thus the lemma holds.   □

Definition 1. For a graph G and a vertex v^* of G, let $G - [v^*]$ denote the graph obtained from G by deleting the vertex v^* and all the edges whose two endvertices are in $N_G[v^*]$, i.e., $V(G - [v^*]) = V(G) - \{v^*\}$ and $E(G - [v^*]) = E(G) - \{xy \in E(G) \mid x, y \in N_G[v^*]\}$.

Lemma 2.2. *Let G be a diamond-free graph. Then the graph $G - [v^*]$ is diamond-free.*

Proof. Since G is diamond-free, $N_G[v^*]$ consists of maximal cliques $K_{(1)}, \ldots, K_{(t)}$ of G that share the vertex v^*, where $t = \theta_V(N_G(v^*))$. Let $G_0 := G$ and let $G_i := G_{i-1} - E(K_{(i)})$ for $i = 1, \ldots, t$. Note that $K_{(i)}$ is also a maximal clique in G_{i-1}. By Lemma 2.1, if G_{i-1} is diamond-free, then G_i is diamond-free. Therefore G_t is diamond-free. Note that $G - [v^*] = G_t - \{v^*\}$ by the definition of $G - [v^*]$, and that the vertex v^* is isolated in G_t. Thus $G - [v^*] (= G_t - \{v^*\})$ is diamond-free.

Lemma 2.3. *Let G be a diamond-free graph and let v^* be a vertex of G.*

- *If $k(G - [v^*]) \geq 1$, then $k(G) \leq k(G - [v^*]) + \theta_V(N_G(v^*)) - 1$.*
- *If $k(G - [v^*]) = 0$, then $k(G) \leq \theta_V(N_G(v^*))$.*

Proof. Let $G^* := G - [v^*]$. Let D^* be an acyclic digraph such that $C(D^*) = G^* \cup I_{k(G^*)}$. Let $m := \theta_V(N_G(v^*))$. Since G is diamond-free, $N_G(v^*)$ is the disjoint union of m cliques $C_1, \ldots, C_m$. Note that $\{C_1, \ldots, C_m\}$ is the minimum vertex clique cover of $N_G(v^*)$. First, consider the case where $k(G^*) \geq 1$. Then we can take a vertex $a \in I_{k(G^*)}$. Let D be a digraph defined by $V(D) = (V(D^*) \backslash \{a\}) \cup \{v^*\} \cup \{a_1, \ldots, a_m\}$ and $A(D) = (A(D^*) - \{(v, a) \mid v \in N_{D^*}^-(a)\}) \cup \{(v, v^*) \mid v \in N_{D^*}^-(a)\} \cup \bigcup_{i=1}^m \{(v, a_i) \mid v \in C_i \cup \{v^*\}\}$. Then D is acyclic and $C(D)$ is G together with $k(G^*) + m - 1$ isolated vertices. Thus $k(G) \leq k(G^*) + m - 1$.

Second, consider the case where $k(G^*) = 0$. Let D be a digraph defined by $V(D) = V(D^*) \cup \{v^*\} \cup \{z_1, \ldots, z_m\}$ and $A(D) = A(D^*) \cup \bigcup_{i=1}^m \{(v, z_i) \mid v \in C_i \cup \{v^*\}\}$. Then D is acyclic and $C(D)$ is G together with m isolated vertices. Thus $k(G) \leq m$.   □

Definition 2. A vertex v in a graph G is said to be *essential* if $\theta_V(N_G(v)) \geq 3$, and is said to be *non-essential* if $\theta_V(N_G(v)) \leq 2$. We denote by $V^\circ(G)$ the set of essential vertices in G, i.e., $V^\circ(G) := \{v \in V(G) \mid \theta_V(N_G(v)) \geq 3\}$.

For a vertex v in a graph G, let $N_G^\circ(v)$ be the set of neighbors of v in G that are essential, i.e., $N_G^\circ(v) := N_G(v) \cap V^\circ(G)$. An *essential clique* of a graph G is a maximal clique of G in which all the vertices are essential.

Lemma 2.4. *Let G be a diamond-free graph and let v^* be a vertex of G. Let v be a neighbor of v^*. Then, v is non-essential in G if and only if v is simplicial in $G - [v^*]$.*

Proof. Since $v \in N_G(v^*)$ and G is diamond-free, we have $\theta_V(N_{G-[v^*]}(v)) = \theta_V(N_G(v)) - 1$. Then, $\theta_V(N_G(v)) \leq 2$ if and only if $\theta_V(N_{G-[v^*]}(v)) \leq 1$. Thus, v is non-essential in G if and only if v is simplicial in $G - [v^*]$.   □

Definition 3. For a graph G, let

$$\mu(G) := \sum_{v \in V_{\mathrm{ns}}(G)} (\theta_V(N_G(v)) - 2) = \sum_{v \in V^\circ(G)} (\theta_V(N_G(v)) - 2),$$

where $V_{\mathrm{ns}}(G)$ denotes the set of non-simplicial vertices in G.

Lemma 2.5. *Let G be a diamond-free graph and let v^* be a vertex of G. Then*

$$\mu(G - [v^*]) = \mu(G) - |N_G^\circ(v^*)|.$$

Proof. For $v \in N_G(v^*)$, it follows from Lemma 2.4 that v is essential in G if and only if v is non-simplicial in $G - [v^*]$. Therefore, $V_{\mathrm{ns}}(G-[v^*]) \cap N_G(v^*) = N_G^\circ(v^*)$. For $v \in V(G) \backslash N_G[v^*]$, $v \in V_{\mathrm{ns}}(G-[v^*])$ if and only if $v \in V_{\mathrm{ns}}(G)$ by the definition of $G - [v^*]$. Therefore, $V_{\mathrm{ns}}(G - [v^*]) \backslash N_G[v^*] = V_{\mathrm{ns}}(G) \backslash N_G[v^*]$. Thus it follows that $V_{\mathrm{ns}}(G - [v^*])$ is the disjoint union of $N_G^\circ(v^*)$ and $V_{\mathrm{ns}}(G) \backslash N_G[v^*]$.

For any vertex $v \in N_G^\circ(v^*) \subseteq N_G(v^*)$, we have $\theta_V(N_{G-[v^*]}(v)) = \theta_V(N_G(v)) - 1$ since G is diamond-free. For any vertex $v \in V_{\mathrm{ns}}(G) \backslash N_G[v^*]$, we have $\theta_V(N_{G-[v^*]}(v)) = \theta_V(N_G(v))$. Thus it follows from the definition of μ that $\mu(G - [v^*]) = \mu(G) - |N_G^\circ(v^*)|$.   □

3   Proof of Theorem 1.4

We are now ready to give a proof for Theorem 1.4, that is, we show

$$k(G) \leq \begin{cases} 1 + \frac{1}{2}\mu(G) & \text{if } G \text{ has an isolated vertex,} \\ \frac{3}{2} + \frac{1}{2}\mu(G) & \text{if } G \text{ has a simplicial vertex,} \\ 2 + \frac{1}{2}\mu(G) & \text{otherwise.} \end{cases}$$

Proof of Theorem 1.4. We show by induction on the number $|V(G)| + |E(G)|$ of a graph G. If a graph has one vertex, then the inequality trivially holds. Suppose that the inequality holds for any graph G such that $|V(G)| + |E(G)| \leq m$. Take a graph G with $|V(G)| + |E(G)| = m + 1$.

(**Case 1**) G has an isolated vertex v^*.

Let $G^* := G - v^*$. Then G^* is diamond-free and $\frac{1}{2}\mu(G^*) = \frac{1}{2}\mu(G)$. If $k(G^*) = 0$, then $k(G) = 0$ by Lemma 2.3. Therefore, $k(G) = 0 \leq \frac{1}{2}\mu(G) + 1$. If $k(G^*) \geq 1$, then $k(G) \leq k(G^*) - 1$ by Lemma 2.3. By the induction hypothesis, we have $k(G) \leq k(G^*) - 1 \leq \frac{1}{2}\mu(G^*) + 2 - 1 = \frac{1}{2}\mu(G) + 1$.

(**Case 2**) G has no isolated vertex but G has a simplicail vertex v^*.

Let $G^* := G - [v^*]$. By Lemma 2.2, G^* is diamond-free. By Lemma 2.5, $\frac{1}{2}\mu(G^*) = \frac{1}{2}\mu(G) - \frac{1}{2}|N_G^\circ(v^*)|$. If $k(G^*) = 0$, then $k(G) \leq 1$ by Lemma 2.3. Therefore, $k(G) \leq 1 \leq \frac{1}{2}\mu(G) + \frac{3}{2}$. If $k(G^*) \geq 1$, then $k(G) \leq k(G^*)$ by Lemma 2.3.

(**Case 2-1**) $N_G^\circ(v^*) \neq \emptyset$.

By the induction hypothesis and Lemma 2.5, $k(G) \leq k(G^*) \leq \frac{1}{2}\mu(G^*) + 2 = \frac{1}{2}\mu(G) - \frac{1}{2}|N_G^\circ(v^*)| + 2 \leq \frac{1}{2}\mu(G) + \frac{3}{2}$.

(**Case 2-2**) $N_G^\circ(v^*) = \emptyset$.

By Lemma 2.4, any vertex in $N_G(v^*)$ is a simplicial vertex in G^*. By the induction hypothesis, we obtain $k(G) \leq k(G^*) \leq \frac{1}{2}\mu(G^*) + \frac{3}{2} = \frac{1}{2}\mu(G) + \frac{3}{2}$.

(**Case 3**) G has no simplicial vertex.

In this case, $\theta_V(N_G(v)) \geq 2$ holds for any vertex v of G. If G has no essential vertex, then $\frac{1}{2}\mu(G) = 0$ and G is a quasi-line graph. By Theorem 1.3, we have $k(G) \leq 2$. Now we suppose that G has an essential vertex w^*.

(**Case 3-1**) w^* is not contained in any essential clique of G.

Since any maximal clique of G containing w^* contains a vertex v such that $\theta_V(N_G(v)) = 2$, we can take a vertex v^* such that $v^* \in N_G(w^*)$ and $\theta_V(N_G(v^*)) = 2$. Let $G^* := G - [v^*]$. By Lemma 2.5, $\frac{1}{2}\mu(G^*) = \frac{1}{2}\mu(G) - \frac{1}{2}|N_G^\circ(v^*)|$. If $k(G^*) = 0$, then $k(G) \leq 2$ by Lemma 2.3. Therefore, $k(G) \leq 2 \leq \frac{1}{2}\mu(G) + 2$. If $k(G^*) \geq 1$, then $k(G) \leq k(G^*) + 1$ by Lemma 2.3. Note that $|N_G^\circ(v^*)| \geq 1$ since $w^* \in N_G^\circ(v^*)$.

(**Case 3-1-1**) $|N_G^\circ(v^*)| \geq 2$.

By the induction hypothesis, $k(G) \leq k(G^*) + 1 \leq \frac{1}{2}\mu(G^*) + 2 + 1 = \frac{1}{2}\mu(G) - \frac{1}{2}|N_G^\circ(v^*)| + 2 + 1 \leq \frac{1}{2}\mu(G) + 2$.

(**Case 3-1-2**) $|N_G^\circ(v^*)| = 1$.

From the facts that any vertex in $N_G(v^*) \setminus N_G^\circ(v^*)$ is a simplicial vertex of G^* and that $|N_G(v^*)| \geq 2$, G^* has a simplicial vertex. Then by the induction hypothesis, $k(G) \leq k(G^*) + 1 \leq \frac{1}{2}\mu(G^*) + \frac{3}{2} + 1 = \frac{1}{2}\mu(G) - \frac{1}{2}|N_G^\circ(v^*)| + \frac{3}{2} + 1 = \frac{1}{2}\mu(G) + 2$.

(**Case 3-2**) There exists an essential clique of G containing w^*.

Let K be an essential clique of G containing w^*. Note that $|K| \geq 2$ since w^* is not isolated. Let $G^* := G - E(K)$. Then G^* is diamond-free by Lemma 2.1. It is easy to see that $k(G) \leq k(G^*) + 1$ since we can add one isolated vertex to cover the edges of K. Note that $\theta_V(N_{G^*}(v)) = \theta_V(N_G(v))$ if $v \notin K$ and $\theta_V(N_{G^*}(v)) = \theta_V(N_G(v)) - 1$ if $v \in K$. In addition, $V_{ns}(G) = V_{ns}(G^*)$. Therefore, $\frac{1}{2}\mu(G^*) = \frac{1}{2}\mu(G) - \frac{1}{2}|K|$. By the induction hypothesis, $k(G) \leq k(G^*) + 1 \leq \frac{1}{2}\mu(G^*) + 2 + 1 = \frac{1}{2}\mu(G) - \frac{1}{2}|K| + 2 + 1 \leq \frac{1}{2}\mu(G) + 2$.

Hence the theorem holds. $\qquad\qquad\square$

Acknowledgment. The author is grateful to the anonymous reviewers for careful reading and valuable comments. This work was supported by JSPS KAKENHI Grant Number 15K20885.

References

1. Cohen, J.E.: Interval Graphs and Food Webs: A Finding and a Problem. RAND Corporation Document 17696-PR, Santa Monica (1968)
2. Kim, S.-R., Lee, J.Y., Park, B., Sano, Y.: Competitively tight graphs. Ann. Comb. **17**, 733–741 (2013)
3. Kim, S.-R., Lee, J.Y., Park, B., Sano, Y.: A generalization of Opsut's result on the competition numbers of line graphs. Discret. Appl. Math. **181**, 152–159 (2015)

4. McKay, B.D., Schweitzer, P., Schweitzer, P.: Competition numbers, quasi-line graphs, and holes. SIAM J. Discret. Math. **28**, 77–91 (2014)
5. Opsut, R.J.: On the computation of the competition number of a graph. SIAM J. Algebr. Discret. Methods **3**, 420–428 (1982)
6. Park, B., Sano, Y.: The competition number of a generalized line graph is at most two. Discret. Math. Theor. Comput. Sci. **14**(2), 1–10 (2012)
7. Roberts, F.S.: Food webs, competition graphs, and the boxicity of ecological phase space. In: Alavi, Y., Lick, D.R. (eds.) Theory and Applications of Graphs. Lecture Notes in Mathematics, vol. 642, pp. 477–490. Springer, Heidelberg (1978)
8. Sano, Y.: A generalization of Opsut's lower bounds for the competition number of a graph. Graphs Comb. **29**, 1543–1547 (2013)

On Evasion Games on Graphs

Satoshi Tayu$^{(\boxtimes)}$ and Shuichi Ueno

Department of Information and Communications Engineering,
Tokyo Institute of Technology, Tokyo 152-8550-S3-57, Japan
tayu@eda.ce.titech.ac.jp

Abstract. We consider an evasion game on a connected simple graph. We first show that the pursuit number of a graph G, the smallest k such that k pursuers win the game, is bounded above by the pathwidth of G. We next show that the pursuit number of G is two if and only if the pathwidth of G is one. We also show that for any integer $w \geq 2$, there exists a tree T such that the pursuit number of T is three and the pathwidth of T is w.

1 Introduction

In an evasion game on a connected simple graph, we have k pursuers and an evader. The evader moves invisibly along the edges of the graph. The pursuers must guess the position of the evader. At each round, k pursuers guess at most k vertices. The pursuers win if the current vertex of the evader is contained in the guessed vertices. Otherwise, the evader either stays at its vertex or moves to one of its neighbors. The pursuit number of a graph G, denoted by $\rho(G)$, is the minimum number k such that we have a winning strategy on G for k pursuers. In the active version, the evader is required to move at each round. We denote by $\rho^*(G)$ the pursuit number of a graph G for the active evasion game. We have $\rho(G) \geq \rho^*(G)$ by definition.

We denote the vertex set and the edge set of a graph G by $V(G)$ and $E(G)$, respectively. Let $\mathcal{X} = (X_1, X_2, \ldots, X_r)$ be a sequence of subsets of $V(G)$. The *width* of $\mathcal{X}$ is $\max_{1 \leq i \leq r} |X_i| - 1$. $\mathcal{X}$ is called a *path-decomposition* of G if the following conditions are satisfied:

(i) $\bigcup_{1 \leq i \leq r} X_i = V(G)$;
(ii) for any edge $(u, v) \in E(G)$, there exists an i such that $u, v \in X_i$;
(iii) for all l, m, and n with $1 \leq l \leq m \leq n \leq r$, $X_l \cap X_n \subseteq X_m$.

The *pathwidth* of G, denoted by $\mathrm{pw}(G)$, is the minimum width over all path-decompositions of G [5].

A connected graph of pathwidth one is called a *caterpillar*, which is a nontrivial tree that contains no 2-claw as a subtree, where a k-*claw* is a tree obtained from a complete bipartite graph $K_{1,3}$ by replacing each edge with a path of length k. A 2-*directional orthogonal ray tree (2DORT)* is a tree that contains no 3-claw as a subtree [7]. It is easy to see that the pathwidth of a 2DORT is at most 2. A caterpillar is a 2DORT by definition.

© Springer International Publishing AG 2016
J. Akiyama et al. (Eds.): JCDCGG 2015, LNCS 9943, pp. 253–264, 2016.
DOI: 10.1007/978-3-319-48532-4_23

It has been known that $\rho^*(G) = 1$ if and only if G is a 2DORT [2, 4]. It was recently shown that $\rho^*(G) \leq \text{pw}(G)+1$ for any graph G, and that for any integer $k \geq 2$, there exists a graph G such that $\text{pw}(G) = k$ and $\rho^*(G) = k+1$ [1]. It follows that $\rho^*(T) = O(\log n)$ for any n-vertex tree T, since $\text{pw}(T) = O(\log n)$ for any n-vertex tree T [6]. Very recently, it was shown in [3] that there exists an n-vertex tree T such that $\rho^*(T) = \Omega(\log n)$.

We show the following four theorems.

Theorem 1. *For any graph G, $\rho(G) \leq \text{pw}(G) + 1$. In particular, $\rho(T) = O(\log n)$ for any n-vertex tree T.*

It should be noted that for any integer $k \geq 0$, there exists a graph G such that $\text{pw}(G) = k$ and $\rho(G) = k+1$, since $\rho(G) \geq \rho^*(G)$ for any graph G. Notice also that there exists an n-vertex tree T such that $\rho(T) = \Omega(\log n)$, immediate from a result of [3] mentioned above.

Theorem 2. *$\rho(G) = 2$ if and only if $\text{pw}(G) = 1$.*

It should be noted that $\rho(G) = 1$ if and only if $\text{pw}(G) = 0$ (G has just one vertex), as mentioned in [4].

Theorem 3. *If $\text{pw}(G) = 2$ then $\rho(G) = 3$.*

However, the converse of Theorem 3 does not hold as shown in the following.

Theorem 4. *For any integer $k \geq 2$, there exists a tree T such that $\rho(T) = 3$ and $\text{pw}(T) = k$.*

2   Preliminaries

For a graph G, a k *pursuers' strategy* is a sequence $\mathcal{P} = (P_1, P_2, \ldots, P_r)$ of guessed vertices, where $P_i \subseteq V(G)$ and $|P_i| \leq k$ for any $i \in [r]$, where $[n] = \{1, 2, \ldots, n\}$ for a positive integer n; the pursuers guess the vertices in P_i at the i-th round of the game.

An *evader's strategy* is a sequence $\mathcal{M} = (m_0, m_1, \ldots, m_r)$ of vertices of G such that $m_i = m_{i-1}$ or m_i is adjacent to m_{i-1} for any $i \geq 1$; vertex m_0 is an initial position of the evader, and the evader stays at vertex m_i in the i-th round of the game.

A pursuers' strategy $\mathcal{P} = (P_1, P_2, \ldots, P_r)$ is a *winning strategy* if for any evader's strategy $\mathcal{M} = (m_0, m_1, \ldots, m_r)$, there exists an $i \geq 1$ such that $m_i \in P_i$. The *pursuit number* $\rho(G)$ of G is the minimum k such that there exists a winning strategy for k pursuers on G.

For a pursuer's strategy $\mathcal{P} = (P_1, P_2, \ldots, P_r)$, a vertex $v \in V(G)$ is said to be *contaminated* at the i-th round if there exists an evader's strategy $\mathcal{M} = (m_0, m_1, \ldots, m_r)$ such that $v = m_i$ and $m_j \notin P_j$ for any $j \in [i]$. Otherwise, v is said to be *clear* at the i-th round.

For a vertex $u \in V(G)$, let $N(u) = \{v \mid v \in V(G), (u,v) \in E(G)\} \cup \{u\}$, and for a vertex set $U \subseteq V(G)$, define that $N(U) = \bigcup_{u \in U} N(u)$. For a pursuers' strategy $\mathcal{P} = (P_1, P_2, \ldots, P_r)$, let $\mathcal{D}(\mathcal{P}) = (D_0, D_1, \ldots, D_r)$ be the sequence of *contaminated sets* of vertices for $\mathcal{P}$; D_i is the set of contaminated vertices at the i-th round, where $D_0 = V(G)$. It should be noted that

$$D_i = N(D_{i-1}) - P_i \tag{1}$$

for any $i \in [r]$, and $\mathcal{P}$ is a winning strategy if and only if $D_i = \emptyset$ for some $i \in [r]$.

3   Proof of Theorem 1

We can show that a path-decomposition $\mathcal{X} = (X_1, X_2, \ldots, X_r)$ of G with width k is a winning strategy for $k + 1$ pursuers on G by the same arguments as the proof of $\rho^*(G) \leq \mathrm{pw}(G) + 1$ in [1].

It is shown in [6] that $\mathrm{pw}(T) \leq \log_3(2n + 1)$ for any n-vertex tree T. Thus, we have $\rho(T) = O(\log n)$ for any n-vertex tree T.

4   Proof of Theorems 2 and 3

Lemma 1. *If G contains a cycle then $\rho(G) \geq 3$.*

Proof. It suffices to show that $\rho(C) \geq 3$ for any cycle C. Let $\mathcal{P} = (P_1, P_2, \ldots, P_r)$ be a strategy for two pursuers on C, where $|P_i| \leq 2$ for any $i \in [r]$. We define an evader's strategy $\mathcal{M} = (m_0, m_1, \ldots, m_r)$ as follows. Let m_0 be any vertex in $V(C)$. We recursively define m_i as a vertex in $N(m_{i-1}) - P_i$ for any $i \in [r]$. Notice that $N(m_{i-1}) - P_i \neq \emptyset$, since $|N(m_{i-1})| = 3$ and $|P_i| \leq 2$. Thus, $\mathcal{P}$ is not a winning strategy, since $m_i \notin P_i$ for any $i \in [r]$, and we conclude that $\rho(C) \geq 3$.

Lemma 2. *If G is a 2-claw then $\rho(G) \geq 3$.*

Proof. Let T_2 be a 2-claw shown in Fig. 1. Define that

$$F_j = \{a_j, b_1, b_2, b_3, c\}, \text{ for any } j \in [3], \text{ and}$$
$$F_{j,j'} = \{a_j, a_{j'}, b_j, b_{j'}, c\}, \text{ for any } j \neq j' \in [3].$$

Let $\mathcal{P} = (P_1, P_2, \ldots, P_r)$ be a strategy for two pursuers on T_2, where $|P_i| \leq 2$ for any $i \in [r]$, and $\mathcal{D}(\mathcal{P}) = (D_0, D_1, \ldots, D_r)$ be the sequence of contaminated sets of vertices for $\mathcal{P}$, where $D_0 = V(T_2) = \{a_1, a_2, a_3, b_1, b_2, b_3, c\}$

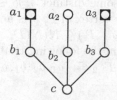

Fig. 1. 2-claw T_2.

Claim 1. *For any $i \in [r]$, $F_j \subseteq N(D_i)$ for some $j \in [3]$ or $F_{j,j'} \subseteq N(D_i)$ for some distinct $j, j' \in [3]$.*

Proof of Claim 1. The proof is by induction on i. We first show that $N(D_1)$ satisfies the claim. Since $D_0 = V(T_2)$, $D_1 = N(D_0) - P_1 = V(T_2) - P_1$ by (1). Let $P_1 = \{u, v\}$. If $(u, v) \notin E(T_2)$ then $N(u) \cap D_1 \neq \emptyset$ and $N(v) \cap D_1 \neq \emptyset$. Thus, $P_1 \subseteq N(D_1)$, and we have $N(D_1) = V(T_2)$, which satisfies the claim. If $(u, v) \in E(T_2)$ then P_1 is $\{a_j, b_j\}$ or $\{b_j, c\}$ for some $j \in [3]$. If $P_1 = \{a_j, b_j\}$ then $c \in D_1$, and so $N(D_1) = V(T_2) - \{a_j\}$, which contains $F_{j',j''}$ for $j' \neq j'' \in [3] - \{j\}$, and we are done. If $P_1 = \{b_j, c\}$ then $P_1 \subseteq N(D_1)$, and we have $N(D_1) = V(T_2)$, which satisfies the claim.

Suppose that Claim 1 holds for $i - 1 \in [r - 1]$, that is, $F_j \subseteq N(D_{i-1})$ for some $j \in [3]$ or $F_{j,j'} \subseteq N(D_{i-1})$ for some distinct $j, j' \in [3]$. We show that Claim 1 also holds for i. We distinguish two cases.

Case 1 $F_j \subseteq N(D_{i-1})$ for some $j \in [3]$: If $|F_j \cap P_i| \leq 1$, then $F_j \subseteq N(F_j - P_i)$, since any vertex of F_j is adjacent with another vertex of F_j. Therefore, $F_j \subseteq N(F_j - P_i) \subseteq N(N(D_{i-1}) - P_i) = N(D_i)$ by (1), and we are done. Thus, we assume in the following that $|F_j \cap P_i| = 2$. Without loss of generality, we assume that $F_1 = \{a_1, b_1, b_2, b_3, c\} \subseteq N(D_{i-1})$. We further distinguish three cases.

Case 1-1 $a_1 \in P_i$: In this case, P_i is $\{a_1, c\}$, $\{a_1, b_1\}$, $\{a_1, b_2\}$, or $\{a_1, b_3\}$. If $P_i = \{a_1, c\}$ then $\{b_1, b_2, b_3\} \subseteq D_i$ by (1). Thus, $N(D_i) = V(T_2)$, which satisfies the claim. If $P_i = \{a_1, b_1\}$ then $\{b_2, b_3, c\} \subseteq D_i$. Thus, $F_{2,3} = \{a_2, b_2, a_3, b_3, c\} \subseteq N(D_i)$, and we are done. If $P_i = \{a_1, b_2\}$ then $\{b_1, b_3, c\} \subseteq D_i$. Thus, $F_{1,3} = \{a_1, b_1, b_2, a_3, b_3, c\} \subseteq N(D_i)$, and we are done. If $P_i = \{a_1, b_3\}$ then $\{b_1, b_2, c\} \subseteq D_i$. Thus, $F_{1,2} = \{a_1, b_1, a_2, b_2, b_3, c\} \subseteq N(D_i)$, and we are done.

Case 1-2 $a_1 \notin P_i$ and $b_1 \in P_i$: In this case, P_i is $\{b_1, b_2\}$, $\{b_1, b_3\}$, or $\{b_1, c\}$. If $P_i = \{b_1, b_2\}$ then $\{a_1, b_3, c\} \subseteq D_i$ and $\{a_1, b_1, b_2, a_3, b_3, c\} \subseteq N(D_i)$ by (1). Thus, we have $F_{1,3} \subseteq N(D_i)$, and we are done. If $P_i = \{b_1, b_3\}$ then $\{a_1, b_2, c\} \subseteq D_i$ and $\{a_1, b_1, a_2, b_2, b_3, c\} \subseteq N(D_i)$. Thus, we have $F_{1,2} \subseteq N(D_i)$, and we are done. If $P_i = \{b_1, c\}$ then $\{a_1, b_2, b_3\} \subseteq D_i$ and $N(D_i) = V(T_2)$, which satisfies the claim.

Case 1-3 $a_1, b_1 \notin P_i$: In this case, P_i is $\{b_2, b_3\}$, $\{b_2, c\}$, or $\{b_3, c\}$. If $P_i = \{b_2, b_3\}$ then $\{a_1, b_1, c\} \subseteq D_i$ by (1) and $\{a_1, b_1, b_2, b_3, c\} \subseteq N(D_i)$. Thus, $F_1 \subseteq N(D_i)$, and we are done. If $P_i = \{b_2, c\}$ then $\{a_1, b_1, b_3\} \subseteq D_i$ and $\{a_1, b_1, b_2, a_3, b_3, c\} \subseteq N(D_i)$. Thus, $F_1 \subseteq N(D_i)$, and we are done. If $P_i = \{b_3, c\}$ then $\{a_1, b_1, b_2\} \subseteq D_i$ and $\{a_1, b_1, a_2, b_2, b_3, c\} \subseteq N(D_i)$. Thus, $F_1 \subseteq N(D_i)$, and we are done.

Case 2 $F_{j,j'} \subseteq N(D_{i-1})$ for some distinct $j, j' \in [3]$: If $|F_{j,j'} \cap P_i| \leq 1$, then $F_{j,j'} \subseteq N(F_{j,j'} - P_i)$, since any vertex of $F_{j,j'}$ is adjacent with another vertex of $F_{j,j'}$. Therefore, $F_{j,j'} \subseteq N(F_{j,j'} - P_i) \subseteq N(N(D_{i-1}) - P_i) = N(D_i)$ by (1), and we are done. Thus, we assume in the following that $|F_{j,j'} \cap P_i| = 2$. Without loss of generality, we assume that $F_{1,2} = \{a_1, a_2, b_1, b_2, c\} \subseteq N(D_{i-1})$. We further distinguish four cases.

Case 2-1 $P_i = \{a_1, a_2\}$: In this case, $\{b_1, b_2, c\} \subseteq D_i$ by (1) and $\{a_1, b_1, a_2, b_2, b_3, c\} \subseteq N(D_i)$. Thus, $F_{1,2} \subseteq N(D_i)$, and we are done.

Case 2-2 $a_1 \in P_i$ and $a_2 \notin P_i$: In this case, P_i is $\{a_1, b_1\}$, $\{a_1, b_2\}$, or $\{a_1, c\}$. If $P_i = \{a_1, b_1\}$ then $\{a_2, b_2, c\} \subseteq D_i$ and $\{b_1, a_2, b_2, b_3, c\} \subseteq N(D_i)$. Thus, $F_2 \subseteq N(D_i)$, and we are done. If $P_i = \{a_1, b_2\}$ then $\{b_1, a_2, c\} \subseteq D_i$ and $\{a_1, b_1, a_2, b_2, b_3, c\} \subseteq N(D_i)$. Thus, $F_{1,2} \subseteq N(D_i)$, and we are done. If $P_i = \{a_1, c\}$ then $\{b_1, a_2, b_2\} \subseteq D_i$ and $\{a_1, b_1, a_2, b_2, c\} \subseteq N(D_i)$. Thus, $F_{1,2} \subseteq N(D_i)$, and we are done.

Case 2-3 $a_1 \notin P_i$ and $a_2 \in P_i$: The proof is similar to the proof of Case 2-2, and omitted.

Case 2-4 $a_1, a_2 \notin P_i$, i.e., $P_i \subseteq \{b_1, b_2, c\}$: In this case, P_i is $\{b_1, b_2\}$, $\{b_1, c\}$ or $\{b_2, c\}$. If $P_i = \{b_1, b_2\}$, then $\{a_1, a_2, c\} \subseteq D_i$ by (1) and $\{a_1, b_1, a_2, b_2, b_3, c\} \subseteq N(D_i)$. Thus, $F_{1,2} \subseteq N(D_i)$, and we are done. If $P_i = \{b_1, c\}$ then $\{a_1, a_2, b_2\} \subseteq D_i$ and $\{a_1, b_1, a_2, b_2, c\} \subseteq N(D_i)$. Thus, $F_{1,2} \subseteq N(D_i)$, and we are done. If $P_i = \{b_2, c\}$ then $\{a_1, b_1, a_2\} \subseteq D_i$ and $\{a_1, b_1, a_2, b_2, c\} \subseteq N(D_i)$. Thus, $F_{1,2} \subseteq N(D_i)$, and we are done.

This completes the proof of Claim 1.   □

Claim 2. $\mathcal{P}$ *is not a winning strategy.*

Proof of Claim 2. From Claim 1, $N(D_i) \geq 5$ for any $i \in [r]$. Therefore, $|D_i| = |N(D_{i-1}) - P_i| \geq 5 - 2$, which means that $D_i \neq \emptyset$ for any $i \in [r]$. Thus, we have the claim.   □

From Claim 2, we conclude that $\rho(T_2) \geq 3$, and we have Lemma 2.   □

If $\rho(G) \leq 2$ then G is a tree by Lemma 1, and G contains no 2-claw by Lemma 2, that is, G is a caterpillar. Thus, if $\rho(G) \leq 2$ then $pw(G) \leq 1$. On the other hand, if $pw(G) \leq 1$ then $\rho(G) \leq 2$ by Theorem 1. Thus, we have the following.

Lemma 3. $\rho(G) \leq 2$ *if and only if* $pw(G) \leq 1$.

Since $\rho(G) = 1$ if and only if $pw(G) = 0$, it follows from Lemma 3 that $\rho(G) = 2$ if and only if $pw(G) = 1$, and we obtain Theorem 2.

If $pw(G) = 2$ then $\rho(G) \geq 3$ by Lemma 3, and $\rho(G) \leq 3$ by Theorem 1. Thus, $\rho(G) = 3$ if $pw(G) = 2$, and we obtain Theorem 3.

5   Proof of Theorem 4

We need some preliminaries. For a graph G and $U \subseteq V(G)$, we denote by $G - U$ the graph obtained from G by deleting all vertices of U and all edges incident to a vertex of U. The following is shown in [8].

Theorem 5. *Let G be a connected graph and $k \geq 1$ be an integer. If G has a vertex v such that $G - \{v\}$ has at least three connected components with pathwidth $k - 1$ or more, then $pw(G) \geq k$.*

Lemma 4. *The pathwidth of a tree T is at most $k \geq 1$ if and only if there exists a path Q in T such that the pathwidth of every connected component of $T - V(Q)$ is at most $k - 1$.*

Proof. We first show the necessity. Let T be a tree of pathwidth at most k, and $\mathcal{X} = (X_1, X_2, \ldots, X_r)$ be a path-decomposition of T with width at most k. Let $u \in X_1$ and $v \in X_r$ be any vertices, and Q be the path connecting u and v in T.

We show that

$$V(Q) \cap X_i \neq \emptyset \tag{2}$$

for any $i \in [r]$. Assume to the contrary that there exists X_j $(1 < j < r)$ with $X_j \cap V(Q) = \emptyset$. Define that $U_1 = (\bigcup_{i \leq j-1} X_i) \cap V(Q)$ and $U_2 = (\bigcup_{i \geq j+1} X_i) \cap V(Q)$. From (iii), we have $U_1 \cap U_2 \subseteq X_j \cap V(Q) = \emptyset$, that is, $U_1 \cap U_2 = \emptyset$. Since Q is a path connecting u and v, Q contains an edge (x, y) with $x \in U_1$ and $y \in U_2$. However, we have no X_i such that $x, y \in X_i$, contradicting to (ii).

If we define $X_i' = X_i - V(Q)$ for any $i \in [r]$, $\mathcal{X}' = (X_1', X_2', \ldots, X_r')$ is a path-decomposition of $T - V(Q)$. Since the width of $\mathcal{X}'$ is at most $k-1$ by (2), we conclude that every connected component of $T - V(Q)$ has pathwidth at most $k-1$. This completes the proof of the necessity.

We next show the sufficiency. Let Q be a path in T such that every connected component of $T - V(Q)$ has pathwidth at most $k-1$. Let $V(Q) = \{v_1, v_2, \ldots, v_p\}$ and $E(Q) = \{(v_i, v_{i+1}) \mid i \in [p-1]\}$. Let $C_1, C_2, \ldots C_q$ be the connected components of $T - V(Q)$ such that if C_i contains a vertex adjacent to v_j then C_{i+1} contains a vertex adjacent to $v_{j'}$ for some $j' \geq j$. For any $i \in [q]$, let $\mathcal{X}^i = (X_1^i, X_2^i, \ldots, X_{r_i}^i)$ be a path-decomposition of C_i of width at most $k-1$, that is, $|X_j^i| \leq k$ for any $j \in [r_i]$. Let $J(i)$ be an integer such that C_i contains a vertex adjacent to $v_{J(i)} \in V(Q)$. Since T is a tree, $J(i)$ is uniquely determined.

A path-decomposition of T of width at most k is constructed as follows. For any $i \in [q]$ and $j \in [r_i]$, let $Y_j^i = X_j^i \cup \{v_{J(i)}\}$ and $\mathcal{Y}^i = (Y_1^i, Y_2^i, \ldots, Y_{r_i}^i)$. If $J(1) \geq 2$, $\mathcal{Z}^0$ is defined as an empty sequence. Otherwise, we define $\mathcal{Z}^0 = (Z_1^0, Z_2^0, \ldots, Z_{J(1)-1}^0)$, where $Z_j^0 = \{v_j, v_{j+1}\}$ for any $j \in [J(1) - 1]$. For any $i \in [q-1]$ and $l \in [J(i+1) - J(i)]$, define that $Z_l^i = \{v_{J(i)+l-1}, v_{J(i)+l}\}$ and $\mathcal{Z}^i = (Z_1^i, Z_2^i, \ldots, Z_{J(i+1)-J(i)}^i)$, where $\mathcal{Z}^i$ is an empty sequence if $J(i+1) = J(i)$. If $J(q) = p$ then $\mathcal{Z}^q$ is defined as an empty sequence. Otherwise, define that $Z_j^q = \{v_{J(q)+j-1}, v_{J(q)+j}\}$ for any $j \in [p - J(q)]$ and $\mathcal{Z}^q = (Z_1^q, Z_2^q, \ldots, Z_{p-J(q)}^q)$. Define that $\mathcal{X}' = (\mathcal{Z}^0, \mathcal{Y}^1, \mathcal{Z}^1, \mathcal{Y}^2, \ldots, \mathcal{Y}^q, \mathcal{Z}^q)$. We denote $\mathcal{X}'$ by $(X_1', X_2', \ldots, X_{r'}')$.

We show that $\mathcal{X}'$ is a path-decomposition for T of width at most k. We first show that $\mathcal{X}'$ satisfies (i). Any vertex of Q is contained in some Z_j^i by definition. Since $\mathcal{X}^i$ is a path-decomposition of C_i, we have $V(C_i) = \bigcup_{j \in [r_i]} X_j^i \subseteq \bigcup_{j \in [r_i]} Y_j^i$. Thus, we conclude that $V(T) = \bigcup_{i \in [r']} X_i'$. Thus, $\mathcal{X}'$ satisfies (i).

We next show that $\mathcal{X}'$ satisfies (ii). We distinguish 3 cases. 1) $(u, v) \in E(C_i)$ for some $i \in [q]$: Since $\mathcal{X}^i$ is a path-decomposition of C_i, $u, v \in X_j^i$ for some $j \in [r_i]$. Thus, we conclude that $u, v \in X_j^i \subset Y_j^i = X_l'$ for some l. 2) $(u, v) \in E(Q)$: Since $u, v \in Z_j^i$ for some i, j, we conclude that $u, v \in X_l'$ for some l. 3) $(u, v) \in E(T)$ such that $u \in V(C_i)$ for some $i \in [q]$ and $v = v_{J(i)} \in V(Q)$: We have $u, v \in Y_j^i = X_l'$ for some l by definition. Thus, $\mathcal{X}'$ satisfies (ii).

We now show that $\mathcal{X}'$ satisfies (iii). Let l, m, and n be arbitrary integers with $1 \leq l \leq m \leq n \leq r'$. If $l = n$, $X_m' = X_l' = X_n'$, and we are done. Assume that $l \leq n - 1$. Let $y(i)$ be an integer such that $X'_{y(i)+1} = Y^i_1$. If $y(i) + 1 \leq l < n \leq y(i) + r_i$ for some $i \in [p]$, $X'_l \cap X'_n = Y^i_{l-y(i)} \cap Y^i_{n-y(i)} = (X^i_{l-y(i)} \cup \{v_{J(i)}\}) \cap (X^i_{n-y(i)} \cup \{v_{J(i)}\}) = (X^i_{l-y(i)} \cap X^i_{n-y(i)}) \cup \{v_{J(i)}\} \subseteq X^i_{m-y(i)} \cup \{v_{J(i)}\} = X'_m$, since $\mathcal{X}^i = (X^i_1, X^i_2, \ldots, X^i_{r_i})$ is a path-decomposition of C_i. Thus, we have $X'_l \cap X'_n \subseteq X'_m$. Otherwise, $X'_l \cap X'_n$ contains only vertices of Q. Since any vertex in Q appears only in consecutive subsets in $\mathcal{X}'$, we have $X'_l \cap X'_n \subseteq X'_m$. Therefore, $\mathcal{X}'$ satisfies (iii).

Thus, $\mathcal{X}'$ is a path-decomposition of T. Since $|X'_i| \leq k + 1$ for any $i \in [r']$ by definition, the width of $\mathcal{X}'$ is at most k, and we conclude that $\mathrm{pw}(T) \leq k$. This completes the proof of the sufficiency.   □

Now, we are ready to prove Theorem 4. We prove the theorem by induction on k. The following lemma is the basis of the induction. For a graph G and $x, y \in V(G)$, a winning strategy $\mathcal{P} = (P_1, P_2, \ldots, P_r)$ on G is called an (x, y)-winning strategy if the following conditions are satisfied:

- $x \in P_i$ if and only if $i = 1$, and
- $y \in P_i$ if and only if $i = r$.

Lemma 5. *For the 2-claw T_2 shown in Fig. 1, $\rho(T_2) = 3$ and $\mathrm{pw}(T_2) = 2$. Moreover, there exists an (x, y)-winning strategy for three pursuers on T_2 for some $x, y \in V(T_2)$.*

Proof. By Lemma 3, $\rho(T_2) \geq 3$. We show that $\mathcal{P} = (\{a_1, b_1, c\}, \{a_2, b_2, c\}, \{a_3, b_3, c\})$ is an (a_1, a_3)-winning strategy for three pursuers on T_2. Let $\mathcal{D}(\mathcal{P}) = (D_0, D_1, D_2, D_3)$ be the sequence of contaminated sets of vertices for $\mathcal{P}$, where $D_0 = V(T_2)$. By (1), $D_1 = N(D_0) - P_1 = \{a_2, b_2, a_3, b_3\}$, $D_2 = N(D_1) - P_2 = \{a_3, b_3\}$, and $D_3 = N(D_2) - P_3 = \emptyset$. Thus, $\mathcal{P}$ is an (a_1, a_3)-winning strategy for three pursuers on T_2, and we conclude that $\rho(T_2) = 3$.

From Theorem 5 and Lemma 4, we have $\mathrm{pw}(T_2) = 2$, since the pathwidth of every connected components of $T_2 - \{c\}$ is 1. This completes the proof of the lemma.   □

We need some more preliminaries to show the induction step. Let G be a graph, $\mathcal{P} = (P_1, P_2, \ldots, P_r)$ be a pursuers' strategy on G, and $\mathcal{D}(\mathcal{P}) = (D_0, D_1, \ldots, D_r)$ be a sequence of contaminated sets of vertices for $\mathcal{P}$. From (1), we have the following.

Lemma 6. *For any $i \in [r - 1]$, if $N(D_i) - D_i \subseteq P_{i+1}$, then $D_{i+1} = D_i - P_{i+1}$.*

For any $U \subseteq V(G)$, define that $N^1(U) = N(U)$, and $N^{i+1}(U) = N(N^i(U))$ for any $i \geq 1$. From (1) we have the following.

$$D_{i+k} \subseteq N^k(D_i) - P_{i+k} \qquad (3)$$

for any $i \in [r - 1]$ and $k \in [r - i]$. For any two vertices $u, v \in V(G)$, we denote by $\mathrm{dist}_G(u, v)$ the distance between vertices u and v in G. From (3), we have the following.

Lemma 7. *If* $\mathrm{dist}_G(u, v) \geq k+1$ *for any* $u \in U$ *and* $v \in D_i$*, then* $U \cap D_{i+k} = \emptyset$.

For a sequence $\mathcal{X} = (X_1, X_2, \ldots, X_r)$, r is called the *length* of $\mathcal{X}$ and denoted by $|\mathcal{X}|$. For sequences $\mathcal{X}^i = (X_1^i, X_2^i, \ldots, X_{r_i}^i)$ for $i \in [n]$, $(\mathcal{X}^1, \mathcal{X}^2, \ldots, \mathcal{X}^n)$ is a sequence obtained by concatenating $\mathcal{X}^1, \mathcal{X}^2, \ldots, \mathcal{X}^n$, that is, $(\mathcal{X}^1, \mathcal{X}^2, \ldots, \mathcal{X}^n) = (X_1^1, X_2^1, \ldots, X_{r_1}^1, X_1^2, X_2^2, \ldots, X_{r_2}^2, \ldots, X_1^n, X_2^n, \ldots, X_{r_n}^n)$. Notice that

$$|(\mathcal{X}^1, \mathcal{X}^2, \ldots, \mathcal{X}^n)| = \sum_{i=1}^n |\mathcal{X}^i| = \sum_{i=1}^n r_i.$$

Now, we are ready to show the induction step.

Lemma 8. *Let* T_{k-1} *($k \geq 3$) be a tree with* $\rho(T_{k-1}) = 3$ *and* $\mathrm{pw}(T_{k-1}) = k - 1$. *Assume that there exists an* (x, y)*-winning strategy for three pursuers on* T_{k-1} *for some* $x, y \in V(T_{k-1})$. *Then, we can construct from three copies of* T_{k-1} *a tree* T_k *with* $\rho(T_k) = 3$ *and* $\mathrm{pw}(T_k) = k$. *Moreover, there exists an* (x, y)*-winning strategy for three pursuers on* T_k *for some* $x, y \in V(T_k)$.

Proof. Let T_{k-1} be a tree with $\rho(T_{k-1}) = 3$ and $\mathrm{pw}(T_{k-1}) = k - 1$, and $\mathcal{P} = (P_1, P_2, \ldots, P_r)$ be an (x, y)-winning strategy for three pursuers on T_{k-1}. Let T_{k-1}^i be a copy of T_{k-1} for $i \in [3]$, and $v^i \in V(T_{k-1}^i)$ be the vertex corresponding to a vertex $v \in V(T_{k-1})$. Let $\mathcal{P}^i = (P_1^i, P_2^i, \ldots, P_r^i)$ be an (x^i, y^i)-winning strategy corresponding to $\mathcal{P}$ for $i \in [3]$. Let Q be a path with $V(Q) = \{q_i \mid i \in [r]\}$ and $E(Q) = \{(q_i, q_{i+1}) \mid i \in [r-1]\}$.

Define that T_k is a tree obtained from T_{k-1}^1, T_{k-1}^2, T_{k-1}^3, and Q by adding three edges (q_1, y^1), (q_r, y^2), and (q_r, x^3) (See Fig. 2).

Since the pathwidth of any connected component of $T_k - \{q_r\}$ is at least $\mathrm{pw}(T_{k-1})$, we have $\mathrm{pw}(T_k) \geq \mathrm{pw}(T_{k-1}) + 1 = k$ by Theorem 5. On the other hand, since T_{k-1}^1, T_{k-1}^2, and T_{k-1}^3 are the connected components of $T_k - V(Q)$, we have $\mathrm{pw}(T_k) \leq \mathrm{pw}(T_{k-1}) + 1 = k$ by Lemma 4. Thus, we have $\mathrm{pw}(T_k) = k$.

We have $\rho(T_k) \geq 3$, since $\rho(T_{k-1}) = 3$. We will show an (x^1, y^3)-winning strategy for three pursuers on T_k, which means that $\rho(T_k) = 3$. Let $h = \lceil r/2 \rceil$,

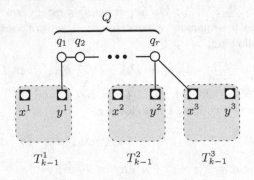

Fig. 2. Tree T_k.

and $\mathcal{R} = (R_1, R_2, \ldots, R_h)$ and $\mathcal{S} = (S_1, S_2, \ldots, S_{r+1})$ be sequences of subsets of $V(T_k)$ defined as follows.

$$R_j = \begin{cases} \{y_1, q_1, q_2\} & \text{if } j = 1, \\ \{q_{2j-2}, q_{2j-1}, q_{2j}\} & \text{if } 2 \le j \le h-1, \\ \{q_{r-2}, q_{r-1}, q_r\} & \text{if } j = h, \end{cases} \tag{4}$$

$$S_j = \begin{cases} \{y^1, y^2, q_1\} & \text{if } j = 1, \\ \{y^2, q_{j-1}, q_j\} & \text{if } 2 \le j \le r, \\ \{q_{r-1}, q_r, x^3\} & \text{if } j = r+1. \end{cases} \tag{5}$$

Define that $\mathcal{P}' = (\mathcal{P}^1, \mathcal{R}, \mathcal{P}^2, \mathcal{S}, \mathcal{P}^3)$. Notice that $|\mathcal{P}'| = |\mathcal{P}^1| + |\mathcal{R}| + |\mathcal{P}^2| + |\mathcal{S}| + |\mathcal{P}^3| = 4r + h + 1$. We now show the following.

Claim 3. $\mathcal{P}'$ is an (x^1, y^3)-winning strategy for three pursuers on T_k.

Proof of Claim 3. Let $r' = |\mathcal{P}'| = 4r + h + 1$ and $\mathcal{D}(\mathcal{P}') = (D_0, D_1, \ldots, D_{r'})$ be the sequence of contaminated sets of vertices for $\mathcal{P}'$. Since $\mathcal{P}^1$ is an (x^1, y^1)-winning strategy on T_{k-1}^1 and $T_{k-1}^1 - \{y^1\}$ is a connected component of $T_k - \{y^1\}$, we have

$$D_r = V(T_k) - V(T_{k-1}^1). \tag{6}$$

Similarly, by noting $|\mathcal{P}^1| + |\mathcal{R}| + |\mathcal{P}^2| = 2r + h$, we also have

$$D_{2r+h} \cap V(T_{k-1}^2) = \emptyset, \tag{7}$$

since $\mathcal{P}^2$ is an (x^2, y^2)-winning strategy on T_{k-1}^2 and $T_{k-1}^2 - \{y^2\}$ is a connected component of $T_k - \{y^2\}$.

(I). $D_{r+i} = V(T_k) - V(T_{k-1}^1) - \{q_j \mid j \in [2i]\}$ for any i with $0 \le i \le h - 1$.

Proof of (I). We show (I) by induction on i. From (6), (I) holds for $i = 0$, since $\{q_j \mid j \in [2i]\} = \emptyset$ if $i = 0$. Let $i \ge 1$ and assume that (I) holds for $i - 1$, that is, $D_{r+i-1} = V(T_k) - V(T_{k-1}^1) - \{q_1, q_2, \ldots, q_{2i-2}\}$. It follows that $N(D_{r+i-1}) - D_{r+i-1} = \{q_{2i-2}\}$, where we assume that $q_{2i-2} = y^1$ if $i = 1$. Therefore, $N(D_{r+i-1}) - D_{r+i-1} \subseteq R_i$ by (4). Thus from Lemma 6, $D_{r+i} = D_{r+i-1} - R_i = V(T_k) - V(T_{k-1}^1) - \{q_j \mid j \in [2i]\}$, and (I) holds for i. $\square$

From (I), we have $D_{r+h-1} = V(T_k) - V(T_{k-1}^1) - \{q_i \mid i \in [2h-2]\}$. Therefore, $N(D_{r+h-1}) - D_{r+h-1} = \{q_{2h-2}\} \subseteq R_h$, and we have

$$D_{r+h} = V(T_k) - V(T_{k-1}^1) - V(Q) \tag{8}$$

by Lemma 6. From Lemma 7 and (8), $D_{2r+h} \cap V(T_{k-1}^1) = \emptyset$, i.e., $D_{2r+h} \subseteq V(T_k) - V(T_{k-1}^1)$. Thus from (7), we have

$$D_{2r+h} \subseteq V(T_k) - (V(T_{k-1}^1) \cup V(T_{k-1}^2)). \tag{9}$$

Let $\mathcal{P}' = (P_1', P_2', \ldots, P_{r'}')$, and $\mathcal{M} = (m_0, m_1, \ldots, m_{r'})$ be an evader's strategy such that $m_i = x^3$ for $i \leq r + h$, and $m_i = q_{2r+h+1-i}$ for $r + h + 1 \leq i \leq 2r + h$. Then, $m_i \notin P_i'$ for any $i \in [2r + h]$. Therefore, $q_1 \in D_{2r+h}$. Similarly, we can prove that $q_i \in D_{2r+h}$ for any $i \in [2r + h]$, and thus, $V(Q) \subseteq D_{2r+h}$. Since $V(T_{k-1}^3) \subseteq D_{2r+h}$, we have

$$V(T_k) - (V(T_{k-1}^1) \cup V(T_{k-1}^2)) = V(Q) \cup V(T_{k-1}^3)$$
$$\subseteq D_{2r+h}. \tag{10}$$

Thus, from (9) and (10), we have

$$D_{2r+h} = V(T_k) - (V(T_{k-1}^1) \cup V(T_{k-1}^2)). \tag{11}$$

(II). $D_{2r+h+i} = V(T_k) - V(T_{k-1}^1) - V(T_{k-1}^2) - \{q_j \mid j \in [i]\}$ for any $i \in [r]$.

Proof of (II). From (11), $N(D_{2r+h}) \cap (V(T_{k-1}^1) \cup V(T_{k-1}^2)) = \{y^1, y^2\}$. Thus from Lemma 6 and (5), we have

$$D_{2r+h+1} = V(T_k) - V(T_{k-1}^1) - V(T_{k-1}^2) - \{q_1\}. \tag{12}$$

We now show that

$$D_{2r+h+i} = V(T_k) - V(T_{k-1}^1) - V(T_{k-1}^2) - \{q_j \mid j \in [i]\} \tag{13}$$

by induction on i. Clearly, (13) holds for $i = 1$ by (12). Assume that (13) holds for $i - 1$ with $i \geq 2$, that is, $D_{2r+h+i-1} = V(T_k) - V(T_{k-1}^1) - V(T_{k-1}^2) - \{q_j \mid j \in [i-1]\}$, and we will show that (13) also holds for i. By induction hypothesis, $N(D_{2r+h+i-1}) - D_{2r+h+i-1} \subseteq \{q_{i-1}, y^2\}$. Thus from Lemma 6 and (5), (13) holds for i. This completes the proof of (II). □

From (5) and (II), we have

$$D_{3r+h+1} = V(T_k) - V(T_{k-1}^1) - V(T_{k-1}^2) - V(Q) - \{x^3\}$$
$$= V(T_{k-1}^3) - \{x^3\}. \tag{14}$$

Therefore, we have $D_{4r+h+1} = \emptyset$, since $\mathcal{P}^3$ is an (x^3, y^3)-winning strategy on T_{k-1}^3 and $T_{k-1}^3 - \{x^3\}$ is a connected component of $T_k - \{x^3\}$. Since $x^1 \in P_i'$ if and only if $i = 1$, and $y^3 \in P_i'$ if and only if $i = r'$, we conclude that $\mathcal{P}'$ is an (x^1, y^3)-winning strategy on T_k, and we have Claim 3. □

This completes the proof of the lemma. □

From Lemmas 5 and 8, we have Theorem 4.

6   Active Pursuit Number of T_k

We show the following for tree T_k defined in the previous section.

Theorem 6. *For any $k \geq 3$, $\rho^*(T_k) = 2$.*

Proof. For a bipartite graph G with a bipartition (B_0, B_1) and $P \subseteq V(G)$, define that $\mathcal{B}_G(P) = \max\{|P \cap B_0|, |P \cap B_1|\}$. For a pursuers strategy $\mathcal{P} = (P_1, P_2, \ldots, P_r)$ on G, $\mathcal{B}_G(\mathcal{P}) = \max\{\mathcal{B}_G(P_i) \mid i \in [r]\}$.

Lemma 9. *For a bipartite graph G, if there exists a winning strategy $\mathcal{P} = (P_1, P_2, \ldots, P_r)$ for the general evasion game on G with $\mathcal{B}_G(\mathcal{P}) \leq l$, then $\rho^*(G) \leq l$.*

Proof of Lemma 9. Let $\mathcal{P} = (P_1, P_2, \ldots, P_r)$ be a winning strategy for the general evasion game on G satisfying $\mathcal{B}_G(\mathcal{P}) \leq l$. Without loss of generality, we assume that r is odd, since otherwise, $\mathcal{P}' = (P_1, P_2, \ldots, P_r, \emptyset)$ is also a winning strategy of odd length on G satisfying $\mathcal{B}_G(\mathcal{P}') \leq l$.

Let (B_0, B_1) be a bipartition of G. Define that $W_i = P_i \cap B_{i \bmod 2}$, i.e.,

$$W_i = \begin{cases} P_i \cap B_0 & \text{if } i \text{ is even, and} \\ P_i \cap B_1 & \text{if } i \text{ is odd,} \end{cases}$$

for any $i \in [r]$. Define also that $W_i = W_{i-r}$ for $r + 1 \leq i \leq 2r$, and $\mathcal{W}^* = (W_1, W_2, \ldots, W_{2r})$. We will show that pursuers' strategy $\mathcal{W}^*$ is a winning strategy on G for the active evasion game.

Let $\mathcal{M}^* = (m_0, m_1, \ldots, m_{2r})$ be any evader's strategy on G for the active evasion game. From the definition of the active evasion game, the evader must move at each round and we have

$$m_i \in B_0 \Leftrightarrow m_{i-1} \in B_1 \text{ for any } i \in [2r]. \tag{15}$$

Since r is odd, we also have

$$m_0 \in B_0 \Leftrightarrow m_r \in B_1. \tag{16}$$

Define that

$$\mathcal{M}^L = (m_0, m_1, \ldots, m_r) \text{ and}$$
$$\mathcal{M}^R = (m_r, m_{r+1} \ldots, m_{2r}).$$

It should be noted that $\mathcal{M}^L$ and $\mathcal{M}^R$ both can be considered as evader's strategies of r rounds for the general evasion game on G. Since $\mathcal{P}$ is a winning strategy on G for the general evasion game, there exist integers α and β with $1 \leq \alpha \leq r < \beta \leq 2r$ such that

$$m_\alpha \in P_\alpha, \text{ and} \tag{17}$$
$$m_\beta \in P_{\beta-r}. \tag{18}$$

We now show that there exists an integer $i \in [2r]$ such that $m_i \in W_i$. We distinguish two cases.

Case 1 $m_0 \in B_0$: From (15) and $m_0 \in B_0$, we have

$$m_i \in B_{i \bmod 2} \tag{19}$$

for any $i \in [r]$. Thus from (17) and (19), we have $m_\alpha \in P_\alpha \cap B_{\alpha \bmod 2}$, i.e., $m_\alpha \in W_\alpha$.

Case 2 $m_0 \in B_1$: From (15) and (16),

$$m_{r+i} \in B_{i \bmod 2} \tag{20}$$

for any $i \in [r]$. Let $\beta' = \beta - r$. From (18) and (20), we have $m_\beta \in P_{\beta'} \cap B_{\beta' \bmod 2} = W_{\beta'}$.

Thus, $\mathcal{W}^*$ is a winning strategy for the active evasion game on G. Since $\mathcal{B}_G(\mathcal{P}) \leq l$, we have $|W_i| \leq l$ for any $i \in [2r]$. Thus, $\rho^*(G) \leq l$, and we have the lemma. $\square$

If $\mathcal{P} = (P_1, P_2, \ldots, P_r)$ is the (x, y)-winning strategy for three pursuers on T_k defined in the previous section then $\mathcal{B}_G(\mathcal{P}) \leq 2$, since $|P_i| = 3$ and every P_i contains a pair of adjacent vertices for any $i \in [r]$. Thus, we have $\rho^*(T_k) \leq 2$ for any $k \geq 2$ by Lemma 9. Since T_k contains a 3-claw if $k \geq 3$, we have $\rho^*(T_k) \geq 2$ if $k \geq 3$, and we conclude that $\rho^*(T_k) = 2$ for any $k \geq 3$. (Notice that $\rho^*(T_2) = 1$, since T_2 is a 2DORT.) This completes the proof of the theorem. $\square$

7   Concluding Remarks

Since it is well-known that the longest path in a tree can be found in linear time, caterpillars and 2DORTs can be recognized in linear time [7]. Therefore, we can decide in linear time whether $\rho^*(G) = 1$ and $\rho(G) = 2$. The complexity of computing $\rho(G)$ and $\rho^*(G)$ is open.

Acknowledgements. The research was partially supported by JSPS KAKENHI Grant Number 26330007.

References

1. Abramovskaya, T.V., Fomin, F.V., Golovach, P.A., Pilipczuk, M.: How to hunt an invisible rabbit on a graph. Eur. J. Comb. **52**, 12–26 (2016)
2. Britnell, J.R., Wildon, M.: Finding a princess in a palace: a pursuit evasion problem. Electron. J. Comb. **20**, 25 (2013)
3. Gruslys, V., Mèrouéh, A.: Catching a mouse on a tree. arXiv.org/abs/1502.06591 (2015)
4. Haslegrave, J.: An evasion game on a graph. Discrete Math. **314**, 1–5 (2014)
5. Robertson, N., Seymour, P.: Graph minors I. Excluding a forest. J. Comb. Theor. Series B **35**, 39–61 (1983)
6. Scheffler, P.: Optimal embedding of a tree into an interval graph in linear time. Ann. Discret. Math. **51**, 278–291 (1992)
7. Shrestha, A.M.S., Tayu, S., Ueno, S.: On orthogonal ray graphs. Discrete Appl. Math. **158**, 1650–1659 (2010)
8. Takahashi, A., Ueno, S., Kajitani, Y.: Minimal acyclic forbidden minors for the family of graphs with bounded path-width. Discrete Math. **127**, 293–304 (1994)

Sudoku Colorings of a 16-Cell Pre-fractal

Hideki Tsuiki[✉] and Yasuyuki Tsukamoto

Graduate School of Human and Environmental Studies,
Kyoto University, Kyoto, Japan
{tsuiki,tsukamoto}@i.h.kyoto-u.ac.jp

Abstract. We study coloring problems of the third-level approximation of a 16-cell fractal. This four-dimensional object is projected to a cube in eight different ways, after which it forms an $8 \times 8 \times 8$ grid of cubes. On each such grid, we can consider two Sudoku-like colorings. Our question is whether it is possible to assign colors to the 8^3 pieces of this pre-fractal object in such a manner that all of its eight cubic projections form Sudoku-like colorings. We analyzed this problem and its variants and constructed solution patterns to the cases they exist. We also enumerated the number of solutions with computer programs for some of the cases.

1 Introduction

An imaginary cube is a three-dimensional object that has square projections in three orthogonal directions, just as a cube has [3,4]. Among imaginary cubes, a hexagonal bipyramid imaginary cube (simply called an H, Fig. 1) is a double imaginary cube, i.e., it is an imaginary cube of two different cubes. Therefore, it has square projections in six ways. In addition, from an H, a double imaginary cube fractal with the similarity dimension two is generated. When the first author designed a sculpture based on the second-level approximation of this fractal, 81 pieces were colored with nine colors so that the colors form a Sudoku solution pattern in each of the six square projections, which form 9×9 grids (Fig. 2) [1]. As the upper-middle picture of Fig. 2 indicates, this coloring pattern is based on simple rules. In [2], he studied this Sudoku coloring problem and showed that it has 140 solutions modulo change of colors and 30 solutions modulo change of colors and congruences of the object. This calculation was first done with a computer program and then performed manually, i.e., it was shown mathematically as a proof. Tsuiki and Yokota also studied this Sudoku coloring problem only for three orthogonal square projections, and they enumerated their solutions using computer programs [5].

In this paper, we report our study of Sudoku colorings of the third level approximation of the 16-cell fractal. This four-dimensional object is projected to a cube in eight different ways, after which it forms an $8 \times 8 \times 8$ grid of cubes. On each such grid of cubes, we can consider two Sudoku-like coloring problems indicated in Fig. 3(a,b). Our question is whether it is possible to assign colors to the 8^3 pieces of this pre-fractal object in such a manner that all of its eight cubic projections form Sudoku-like colorings. We analyzed this problem and

© Springer International Publishing AG 2016
J. Akiyama et al. (Eds.): JCDCGG 2015, LNCS 9943, pp. 265–276, 2016.
DOI: 10.1007/978-3-319-48532-4_24

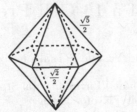
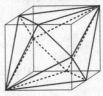

Fig. 1. H (Hexagonal bipyramid imaginary cube). This object is projected to a square in six different ways.

its variants and constructed solution patterns to the cases they exist. We also enumerated the number of solutions with computer programs for some of the cases.

In the next section, we explain a 16-cell and its pre-fractals, and explain Puzzle A and Puzzle B which are the two Sudoku-like coloring problems which we study in this paper. In Sect. 3, we study properties of cubic projections of a 16-cell. We study Puzzle A and its variants in Sect. 4, and Puzzle B and its variants in Sect. 5.

Fig. 2. Fractal Sudoku Sculpture [1,2] (reassembled by Y. Tsukamoto in 2013) (Color figure online).

2   A 16-Cell and Sudoku-like Coloring Problems

A 16-cell is a four-dimensional regular polytope with eight vertices and sixteen regular tetrahedron facets. We first review properties of this object and (pre-) fractals generated by it based on [4]. Then, we explain our Sudoku-like coloring problems.

A 16-cell is a four-dimensional counterpart of a regular octahedron in that it is a cross-polytope. That is, $V = \{(\pm 2, 0, 0, 0), (0, \pm 2, 0, 0), (0, 0, \pm 2, 0),$ $(0, 0, 0, \pm 2)\}$ is the set of vertices of a 16-cell. A 16-cell is also obtained by selecting eight non-adjacent vertices of a hypercube. That is, let V_1 and V_2 be the subsets of $(\pm 1, \pm 1, \pm 1, \pm 1)$ with even and odd number of $+1$ coordinates, respectively. Then, V_1 and V_2 are sets of vertices of 16-cells. If these two 16-cells are projected along each of the four axis of coordinates, then we have cubes. Therefore, a 16-cell is an imaginary 4-cube. Here, an *imaginary n-cube* is an n-dimensional object that has $(n-1)$-dimensional hypercube projections in n orthogonal directions just as an n-dimensional hypercube has. One can see that these cubic projections are projections from four pairs of facets of a 16-cell. Therefore, by symmetry, it also has cubic projections from the other four pairs of facets. Thus, a 16-cel has cubic projections in eight directions and the eight directions are divided into two sets of four mutually orthogonal directions. We call such an object a *double imaginary 4-cube*.

More interestingly, not only a 16-cell but also a fractal based on a 16-cell is a double imaginary 4-cube. Let F_i $(1 \le i \le 8)$ be the homothetic transformations with centers at the eight vertices of a 16-cell S and with scales $1/2$. Here, a homothetic transformation is a similitude that performs no rotations. We define a map G on the space $\mathcal{H}^4$ of non-empty compact subsets of $\mathbb{R}^4$ as $G(X) = \bigcup_{i=1}^{8} F_i(X)$. Since G is a contraction map in $\mathcal{H}^4$, the sequence $S_0 = S, S_1 = G(S_0), S_2 = G(S_1), S_3 = G(S_2), \ldots$ converges to the unique fixedpoint S_∞ of G, which is called the fractal generated by the iterative function system $\{F_i \mid 1 \le i \le 8\}$. One can easily see that S_∞ has the similarity dimension 3, and it is also a double imaginary 4-cube. In addition, not only S_∞ but also S_n, which is a n-th level approximation of S_∞, is a double imaginary hypercube for every $n \ge 1$. We call such an approximation of a fractal a *pre-fractal*.

Since each cubic projection image of S_n consists of a set of 8^n cubes forming a $2^n \times 2^n \times 2^n$ grid, if there is a coloring notion on a $2^n \times 2^n \times 2^n$ grid of cubes, then we have a corresponding coloring notion on the pre-fractal S_n that all the eight cubic projections satisfy the coloring. We consider the case $n = 3$ and consider two Sudoku-like coloring puzzles on a $8 \times 8 \times 8$-grid of cubes.

Puzzle A_0: Assign 64 colors to an $8 \times 8 \times 8$ grid of cubes so that each 8×8-plane (3×8 exist) and each $4 \times 4 \times 4$-block (8 exist) contains all 64 colors (Fig. 3(a)).

Puzzle B_0: Assign 8 colors to an $8 \times 8 \times 8$ grid of cubes so that each 8-sequence (3×64 exist) and each $2 \times 2 \times 2$-block (64 exist) contains all 8 colors (Fig. 3(b)).

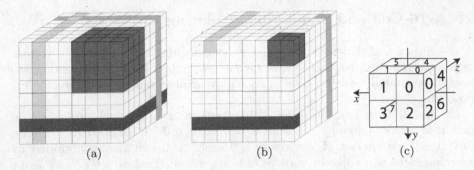

Fig. 3. (a) Sets of pieces with different colors in Puzzle A_0, (b) that of Puzzle B_0, (c) address of cubes in a lattice (Color figure online).

For each of them, there is a Sudoku-like coloring Puzzle A (resp. Puzzle B) of S_3 to assign 64 (resp. 8) colors to the components of S_3 so that each of the eight cubic projections is a solution of Puzzle A_0 (resp. Puzzle B_0). We can also consider four projection variants of these puzzles. That is, fix a set of four orthogonal cubic projections of a 16-cell and consider the condition that each of them is a solution of Puzzle A_0 (or Puzzle B_0). We call them Puzzle A_S and Puzzle B_S, respectively.

Remark: In [3,4], it is shown that (1) all the convex double imaginary 3-cubes are variants of H, (2) a 16-cell is the only convex double imaginary 4-cube, (3) there is no double imaginary n-cube for $n \geq 5$, and (4) H is the only convex double imaginary 3-cube from which one can generate a double imaginary 3-cube fractal with the similarity dimension 2. Therefore, a 16-cell fractal is the only object in three- and higher-dimensional spaces on which one can consider a coloring problem similar to the one on the H pre-fractal.

3   Projections of a 16-Cell Pre-fractal

We study how different cubic projections of S_3 are related.

We first study how vertices of a 16-cell are mapped by cubic projections. Let $v_0 = (2,0,0,0)$, $v_1 = (0,2,0,0)$, $v_2 = (0,0,2,0)$, $v_3 = (0,0,0,2)$ and let S be the 16-cell with the set of vertices $\{\pm v_0, \pm v_1, \pm v_2, \pm v_3\}$. We consider a cube C with vertices $(\pm 1, \pm 1, \pm 1)$ and assign a number in $D = \{i \mid 0 \leq i \leq 7\}$ to the vertices of C so that (x, y, z) is given the number $b(z)b(y)b(x)$ in binary notation with $b(-1) = 0$ and $b(1) = 1$ (c.f. Fig. 3(c)). We sometime use binary notation for elements of D. We define $\mathrm{inv}(i) = 7 - i$ so that i and $7 - i$ specify space diagonal vertices.

For each tuple $(a_0, a_1, a_2, a_3) \in \{-1, 1\}^4$, there is a regular tetrahedron facet F of S with the set of vertices $\{a_0 v_0, a_1 v_1, a_2 v_2, a_3 v_3\}$. S is projected to a cube when it is projected from F, that is, projected along the vector $a_0 v_0 + a_1 v_1 + a_2 v_2 + a_3 v_3$. We fix $a_0 = 1$ and denote by $P_{(a_1, a_2, a_3)}$ this projection. By $P_{(a_1, a_2, a_3)}$, the four space diagonals of S are projected to the four space diagonals of a cube. We transfer cubes obtained by projections to the cube C through rotations and reflections so that v_0 is mapped to vertex 0 and the space diagonal between $\pm v_i$ is mapped to the space diagonal between the vertices i and $\text{inv}(i)$. We redefine this map from S to C as the projection $P_{(a_1,a_2,a_3)}$.

By $P_{(a_1,a_2,a_3)}$, the two regular tetrahedron facets with the vertices $(v_0, a_1v_1, a_2v_2, a_3v_3)$ and $(-v_0, -a_1v_1, -a_2v_2, -a_3v_3)$ preserve their shapes and these lists of vertices are mapped to lists of vertices of regular tetrahedrons in C, that is, vertices $(0, 6, 5, 3)$ and $(7, 1, 2, 4)$ in C. Since v_0 and $-v_0$ are mapped to vertices 0 and 7, respectively, it determines how $P_{(a_1,a_2,a_3)}$ maps vertices of S to vertices of C. That is, v_0 is always mapped to the vertex 0 and v_i ($i = 1, 2, 3$) is mapped to the vertex 2^{i-1} if $a_i = -1$ and to $\text{inv}(2^{i-1})$ if $a_i = 1$.

Instead of studying colorings on S_3, we consider colorings of the $8 \times 8 \times 8$ grid of cubes obtained by projection $P_{(1,1,1)}$. In order to express the constraints caused by other projections, it is important to know how the same piece of S_3 is mapped by different projections. For this, we first study action of $P_{(a_1,a_2,a_3)} \circ P_{(1,1,1)}^{-1}$ on vertices of C, which can be expressed as a permutation on D. The above observation shows that this action is generated by the three transpositions $\alpha = (1, 6)$, $\beta = (2, 5)$, and $\gamma = (3, 4)$. We denote by U the subgroup of the symmetric group S_8 generated by these three transpositions. The order of U is 8 and it is isomorphic to the group $2 \times 2 \times 2$.

Among the eight projections, the projection lines of $P_{(1,1,1)}$, $P_{(-1,-1,1)}$, $P_{(1,-1,-1)}$, and $P_{(-1,1,-1)}$ are mutually orthogonal. One can see that $P \circ P_{(1,1,1)}^{-1}$ for P these four projections cause the identity permutation, $\alpha\beta$, $\beta\gamma$, and $\gamma\alpha$, respectively. They are even permutations and they form the Klein four-group. We denote by U_S this subgroup of U.

Now, we study how each piece of S_3 is mapped to a piece of an $8 \times 8 \times 8$ grid of cubes. As in Fig. 3(c), we assign numbers in D to a $2 \times 2 \times 2$ grid of cubes. We give addresses (i, j, k) for $i, j, k \in D$ to an $8 \times 8 \times 8$ grid of cubes so that i specifies the big block, j specifies the small block, and k specifies the address in the small block. Therefore, a piece of S_3 that is mapped to the cube (i, j, k) by the projection $P_{(1,1,1)}$ is mapped by $P_{(a_1,a_2,a_3)}$ to the cube $(\delta(i), \delta(j), \delta(k))$. Here, $\delta \in U$ is $\alpha^{b_1}\beta^{b_2}\gamma^{b_3}$ where b_i is 0 or 1 depending on whether a_i is 1 or -1.

4 Solutions of Puzzle A

Based on the observation in the previous section, we formalize Puzzle A as a three-dimensional puzzle on a cube.

Let $c : D \times D \times D \to D \times D$ be a coloring of an $8 \times 8 \times 8$ grid of cubes with $D \times D$. The condition that all of the $4 \times 4 \times 4$-blocks contain all the 64 colors can be expressed as follows.

For each $i \in D$, the cardinality of $\{c(i, j, k) \mid j, k \in D\}$ is 64.   (1)

Let $\widetilde{\mathcal{F}}_0 = \{\mathcal{F}_1, \mathcal{F}_2, \mathcal{F}_3\}$ for

$$\mathcal{F}_1 = \{\{0,1,2,3\}, \{4,5,6,7\}\} \, (= \{\{b_z b_y b_x \mid b_z = 0\}, \{b_z b_y b_x \mid b_z = 1\}\}),$$
$$\mathcal{F}_2 = \{\{0,1,4,5\}, \{2,3,6,7\}\} \, (= \{\{b_z b_y b_x \mid b_y = 0\}, \{b_z b_y b_x \mid b_y = 1\}\}),$$
$$\mathcal{F}_3 = \{\{0,2,4,6\}, \{1,3,5,7\}\} \, (= \{\{b_z b_y b_x \mid b_x = 0\}, \{b_z b_y b_x \mid b_x = 1\}\}).$$

The condition of Puzzle A_0 that all of the 8×8-planes contain all the 64 colors can be expressed as the requirement that the following condition holds for every $\mathcal{F} \in \widetilde{\mathcal{F}}_0$.

For each $(F_1, F_2, F_3) \in \mathcal{F} \times \mathcal{F} \times \mathcal{F}$,

the cardinality of $\{c(i,j,k) \mid i \in F_1, j \in F_2, k \in F_3\}$ is 64. $\qquad$ (2)

For Puzzle A_S, we have the condition that $\mathcal{F} = \delta(\mathcal{F}')$ for $\mathcal{F}' \in \widetilde{\mathcal{F}}_0$ and $\delta \in U_S$ also satisfy (2). Here, $\delta(\{F_1, F_2\}) = \{\delta(F_1), \delta(F_2)\}$ and $\delta(\{i,j,k,l\}) = \{\delta(i), \delta(j), \delta(k), \delta(l)\}$. One can see that $\alpha(\beta(\mathcal{F}_1)) = \mathcal{F}_4$ for

$$\mathcal{F}_4 = \{\{0,3,5,6\}, \{1,2,4,7\}\}. \qquad (3)$$

In addition, the set $\widetilde{\mathcal{F}}_S = \{\mathcal{F}_1, \mathcal{F}_2, \mathcal{F}_3, \mathcal{F}_4\}$ is closed under the action of U_S. Therefore, we can restate this condition to the requirement that (2) is satisfied for every $\mathcal{F} \in \widetilde{\mathcal{F}}_S$. Note that, when $\mathcal{F} = \mathcal{F}_4$, the pieces for the cases $F_1 = F_2 = F_3 = \{0,3,5,6\}$ and $F_1 = F_2 = F_3 = \{1,2,4,7\}$ are third-level cubic approximations of the Sierpinski Tetrahedron.

For Puzzle A, we have the condition that $\mathcal{F} = \delta(\mathcal{F}')$ satisfies (2) for $\mathcal{F}' \in \widetilde{\mathcal{F}}_0$ and $\delta \in U$. In this case, $\mathcal{F}$ ranges over all the eight divisions of D into two sets that do not contain i and $\text{inv}(i)$ for every $i \in D$. The cardinality of this set is 8 and we denote this set by $\widetilde{\mathcal{F}}$. We summarize these results.

Proposition 1. *Let $c : D \times D \times D \to D \times D$ be a coloring.*

(a) *c is a solution of Puzzle A_0 if and only if (1) is satisfied and (2) is satisfied for $\mathcal{F} \in \widetilde{\mathcal{F}}_0$.*

(b) *c is a solution of Puzzle A_S if and only if (1) is satisfied and (2) is satisfied for $\mathcal{F} \in \widetilde{\mathcal{F}}_S$.*

(c) *c is a solution of Puzzle A if and only if (1) is satisfied and (2) is satisfied for $\mathcal{F} \in \widetilde{\mathcal{F}}$.*

Our goal is to see whether there exists a solution to these puzzles and to present a solution if it exists.

Theorem 2. *The following is a solution of Puzzle A (and therefore is a solution of Puzzle A_0 and Puzzle A_S).*

$$
\begin{array}{ll}
c(i,j,k) = (j,k) & (i = 0, 7) \\
c(i,j,k) = (\text{inv}(j), k) & (i = 1, 6) \\
c(i,j,k) = (j, \text{inv}(k)) & (i = 2, 5) \\
c(i,j,k) = (\text{inv}(j), \text{inv}(k)) & (i = 3, 4)
\end{array}
$$

Proof. Condition (1) is obviously satisfied. Let $\mathcal{F} \in \widetilde{\mathcal{F}}$ and $F_1, F_2, F_3 \in \mathcal{F}$. We show that the cardinality of $\{c(i,j,k) \mid i \in F_1, j \in F_2, k \in F_3\}$ is 64. It holds because F_1 contains one element of each of $\{0,7\}$, $\{1,6\}$, $\{2,5\}$, $\{3,4\}$, and the four sets $\{(j,k) \mid j \in F_2, k \in F_3\}$, $\{(\text{inv}(j),k) \mid j \in F_2, k \in F_3\}$, $\{(j,\text{inv}(k)) \mid j \in F_2, k \in F_3\}$, $\{(\text{inv}(j),\text{inv}(k)) \mid j \in F_2, k \in F_3\}$ are all disjoint. $\qquad\Box$

We formalized the condition of Puzzle A as a conjunctive normal form Boolean formula and put it into a SAT solver miniSAT version 2.2.0 to obtain some more solutions. Enumeration of all of the solutions of each puzzle is an open problem.

5   Solutions of Puzzle B

We study Puzzle B and its variants. Let $c : D \times D \times D \to D$ be a coloring of an $8 \times 8 \times 8$ grid of cubes with D. The condition that each $2 \times 2 \times 2$-block contains all the 8 colors can be expressed as follows.

For each $(i,j) \in D \times D$, the cardinality of $\{c(i,j,k) \mid k \in D\}$ is 8.   (4)

Let $\widetilde{\mathcal{L}}_0 = \{\mathcal{L}_1, \mathcal{L}_2, \mathcal{L}_3\}$ for

$$\mathcal{L}_1 = \{\{0,1\}, \{2,3\}, \{4,5\}, \{6,7\}\},$$
$$\mathcal{L}_2 = \{\{0,2\}, \{1,3\}, \{4,6\}, \{5,7\}\},$$
$$\mathcal{L}_3 = \{\{0,4\}, \{1,5\}, \{2,6\}, \{3,7\}\}.$$

See Fig. 4 for the meaning of $\mathcal{L}_1$. The condition of Puzzle B_0 that all the 64×3 sequences contain all the 8 colors can be expressed by stating that the following condition holds for every $\mathcal{L} \in \widetilde{\mathcal{L}}_0$.

For each $(L_1, L_2, L_3) \in \mathcal{L} \times \mathcal{L} \times \mathcal{L}$,
the cardinality of $\{c(i,j,k) \mid i \in L_1, j \in L_2, k \in L_3\}$ is 8.   (5)

For Puzzle B_S, we have the condition that $\mathcal{L} = \delta(\mathcal{L}')$ satisfies (5) for $\mathcal{L}' \in \widetilde{\mathcal{L}}_0$ and $\delta \in U_S$. Here, $\delta(\{L_1, L_2, L_3, L_4\}) = \{\delta(L_1), \delta(L_2), \delta(L_3), \delta(L_4)\}$ and $\delta(\{i,j\}) = \{\delta(i), \delta(j)\}$. We define $\widetilde{\mathcal{L}}_S = \{\mathcal{L}_1, \mathcal{L}_2, \mathcal{L}_3, \mathcal{L}_4, \mathcal{L}_5, \mathcal{L}_6\}$ for

$$\mathcal{L}_4 = \{\{0,6\}, \{2,4\}, \{1,7\}, \{3,5\}\},$$
$$\mathcal{L}_5 = \{\{0,5\}, \{1,4\}, \{2,7\}, \{3,6\}\},$$
$$\mathcal{L}_6 = \{\{0,3\}, \{1,2\}, \{4,7\}, \{5,6\}\}.$$

See Fig. 4 for the meaning of $\mathcal{L}_4$. We have $\alpha(\beta(\mathcal{L}_1)) = \mathcal{L}_4$, $\beta(\gamma(\mathcal{L}_2)) = \mathcal{L}_5$, and $\gamma(\alpha(\mathcal{L}_3)) = \mathcal{L}_6$. In addition, $\widetilde{\mathcal{L}}_S$ is closed under the action of U_S. Therefore, the condition can be restated as the requirement that (5) is satisfied for $\mathcal{L} \in \widetilde{\mathcal{L}}_S$.

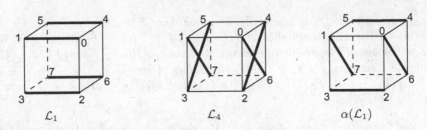

Fig. 4. $\mathcal{L}_1 \in \widetilde{\mathcal{L}}_0$, $\mathcal{L}_4 \in \widetilde{\mathcal{L}}_S$ and $\alpha(\mathcal{L}_1) \in \widetilde{\mathcal{L}}$ considered as relations between vertices of a cube.

The condition that (5) is satisfied for $\mathcal{L}_4$, $\mathcal{L}_5$, and $\mathcal{L}_6$ says that on each of the 24 8×8-planes, different D-colors are assigned to those cubes with the same color in Fig. 5.

Fig. 5. Conditions of Puzzle B_S expressed as colors. (Color figure online)

For Puzzle B, we have the condition that $\mathcal{L} = \delta(\mathcal{L}')$ satisfies (2) for $\mathcal{L}' \in \widetilde{\mathcal{L}}_0$ and $\delta \in U$. In this case, $\mathcal{L}$ ranges over all the 12 divisions of D into four pairs that do not contain i and $\mathrm{inv}(i)$ for $i \in D$ and that if i and j are paired, then $\mathrm{inv}(i)$ and $\mathrm{inv}(j)$ are also paired. We will denote this set by $\widetilde{\mathcal{L}}$. One can see by Fig. 4 that most of the sets of eight cubes which ought to have different colors by condition (5) for $\mathcal{L} = \alpha(\mathcal{L}_1)$ are not on 8×8-planes.

We summarize these results.

Proposition 3. *Let* $c : D \times D \times D \to D$ *be a coloring.*

(a) *c is a solution of Puzzle* B_0 *if and only if (4) is satisfied and (5) is satisfied for* $\mathcal{L} \in \widetilde{\mathcal{L}}_0$.

(b) *c is a solution of Puzzle* B_S *if and only if (4) is satisfied and (5) is satisfied for* $\mathcal{L} \in \widetilde{\mathcal{L}}_S$.

(c) *c is a solution of Puzzle* B *if and only if (4) is satisfied and (5) is satisfied for* $\mathcal{L} \in \widetilde{\mathcal{L}}$.

We obtained the result that Puzzle B has no solution using a computer program, and this fact was verified using the SAT solver miniSAT version 2.2.0.

6   Constructions of Solutions of Puzzle B_0 and Puzzle B_S

We consider D as a linear space over the finite field $F_2 = \{0,1\}$ and use $\oplus$ for addition in D, which is the bitwise "exclusive or" operation. We construct solutions of Puzzle B_0 and Puzzle B_S by considering the address space $D \times D \times D$ and the color space D as linear spaces and restricting the coloring function $c : D \times D \times D \to D$ to linear functions. Thus,

$$c(i,j,k) = c(i,0,0) \oplus c(0,j,0) \oplus c(0,0,k). \tag{6}$$

We call a $2 \times 2 \times 2$ grid of cubes a unit cube and give address to the set of unit cubes with $D \times D$. Through change of colors, we fix the coloring of the unit cube at $(0,0)$ as $c(0,0,k) = k$. We define $\varphi(i) = c(i,0,0)$ and $\psi(j) = c(0,j,0)$. Thus, we have

$$c(i,j,k) = \varphi(i) \oplus \psi(j) \oplus k. \tag{7}$$

Note that the coloring $d : D \to D$ of the unit cube at (i,j) is $d(k) = a \oplus k$ for $a = \varphi(i) \oplus \psi(j)$. We list such colorings in Fig. 6. They are rotations and reflections of the coloring of the unit cube at $(0,0)$. When $a = 001, 010, 100$, it is the image of reflection through the yz-, zx-, xy-coordinate plane, respectively; when $a = 110, 101, 011$, it is the image of a 180-degree rotation along x-, y-, z-coordinate axis, respectively, and when $a = 111$, it is the image of the antipodal map. Note that these maps form an Abelian group of order 8.

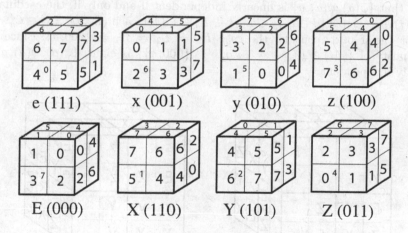

Fig. 6. Coloring $d(k) = a \oplus k$ of the unit cube for $a \in D$. The names E, X, Y, Z, e, x, y, z of the colorings are used in Fig. 7. (Color figure online)

The linear map φ is determined by $\varphi(001), \varphi(010), \varphi(100)$, and ψ is determined by $\psi(001), \psi(010), \psi(100)$. Therefore, the coloring is determined by these six elements of D. We consider conditions on φ and ψ so that (7) forms a solution of Puzzle B_0 and Puzzle B_S. Condition (4) is automatically satisfied. As a part of condition (5) for $\mathcal{L} = \mathcal{L}_1$, it says that $c(i, j, k)$ for $i, j, k \in \{000, 001\}$ are all different and therefore the cardinality of $\{\varphi(i) \oplus \psi(j) \oplus k \mid i, j, k \in \{000, 001\}\}$ is 8. That is, $\{\varphi(001), \psi(001), 001\}$ is linearly independent in the linear space D over F_2. Similarly, condition (5) for $\mathcal{L} = \mathcal{L}_2$ and $\mathcal{L} = \mathcal{L}_3$ imply that $\{\varphi(010), \psi(010), 010\}$ and $\{\varphi(100), \psi(100), 100\}$ are linearly independent in D, respectively. We show in Theorem 4(a) that they form a necessary and sufficient condition for Puzzle B_0.

Theorem 4

(a) *Coloring (7) is a solution of Puzzle B_0 if and only if each of the sets of vectors $\{\varphi(001), \psi(001), 001\}$, $\{\varphi(010), \psi(010), 010\}$, $\{\varphi(100), \psi(100), 100\}$ is linearly independent.*

(b) *Coloring (7) is a solution of Puzzle B_S if and only if, in addition to the three sets of vectors in (a), each of the sets of vectors $\{\varphi(110), \psi(110), 110\}$, $\{\varphi(101), \psi(101), 101\}$, $\{\varphi(011), \psi(011), 011\}$ is linearly independent.*

Note that $\{\varphi(a), \psi(a), a\}$ is linearly independent in D if and only if $\varphi(a), \psi(a), a$ and $\varphi(a)\psi(a)$ are all different.

Proof. Let $a \in \{001, 010, 100, 110, 101, 011\}$. For any pair of elements (b, c) such that $\{a, b, c\}$ is linearly independent, the eight elements $\{000, a, b, c, a \oplus b, b \oplus c, c \oplus a, a \oplus b \oplus c\}$ are all different and the set $\mathcal{L}_a = \{\{000, a\}, \{b, b \oplus a\}, \{c, c \oplus a\}, \{b \oplus c, a \oplus b \oplus c\}\}$ is uniquely determined by a. Since $\mathcal{L}_i (1 \le i \le 6)$ is $\mathcal{L}_a$ for $a = 001, 010, 100, 110, 101, 011$, respectively, in order to show (a) and (b), we prove that $\{\varphi(a), \psi(a), a\}$ is linearly independent if and only if, the cardinality of $\{\varphi(i) \oplus \psi(j) \oplus k \mid i \in L_1, j \in L_2, k \in L_3\}$ is 8 for each $(L_1, L_2, L_3) \in \mathcal{L}_a^3$.

For the if part, consider the case $L_1 = L_2 = L_3 = \{000, a\}$. Since the cardinarity of $\{\varphi(i) \oplus \psi(j) \oplus k \mid i, j, k \in \{000, a\}\}$ is eight, $\{\varphi(a), \psi(a), a\}$ is linearly independent.

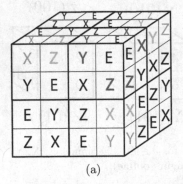
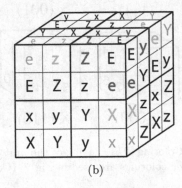

(a)   (b)

Fig. 7. (a)Solution of Puzzle B_0 in Example 5. (b) Solution of Puzzle B_S in Example 6. Meanings of the names E, X, Y, Z, e, x, y, z are given in Fig. 6. Red names are values of φ at $001, 010, 110$ and cyan ones are values of ψ at $001, 010, 110$. They determine green ones as $\varphi(a) \oplus \psi(a)$ for $a \in \{001, 010, 100\}$ and the rest just as a three-dimensional group multiplication table. (Color figure online)

For the only-if part, since $\{\varphi(a), \psi(a), a\}$ is linearly independent, the cardinality of $X = \{\varphi(i) \oplus \psi(j) \oplus k \mid i, j, k \in \{000, a\}\}$ is 8. Let $d_1, d_2, d_3 \in \{000, b, c, b \oplus c\}$. By adding $\varphi(d_1) \oplus \psi(d_2) \oplus d_3$ to each element of X, we have the set $Y = \{\varphi(d_1 \oplus i) \oplus \psi(d_2 \oplus j) \oplus (d_3 \oplus k) \mid i, j, k \in \{000, a\}\}$ whose cardinality is also 8. Let $L_1 = \{d_1, d_1 \oplus a\}$, $L_2 = \{d_2, d_2 \oplus a\}$, $L_3 = \{d_3, d_3 \oplus a\}$. One can see that Y is equal to $\{\varphi(i) \oplus \psi(j) \oplus k \mid i \in L_1, j \in L_2, k \in L_3\}$. Since $\{d, d \oplus a\}$ takes all elements of $\mathcal{L}_a$ if d ranges over $\{000, b, c, b \oplus c\}$, we have the result. $\square$

Example 5. We present a solution of Puzzle B_0 (see Fig. 7(a)). φ and ψ defined as $\varphi(001) = 011$, $\psi(001) = 101$, $\varphi(010) = 110$, $\psi(010) = 011$, $\varphi(100) = 101$, $\psi(100) = 110$ satisfies the condition of Theorem 4(a). This solution consists of only four colorings E, X, Y, Z of the unit cube in Fig. 6, which are the identity map and 180-degree rotations around the three axes. This is not a solution of Puzzle B_S, because $\varphi(110) = 011$, $\psi(110) = 101$, and 110 are linearly dependent.

Example 6. We give a solution of Puzzle B_S. See Fig. 7(b). In order to show its symmetric structure, we present $\mathrm{seq}(i) = (\varphi(i), \psi(i), \varphi(i)\psi(i))$ instead of $(\varphi(i), \psi(i))$ for $i \in \{001, 010, 100\}$.

$$\mathrm{seq}(001) = (100, 011, 111),$$
$$\mathrm{seq}(010) = (110, 111, 001),$$
$$\mathrm{seq}(100) = (111, 010, 101).$$

We can calculate the followings

$$\mathrm{seq}(110) = (001, 101, 100),$$
$$\mathrm{seq}(101) = (011, 001, 010),$$
$$\mathrm{seq}(011) = (010, 100, 110),$$

and see that it satisfies the condition of Theorem 4(b). Using a computer program, we found 480 solutions of Puzzle B_S that satisfy the condition of Theorem 4(b).

Through computer calculation, we have obtained 1148928 solutions of Puzzle B_S modulo change of colors, and this number is verified by a #SAT solver sharpSAT version 1.1 [6]. The enumeration of the solutions of Puzzle B_0 is an open problem.

References

1. Tsuiki, H.: Does it look square? Hexagonal bipyramids, triangular antiprismoids, and their fractals. In: Conferenced Proceedings of Bridges Donostia. Mathematical Connection in Art, Music, and Science, pp. 277–287. Tarquin publications (2007)
2. Tsuiki, H.: SUDOKU colorings of the hexagonal bipyramid fractal. In: Ito, H., Kano, M., Katoh, N., Uno, Y. (eds.) KyotoCGGT 2007. LNCS, vol. 4535, pp. 224–235. Springer, Heidelberg (2008)

3. Tsuiki, H.: Imaginary cubes and their puzzles. Algorithms **5**(2), 273–288 (2012)
4. Tsuiki, H., Tsukamoto, Y.: Imaginary hypercubes. In: Akiyama, J., Ito, H., Sakai, T. (eds.) JCDCGG 2013. LNCS, vol. 8845, pp. 173–184. Springer, Heidelberg (2014)
5. Tsuiki, H., Yokota, Y.: Enumerating 3D-Sudoku solutions over cubic prefractal objects. J. Inf. Process. **20**(3), 667–671 (2012)
6. Thurley, M.: sharpSAT – counting models with advanced component caching and implicit BCP. In: Biere, A., Gomes, C.P. (eds.) SAT 2006. LNCS, vol. 4121, pp. 424–429. Springer, Heidelberg (2006)

The Mathematics of Ferran Hurtado: A Brief Survey

Jorge Urrutia[✉]

Instituto de Matemáticas, Universidad Nacional Autónoma de México,
D.F. Mexico, Mexico
urrutia@matem.unam.mx

Abstract. In this paper, dedicated to our dear friend Prof. Ferran Hurtado, we survey some of the research and open problems arising from his work in collaboration with his many friends and colleagues.

1 Introduction

In this paper, we survey some of the results obtained by our friend and colleague Prof. Ferran Hurtado who left us in October 2014. We have selected some of the topics he studied, and present many of the unsolved problems arising from his work that remain open. Prof. Hurtado had many collaborators, all of whom appreciated his personality, friendship, and his mathematics. Let this paper be a small token of appreciation for all the wisdom and good times he gave all of us. Due to the large extent of the research carried out by Professor Hurtado, it is not possible to cover most of his work.

1.1 The Beginning of Ferran's Mathematics Career

Ferran had a late start in his career as a researcher; he defended his Ph.D. thesis [51] in 1993, when he was already forty-one years old. The thesis was awarded the *Premio Extraordinario de Doctorado UPC 1995*.

The first areas in which Ferran was interested were polygonizations, triangulations, and visibility problems. His first paper, "Poligonizaciones simples," appeared in the *Actas del III Encuentro de Geometría Computacional* held in Zaragoza, Spain in 1992. In 1993 he published three papers, two in conference proceedings, "El número de triangulaciones de un polígono" in the *Actas del IV Encuentro de Geometría Computacional* [53], and his first paper in CCCG, "Looking through a window" [52]. In this paper he studied problems on optimizing the angle of vision of an object, in which the set of positions of the viewer is restricted, say to a line segment. His first journal publication, "Updating polygonizations" [3] also appeared in 1993. In that paper, the authors studied polygonizations that are robust when faced with changes in the positions of

---

J. Urrutia—Research supported in part by SEP-CONACYT of México, Proyecto 80268.

J. Akiyama et al. (Eds.): JCDCGG 2015, LNCS 9943, pp. 277–292, 2016.
DOI: 10.1007/978-3-319-48532-4_25

their vertices. They also studied the problem of finding the maximum distance the vertices of a polygon could be moved away from their position in such a way that the topology on the boundary of the polygon (or its convexity) remains the same. Two follow-up papers appeared in 1994 [5], and 1995. In the last paper they studied the tolerance of arrangements of line segments [6].

His second journal paper, "Ears of triangulations and Catalan numbers" [54] was published in 1996 with Marc Noy. They proved that the number of triangulations of a convex polygon with k leaves is

$$\frac{n}{k}2^{n-2k}\binom{n-4}{2k-4}\mathcal{C}_{k-2}$$

where $\mathcal{C}_n$ is the nth Catalan number.

In his third journal paper, "Hiding points in arrangements of segments" [58] also published in 1996, the following problem was studied: Given a set S of n disjoint line segments, how many points can be placed on the plane in such a way that the line segment joining any two of them intersects an element of S? We say that these points are hidden from each other. They proved that for any set of n disjoint line segments we can hide at least $\sqrt{n}$ points. The proof uses the well known result by Erdős and Szekeres that any sequence of n different numbers contains an increasing or a decreasing subsequence with at least $\sqrt{n}$ elements.

These papers established some of the main topics in which Ferran would work in the years to come; geometric graphs, triangulations, point sets on the plane, and Erdős–Szekeres type problems on sets of points on the plane.

In what follows we will review papers in which Ferran participated, covering the following subjects: triangulations of point sets, geometric graphs and problems on empty polygons in point sets among others. All point sets considered in this paper will be assumed to be in general position.

2   Triangulations

A *triangulation* $\mathcal{T}$ of a point set P is a set of interior disjoint triangles whose vertices are elements of P such that their union is the convex hull of P and no element of $\mathcal{T}$ contains an element of P in its interior. An edge e of a triangulation $\mathcal{T}$ is flippable if it belongs to two triangles of $\mathcal{T}$ whose union is a convex quadrilateral Q; see Fig. 1. Flipping e means deleting e from $\mathcal{T}$ and replacing it by the second diagonal of Q. The vertices of the graph of triangulations of a point set P is the set of triangulations of P, two of which are adjacent if we can obtain one from the other by performing an edge flip. One of Ferran's favourite areas of work was the study of the set of triangulations of point sets and polygons, and related problems. He was interested in studying problems such as

- Finding bounds on the number of triangulations of point sets
- Studying the minimum number of flippable edges that a triangulation may have

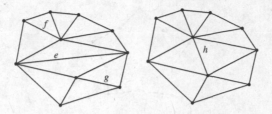

Fig. 1. Flipping edge e. Edges e, f, and g can be flipped simultaneously.

- For colored point sets, the existence of monochromatic empty triangles
- For the graph of triangulations of point sets, parameters such as its connectivity, chromatic number, the existence of hamiltonian cycles, etc.

A well known result of Ajtai et al. [19] states that the number of plane geometric graphs that a point set has is bounded by c^n for a constant c. The best upper bound currently known for c is 87.53; see Sharir and Sheffer [66].

Problems on finding bounds on the number of triangulation $t(P)$ of a polygon were studied in [55]. Sharp bounds on $t(P)$ for simple polygons with k reflex vertices were obtained in the same paper. Let t_i be the number of triangulations of a convex polygon with i vertices. They proved that

$$\left(t_{\lceil(n+k)/(k+1)\rceil}\right)^s \left(t_{\lceil(n+k)/(k+1)\rceil}\right)^{k+1-s} \leq t(P) \leq \sum_{i=0}^{m}(-1)^i \left\| \begin{matrix} \gamma, \overset{u}{..}, \gamma, \gamma+1, \overset{v}{..}, \gamma+1 \\ i \end{matrix} \right\|$$

where s is the residue of dividing $n+k$ by $k+1$, $\gamma = \lfloor k/(n-k)\rfloor$, v is the residue of dividing k by $n-k$, and $u = n-k-v$, and $\left\| \begin{matrix} \alpha_1, \ldots, \alpha_m \\ i \end{matrix} \right\|$ is number of different ways there are of selecting i non-crossing diagonals from a set of reflex diagonals; see [55] for more details. Using the previous result, it follows that the number of triangulations of the point set called *the double circle* [18] is $\sqrt{12}^{2n-\Theta(\log n)}$, which is $o(\mathcal{C}_n)$. The double circle has $2n$ points, n of which are the vertices of a convex polygon, and the remaining n are placed close to the midpoints of the edges of the polygon; see Fig. 2. It is known that the double circle is the only point configuration that minimises the number of triangulations it has for $n \leq 11$. In the same paper [18], they prove that the number of triangulations that any set of n points in the plane has is bounded below by $\Omega(2.33^n)$, $n \geq 1212$. This bound has been improved to $\Omega(2.4317^n)$ in [67].

In [42] it is proved that the so-called double chain has $\Theta^*(8^n)$ triangulation. For some time, it was conjectured that this was the point set configuration that maximised the number of triangulations it has. This conjecture was false, as the double zig-zag chain [16] was proved to have $\Theta^*(\sqrt{72}^n)$ triangulations. The best lower bound on the maximum number of triangulations a point set has, $\Omega(8.65^n)$, was established in [31].

Fig. 2. The double circle with 10 vertices.

2.1   Flipping Edges in Triangulations

The study of edge flipping in triangulations was one of Ferran's favourite subjects of study; see [27]. In [57], it is proved that any triangulation of a point set with n points has at least $\lceil (n-4)/2 \rceil$ edges that can be flipped. The bound is tight. They also show that $O(n+k^2)$ edge flips are always sufficient to transform any triangulation of a polygon Q with k reflex vertices into any other triangulation of Q. In the same paper, they show a polygon with $2n$ vertices and two triangulations of the polygon such that to transform one into the other takes $\Omega(n^2)$ edge flips; see Fig. 3.

In [37], the problem of allowing simultaneous flips is studied. Two flippable edges of a triangulation T can be flipped simultaneously if no triangle of T contains both of them. They prove that any two triangulations of a convex polygon can be transformed into each other using at most $O(\log n)$ parallel flips. They also prove that any two triangulations of a point set or a polygon with n vertices can be transformed into each other using at most a linear number of parallel flips. These bounds are tight. In [68] it was proved that any triangulation of a point set has at most $\frac{n-4}{5}$ edges that can be flipped simultaneously; the bound is tight.

In Ferran et al. [56] the authors introduced a hierarchy on the set of triangulations of a polygon that allowed them to give relatively easy and purely combinatorial proofs of the fact that the graph of triangulations of a convex polygon is Hamiltonian, and that its vertex connectivity is $n-3$. The first result was proved first in Lucas et al. [62]. The second follows from a result proved by Lee [61] that the graph of triangulations of a convex n-gon is the skeleton of a $(n-3)$-polytope. On the other hand, in [10], they prove that the graph of triangulations of polyhedral surfaces is not connected, even for surfaces whose points are in convex position.

Some open problems on flips presented in [27] are the following:

Problem 1. Can every triangulation of a point set in the plane be transformed into a Hamiltonian triangulation by a sequence of $o(n^2)$ edge flips?

Problem 2. Can every triangulation on n points be transformed into a Hamiltonian triangulation by a sequence of $o(n)$ simultaneous edge flips?

Note that in these problems we are working with triangulations of point sets. For triangulations as graphs (that is maximal planar graphs), it is known that

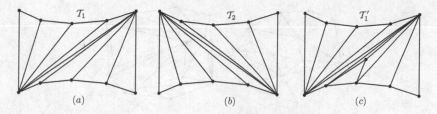

Fig. 3. Two triangulation (a) and (b) with quadratic flip distance.

any triangulation can be transformed into a 4-connected triangulation using a simultaneous edge flip [26].

Suppose that we have two triangulations T_1 and T_2 of a polygon P. If we add a Steiner point u to P that falls in the interior of triangle t_1 of T_1, and a triangle t_2 in T_2, we can obtain two triangulations T_1' and T_2' of $P \cup \{p\}$ by connecting u to the vertices of t_1 and t_2 respectively. In Fig. 3(c), we show the triangulation T_1' obtained by adding a Steiner point p to the triangulation T_1 in Fig. 3(a). It is easy to see that if we add the same point to the triangulation in Fig. 3(b), we can go from triangulation T_1' to T_2' with $4n - 6$ edge flips. This suggests the following new problem.

Take any two triangulations T_1 and T_2 of a polygon P. As above, one at a time, add k Steiner points $p_1, \ldots, p_k$ thus obtaining from T_1 and T_2 two triangulations T_1'' and T_2'' of $P \cup \{p_1, \ldots, p_k\}$.

Problem 3. Given two triangulations T_1 and T_2 of a polygon P, is it always possible to add a sub-linear number of Steiner points $p_1, \ldots, p_k$ in the interior of P such that the flip distance between the resulting triangulations T_1'' and T_2'' of P is linear?

The equivalent question is open for point sets.

2.2 Compatible Triangulations

An intriguing open problem in which Ferran worked was posed in [13]. Let P and P' be two point sets with n points in general position, and assume that their convex hulls have the same number of elements. Two triangulations of P and P' are compatible if there is a mapping of their vertices, edges and faces that preserves adjacencies and incidences.

Conjecture 1. [13] Any two point sets P and P' with n elements such that their convex hulls have the same number of vertices have compatible triangulations.

I believe that this conjecture is false, but, like many others, have failed to disprove it. On the positive side it was proved in [13] that if the number of points in the interior of the convex hulls of P and P' is at most three, then Conjecture 1 is true, and that if the number of points in the interior of the convex hulls of P and P' is k, then if $k - 3$ Steiner points are added to each of P and P', Conjecture 1 holds (Fig. 4).

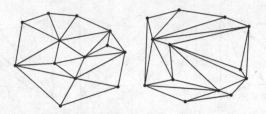

Fig. 4. Two compatible triangulations.

3   Erdős–Szekeres Type Problems on Point Sets

Given a set of points P in the plane, a *k-hole* of P is a subset of points of P with k elements that are the vertices of a convex polygon containing no element of P in its interior. The study of the existence of k-holes in point sets was started by Erdős and Szekeres in [34], where they proved that any point set with n elements always has a logarithmic size subset whose elements are in convex position. In [33], Erdős asked for the existence of k-holes. He asked whether any large enough point set always has a k-hole, for any k. This was disproved by Horton [50], who constructed point sets containing no 7-holes.

In a seminal paper [30], Ferran and several of his co-authors studied the existence of monochromatic 3-holes (called *monochromatic empty triangles* in the following) in families of colored point sets. They proved that any bicolored point set always has an empty monochromatic triangle, and indeed a linear number of them. They also showed that the Horton point sets can be colored with three colors in such a way that no empty monochromatic triangle exists. A question arising from the results in [30] is the following:

Problem 4. Is it true that any bicolored point set contains a superlinear number of empty monochromatic triangles?

This problem was solved in [14] where it is proved that any bichromatic point set with n elements has at least $\Omega(n^{5/4})$ monochromatic triangles. This bound was improved to $\Omega(n^{4/3})$ in [65].

Conjecture 2. [14] Any bichromatic point set has a quadratic number of empty monochromatic triangles.

Another open question posed in [30] is the following:

Conjecture 3. [30] Is it true that any large enough bichromatic point set P always has an empty monochromatic convex quadrilateral?

In Fig. 5 we show a bichromatic point set with 12 points that has no monochromatic 4-hole.

In [17] it was proved that if we allow a hole to be non-convex and P contains at least 5044 points, then Conjecture 3 is true. Recently, it was proved that any

Fig. 5. A point set with no monochromatic 4-hole.

large enough bicolored point set has a monochromatic convex quadrilateral with at most one point in its interior [46].

In [72] Urrutia proved that any sufficiently large 4-colored point set in $\mathbb{R}^3$ has an empty monochromatic tetrahedron. In fact we believe that this is true for any k-colored point set in $\mathbb{R}^3$ that is large enough. The following conjecture by Urrutia would imply Conjecture 3:

Conjecture 4. Any large enough point set in general position in $\mathbb{R}^3$ contains a tetrahedralization with a superlinear number of tetrahedra.

In [17], Aichholzer et al. posed the following problem:

Problem 5. Is it true that any sufficiently large k-colored point set in R^3 in general position contains a convex monochromatic polyhedron that is the union of two interior disjoint tetrahedra that share a face?

They did not specify any specific values of k, but we believe that this is true for any k.

3.1   Measures of Convexity of Point Sets

Convexity is an important area of study in mathematics. Strictly speaking, one might think that convexity has nothing to do with point sets. Nevertheless we can associate some measure of convexity to a point set as follows.

We say that a point set P is in *convex position* if the elements of P are the vertices of a convex polygon. In this sense, we may say that a point set in convex position is *convex*. There are several ways to measure *how convex* a point set is. In [21] some measures of this property are studied. The convex partition number $\mathcal{K}(P)$ of a point set P is the smallest number of disjoint subsets $S_1, \ldots, S_k$ of P such that each S_i is in convex position, and for $1 \leq i < j \leq k$ the convex hulls of S_i and S_j are disjoint; see [70,71]. A polygonization of a point set P is a simple polygon such that its vertices are all of the elements of P. The reflexivity of a polygon [21] is the number of reflex vertices it has. The reflexivity of a point set P, denoted as $\rho(P)$, is the smallest reflexivity of all polygonizations of P. For example if P contains one or two points in the interior of its convex hull, then its reflexivity is one. The point set shown in Fig. 6 has reflexivity 2.

Theorem 1. *[21] The reflexivity of a point set P is at most $\lceil \frac{n_i}{2} \rceil$, where n_i is the number of points of P that belong to the interior of the convex hull of P. The bound is tight.*

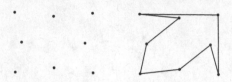

Fig. 6. A point set whose reflexivity is 2.

It is not hard to see that the upper bound is attained with the double circle. It follows easily [21] that $\lfloor \frac{n}{4} \rfloor \leq \rho(P) \leq \lceil \frac{n}{2} \rceil$. The upper bound was improved to $\frac{5}{12}n + O(1)$ in [8].

Let P be a point set. It is easy to see that if we add Steiner points to P placed in the interior of the convex hull of P, the reflexivity of P may decrease or increase. Two interesting open problems are the following:

Conjecture 5. [21] $\rho(P') \geq \rho(P)/2$.

Conjecture 6. [21] $\rho(P) = O(\mathcal{K}(P))$.

3.2 k-convex Polygons and Point Sets

A polygon Q is called k-convex if no line intersects the interior of Q in more than k open intervals. A point set is k-convex if there is a polygonization of P which is k-convex; see [11,12].

Recognizing 2-convex polygons can be done in $O(n \log n)$ time, and recognizing k-convex polygons for $k \geq 4$ takes $O(n^2)$ time [11]. An open problem is that of deciding the complexity of recognizing 3-convex polygons.

Regarding k-convex point sets, we have the following result:

Theorem 2. *[12] Any set P of n points is $O(\sqrt{n})$-convex. The bound is tight.*

In the same paper, it is proved that every set of n points in general position contains a 2-convex subset of size $\Omega(\log^2 n)$.

It is known that to partition the convex hull of a point set P into convex polygons with vertices in P sometimes $n + \sqrt{2(n-3)}$ polygons are necessary [44].

Problem 6. [12] Is it always possible to decompose a given planar point set with a sublinear number of 2-convex polygons?

3.3   Coloring Arrangements of Lines

In [25] the following problem was studied: Given a simple arrangement $\mathcal{A}$ of lines in the plane, what is the minimum number c of colors required so that we can color all lines in a way that no cell of the arrangement is monochromatic? They call c the *chromatic number* of $\mathcal{A}$. They proved that there are arbitrarily large arrangements with chromatic number two, the chromatic clases are the solid and dotted lines; see Fig. 7.

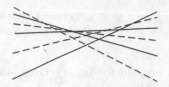

Fig. 7. An arrangement with chromatic number 2.

They also proved that there are arrangements of lines for which $\Omega(\log n/\log\log n) \leq c$, and that for any arrangement of lines $c \leq O(\sqrt{n})$. The upper bound was improved to $\Omega(\sqrt{n/\log n})$ in [9].

The following problem was posed in [25]:

Problem 7. Close the gap between the lower and the upper bound on the chromatic number of arrangments of lines.

4   Packing Trees in Complete Geometric Graphs

Let G be a geometric graph. We say that two geometric graphs H_1 and H_2 can be packed into G if G contains two edge-disjoint subgraphs isomorphic to H_1 and H_2. We will assume that G, H_1 and H_2 have the same number of vertices. The following result was proved by Hedetniemi et al. [49].

Theorem 3. *Let T_1 and T_2 be non-star trees with n vertices. Then there exists a packing of T_1 and T_2 into a complete graph K_n.*

In García et al. [38] the following conjecture was posed:

Conjecture 7. Let T_1 and T_2 be non-star trees with n vertices. Then there exists a simple plane graph G such that T_1 and T_2 can be packed into G.

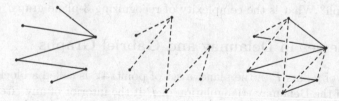

Fig. 8. Packing two isomorphic trees into the same 6-point set. Note that the union of both trees is a planar graph (embedded in the sphere).

García et al. proved their conjecture for the case when T_1 and T_2 are isomorphic, and with the additional constraint that the vertices of T_1 and T_2 are points in convex position; see Fig. 8. We can think of their result as drawing T_1 and T_2 on a coin, with their vertices and some edges of T_1 and T_2 drawn on the rim of the coin, and some edges of T_1 in the top face of the coin, and some edges of T_2 in the bottom face. They also proved:

Theorem 4. *Let T_1 be any tree with n vertices which is different from a star and let T_2 be a path of order n. Then there exists a tight planar packing of T_1 and T_2.*

Conjecture 7 has been verified for binary trees [45], when T_1 is a non-star tree and T_2 is a caterpillar [64], and when T_1 is a non-star and T_2 is obtained from a star by subdividing its edges [32]. It has also been proved for the case when T_1 and T_2 are spider trees [36]. Frati [35] proved that Conjecture 7 is true for trees of diameter at most four, none of which is a star. Görlich [47] proved that two copies of C_n can be packed on a planar graph, for $n = 6$ and $n \geq 8$. She also proved that two copies of some unicycles can also be packed on a planar graph.

In a recent paper, García et al. [39] studied the following problem: Given a point set S, two plane geometric graphs with vertex set S are said to be compatible when their union is a plane geometric graph. They prove:

Theorem 5. *[39] Let S be a point set on the plane and T a plane spanning tree of S. Then there is a spanning tree T' of S compatible with T that has at most $\frac{n-3}{4}$ edges in common with T. Some point sets S have a plane spanning tree T such that any spanning geometric tree of S compatible with T has at least $\frac{n-2}{5}$ edges in common with T.*

Problem 8. Close the gap between the lower and the upper bound of Theorem 5.

More results on biplane geometric graphs can be found in [40,41] where they show among other results that recognizing 2-plane graphs can be done in $O(n \log n)$ time, that every sufficiently large point set admits a 5-connected biplane graph, and that there are arbitrarily large point sets that do not admit any 6-connected biplane graph.

Problem 9. Can we generalise some of the previous results on 2-plane graphs to 3-plane graphs? In particular, what is the maximum number of edges of a 3-plane graph? What is the complexity of recognizing 3-plane graps?

5   Witnesses in Delaunay and Gabriel Graphs

Given a set of points P in the plane, a set of points W is called a blocking set of witnesses of the Delaunay triangulation of P if the interior of any circle passing through two points in P contains an element of W [22]. A similar definition for blocking sets of witnesses for Gabriel graphs and rectangle graphs are defined in [23,24]. In the first case, we want to block circles whose diameters are segments joining pairs of points in P, and in the second, rectangles such that two of their opposite corners belong to P.

In [22] it was proved that any point set P always has a blocking set of witnesses of the Delaunay triangulation of P with at most $2n - 2$ elements. See Fig. 9.

This result was improved in [15] to $\frac{3n}{2}$ for arbitrary point sets P, and to $\frac{5n}{4}$ when the elements of P are in convex position.

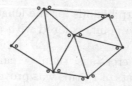

Fig. 9. The non-solid points are a blocking set of witnesses of the Delaunay triangulation of the set of black points.

Conjecture 8. [15] n points are necessary and sufficient to block the Delaunay triangulation of any set of n points in convex position.

In [23] they prove the following result:

Theorem 6. $n - 1$ *witnesses are always sufficient to eliminate all edges of an n-vertex Gabriel graph, while $\frac{3}{4}n - o(n)$ are sometimes necessary.*

Problem 10. Close the gap between the lower and the upper bounds in Theorem 6.

If we want to block the rectangles having two opposite vertices in P, the following essentially tight result is known:

Theorem 7. *[24] Asymptotically $2n - \Theta(\sqrt{n})$ points are necessary to block all rectangles having two opposite corners in P.*

6 Alternating Paths

Let P be a bichromatic set of $2n$ points in the plane whose elements have been colored n red and n blue. An alternating path of P is a simple polygonal path $\mathcal{W}$ all of whose vertices are in P, with each edge in $\mathcal{W}$ having a red and a blue vertex. Alternating paths were first studied in [20] where a quadratic time algorithm was given to decide if a spanning alternating path exists when the elements of P are in convex position. It was proved in [1] that any bichromatic point set has an alternating path that covers at least half of its vertices; see also [4]. Ferran's interest in the problem in this section was evident. This is what he used to call *a poisoned problem.*

The problem of determining the length of the longest alternating path that a bichromatic point set admits is wide open, even for point sets in convex position. In [60] it was proved that any bichromatic point set in convex position has an alternating path with at least $n + c\sqrt{n/\log n}$. This was improved to $n + \Omega(\sqrt{n})$ [48]. The best upper bound known today is asymptotically $\frac{4}{3}n$, proved independently and at about the same time in [1,60]. It is conjectured [60] that this is the real upper bound. We would be happy to prove:

Conjecture 9. Suppose that the elements of P are in convex position. Then there is a constant $c > \frac{1}{2}$ such that P has an alternating path covering cn elements of P.

Surprisingly, there are sharp bounds for the length of the longest alternating path for sets of points with $3n$ points colored with three colors, n of each color. The bound is $2n$ [63].

If we allow at most $n - 1$ crossings, then we can always find an alternating cycle covering all of the elements of P; this was proved by Kano and Kaneko [59]. The bound is best possible. Moreover, if we allow edges to be crossed at most once, then it is always possible to find alternating paths and cycles covering all the elements of P [29]. For double chains [57] (e.g. the set of vertices of the polygon in Fig. 3(a)), Chibuka et al. [28] proved:

Theorem 8. *Let (C_1, C_2) be a double chain with $2n$ points, n blue and n red, and let $|C_i| \geq \frac{1}{5}(|C_1| + |C_2|)$ for $i = 1, 2$. Then (C_1, C_2) has a simple alternating path. Moreover such a path can be found in linear time.*

7 Augmenting the Connectivity of Geometric Graphs

In [2] Abellanas et al. started the study of increasing the connectivity (vertex and edge) of plane geometric graphs by adding edges to them while maintaining planarity. In [2] it is proved that every *geometric* path whose vertices are n points in general position can be augmented to a 2-edge connected plane geometric graph by adding $\frac{n}{2}$ edges, and this bound is the best possible; see Fig. 10. In the same paper it is also also proved that the vertex connectivity of geometric graphs with k cut-vertices can be increased to two by adding k edges. This bound is also best possible.

Fig. 10. The dotted edges increase the edge connectivity of the black geometric path.

Abellanas et al. also proved that every plane geometric tree G with $n \geq 6$ vertices can be completed to a 2-edge connected plane geometric graph by adding at most $\lfloor \frac{2n}{3} \rfloor$ edges. They also conjectured $\frac{n}{2}$ edges would suffice.

This was proved false by Tóth [69], who constructed examples requiring $\frac{17}{3}n - O(1)$ edges. This was improved by García and Tejel [43] to $\frac{6}{11}n - O(1)$; see Fig. 11.

Problem 11. Close the gap between the lower and the upper bound on the number of edges required to complete any geometric tree to a 2-edge connected geometric plane graph.

We believe that the correct upper bound for this problem is close to $\frac{6}{11}n - O(1)$, as García and Tejel's example suggests.

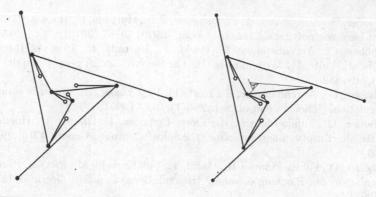

Fig. 11. García and Tejel's recursive construction of trees requiring $\frac{6}{11}n - O(1)$ additional edges to make them 2-edge connected.

8   Final Remarks

We would like to mention that the recently published Computational Geometry Column 61 [7] was dedicated to Prof. Ferran Hurtado. It reviews some results and open problems that appeared in several papers co-authored by Prof. Hurtado. Some (but not all) of these problems also appear here.

References

1. Abellanas, M., García, A., Hurtado, F., Tejel, J.: Caminos alternantes. In: Proceedings of the X Encuentros de Geometría Computacional: Sevilla, Junio 16–17, pp. 7–12 (2003)
2. Abellanas, M., García, A., Hurtado, F., Tejel, J., Urrutia, J.: Augmenting the connectivity of geometric graphs. Comput. Geom. **40**(3), 220–230 (2008)
3. Abellanas, M., García, J., Hernández, G., Hurtado, F., Serra, O., Urrutia, J.: Updating polygonizations. Comput. Graph. Forum **12**(3), 143–152 (1993)
4. Abellanas, M., García, J., Hernández, G., Noy, M., Ramos, P.A.: Bipartite embeddings of trees in the plane. Discrete Appl. Math. **93**(2), 141–148 (1999)
5. Abellanas, M., Hurtado, F., Ramos, P.A.: Tolerance of geometric structures. In: CCCG, pp. 250–255 (1994)
6. Abellanas, M., Hurtado, F., Ramos, P.A.: Tolerancia de arreglos de segmentos. In: VI Encuentros de Geometría Computacional: Barcelona, 5-6-7 de julio de, Departament de Matemàtica Aplicada II, Universitat Politècnica de Catalunya: Actas, pp. 77–84. Departament de Matemàtica Aplicada II (1995)
7. Abrego, B., Dumitrescu, A., Fernández, S., Tóth, C.D.: Computational geometry column 61. ACM SIGACT News **46**(2), 65–77 (2015)
8. Ackerman, E., Aichholzer, O., Keszegh, B.: Improved upper bounds on the reflexivity of point sets. Comput. Geom. **42**(3), 241–249 (2009)
9. Ackerman, E., János, P., Pinchasi, R., Račić, R., Tóth, G.: A note on coloring line arrangements. Electron. J. Combin. **21**(2), Paper p2.23 (2012)
10. Aichholzer, O., Alboul, L.S., Hurtado, F.: On flips in polyhedral surfaces. Int. J. Found. Comput. Sci. **13**(02), 303–311 (2002)

11. Aichholzer, O., Aurenhammer, F., Demaine, E.D., Hurtado, F., Ramos, P., Urrutia, J.: On k-convex polygons. Comput. Geom. **45**(3), 73–87 (2012)
12. Aichholzer, O., Aurenhammer, F., Hackl, T., Hurtado, F., Pilz, A., Ramos, P., Urrutia, J., Valtr, P., Vogtenhuber, B.: On k-convex point sets. Comput. Geom. **47**(8), 809–832 (2014)
13. Aichholzer, O., Aurenhammer, F., Hurtado, F., Krasser, H.: Towards compatible triangulations. Theor. Comput. Sci. **296**(1), 3–13 (2003)
14. Aichholzer, O., Fabila-Monroy, R., Flores-Peñaloza, D., Hackl, T., Huemer, C., Urrutia, J.: Empty monochromatic triangles. Comput. Geom. **42**(9), 934–938 (2009)
15. Aichholzer, O., Fabila-Monroy, R., Hackl, T., Van Kreveld, M., Pilz, A., Ramos, P., Vogtenhuber, B.: Blocking delaunay triangulations. Comput. Geom. **46**(2), 154–159 (2013)
16. Aichholzer, O., Hackl, T., Huemer, C., Hurtado, F., Krasser, H., Vogtenhuber, B.: On the number of plane geometric graphs. Graphs Comb. **23**(1), 67–84 (2007)
17. Aichholzer, O., Hackl, T., Huemer, C., Hurtado, F., Vogtenhuber, B.: Large bichromatic point sets admit empty monochromatic 4-gons. SIAM J. Discrete Math. **23**(4), 2147–2155 (2010)
18. Aichholzer, O., Hurtado, F., Noy, M.: A lower bound on the number of triangulations of planar point sets. Comput. Geom. **29**(2), 135–145 (2004)
19. Ajtai, M., Chvátal, V., Newborn, M.M., Szemerédi, E.: Crossing-free subgraphs. North-Holland Math. Stud. **60**, 9–12 (1982)
20. Akiyama, J., Urrutia, J.: Simple alternating path problem. Discrete Math. **84**(1), 101–103 (1990)
21. Arkin, E.M., Mitchell, J.S.B., Fekete, S.P., Hurtado, F., Noy, M., Sacristán, V., Sethia, S.: On the reflexivity of point sets. In: Aronov, B., Basu, S., Pach, J., Sharir, M. (eds.) Discrete and Computational Geometry, pp. 139–156. Springer, New York (2003)
22. Aronov, B., Dulieu, M., Hurtado, F.: Witness (delaunay) graphs. Comput. Geom. **44**(6), 329–344 (2011)
23. Aronov, B., Dulieu, M., Hurtado, F.: Witness gabriel graphs. Comput. Geom. **46**(7), 894–908 (2013)
24. Aronov, B., Dulieu, M., Hurtado, F.: Witness rectangle graphs. Graphs Comb. **30**(4), 827–846 (2013)
25. Bose, P., Cardinal, J., Collette, S., Hurtado, F., Korman, M., Langerman, S., Taslakian, P.: Coloring and guarding arrangements. Discrete Math. Theor. Comput. Sci. **15**(3), 139–154 (2013)
26. Bose, P., Czyzowicz, J., Gao, Z., Morin, P., Wood, D.R.: Simultaneous diagonal flips in plane triangulations. J. Graph Theory **54**(4), 307–330 (2007)
27. Bose, P., Hurtado, F.: Flips in planar graphs. Comput. Geom. **42**(1), 60–80 (2009)
28. Cibulka, J., Kynčl, J., Mészáros, V., Stolař, R., Valtr, P.: Hamiltonian alternating paths on bicolored double-chains. In: Tollis, I.G., Patrignani, M. (eds.) GD 2008. LNCS, vol. 5417, pp. 181–192. Springer, Heidelberg (2009)
29. Claverol, M., Garijo, D., Hurtado, F., Cuevas, D.L., Seara, C.: The alternating path problem revisited. In: Proceedings XIV Spanish Meeting on Computational Geometry, EGC 2011, Alcalá de Henares, Spain, June 27–30, 2011, pp. 115–118. Universidad de Sevilla (2013)
30. Devillers, O., Hurtado, F., Károlyi, G., Seara, C.: Chromatic variants of the erdos-szekeres theorem on points in convex position. Comput. Geom. **26**(3), 193–208 (2003)

31. Dumitrescu, A., Schulz, A., Sheffer, A., Tóth, C.D.: Bounds on the maximum multiplicity of some common geometric graphs. SIAM J. Discrete Math. **27**(2), 802–826 (2013)
32. Enomoto, H., Kanda, K., Masui, T., Oda, Y., Ota, K.: Private communication
33. Erdös, P.: On some problems of elementary and combinatorial geometry. Annali di Matematica pura ed applicata **103**(1), 99–108 (1975)
34. Erdös, P., Szekeres, G.: A combinatorial problem in geometry. Compos. Math. **2**, 463–470 (1935)
35. Frati, F.: Planar packing of diameter-four trees. In: Proceedings of the XXI Canadian Conference on Computational Geometry: Vancouver, BC, August 17–19, 2009, pp. 95–98 (2009)
36. Frati, F., Geyer, M., Kaufmann, M.: Planar packing of trees and spider trees. Inform. Process. Lett. **109**(6), 301–307 (2009)
37. Galtier, J., Hurtado, F., Noy, M., Perennes, S., Urrutia, J.: Simultaneous edge flipping in triangulations. Int. J. Comput. Geom. Appl. **13**(02), 113–133 (2003)
38. García, A., Hernando, C., Hurtado, F., Noy, M., Tejel, J.: Packing trees into planar graphs. J. Graph Theory **40**(3), 172–181 (2002)
39. García, A., Huemer, C., Hurtado, F., Tejel, J.: Compatible spanning trees. Comput. Geom. **47**(5), 563–584 (2014)
40. García, A., Hurtado, F., Korman, M., Matos, I., Saumell, M., Silveira, R.I., Tejel, J., Tóth, C.D.: Geometric biplane graphs i: maximal graphs. Graphs Comb. **31**(2), 407–425 (2015)
41. García, A., Hurtado, F., Korman, M., Matos, I., Saumell, M., Silveira, R.I., Tejel, J., Tóth, C.D.: Geometric biplane graphs ii: graph augmentation. Graphs Comb. **31**(2), 427–452 (2015)
42. García, A., Noy, M., Tejel, J.: Lower bounds on the number of crossing-free subgraphs of k_n. Comput. Geom. **16**(4), 211–221 (2000)
43. García, A., Tejel, J.: Private communication
44. García-Lopez, J., Nicolás, C.M.: A counterexample about convex partitions. In: IV Jornadas de Matemática Discreta y Algorítmica 2004, Cercedilla, 5–8 Septiembre, 2004, p. 213 (2004)
45. Geyer, M., Hoffmann, M., Kaufmann, M., Kusters, V., Tóth, C.D.: Planar packing of binary trees. In: Dehne, F., Solis-Oba, R., Sack, J.-R. (eds.) WADS 2013. LNCS, vol. 8037, pp. 353–364. Springer, Heidelberg (2013). doi:10.1007/978-3-642-40104-6_31
46. González-Martínez, A.C., Cravioto-Lagos, J., Urrutia, J.: Almost empty monochromatic polygons in planar point sets. In: Actas XVI Spanish Meeting on Computational Geometry, Barcelona, 1–3 de julio, 2015, pp. 81–84 (2015)
47. Görlich, A.: Packing cycles and unicyclic graphs into planar graphs. Demonstr. Math. XLI I(4), 673–679 (2009)
48. Hajnal, P., Mészáros, V.: Note on noncrossing path in colored convex sets. Discr. Math. Theor. Comput. Sci. (2015, accepted)
49. Hedetniemi, S.M., Hedetniemi, S.T., Slater, P.J.: A note on packing two trees into k_n. Ars Combin. **11**, 149–153 (1981)
50. Horton, J.D.: Sets with no empty convex 7-gons. Can. Math. Bull. **26**(4), 482 (1983)
51. Hurtado, F.A.: Problemes geomètrics de visibilitat. Ph.D. thesis, Departamento de Matemática Aplicada i Telemàtica, UPC (1993)
52. Hurtado, F.: Looking through a window. In: Proceedings of the Fifth Canadian Conference on Computational Geometry: University of Waterloo, August 5–9, 1993, pp. 234–239 (1993)

53. Hurtado, F., Marc, N., El número de triangulaciones de un polígono. In: Proceedings of the IV Encuentros de Geometría Computacional: Granada, pp. 1–6 (1993)

54. Hurtado, F., Noy, M.: Ears of triangulations and catalan numbers. Discrete Math. **149**(1), 319–324 (1996)

55. Hurtado, F., Noy, M.: Triangulations, visibility graph and reflex vertices of a simple polygon. Comput. Geom. **6**(6), 355–369 (1996)

56. Hurtado, F., Noy, M.: Graph of triangulations of a convex polygon and tree of triangulations. Comput. Geom. **13**(3), 179–188 (1999)

57. Hurtado, F., Noy, M., Urrutia, J.: Flipping edges in triangulations. Discrete Comput. Geom. **22**(3), 333–346 (1999)

58. Hurtado, F., Serra, O., Urrutia, J.: Hiding points in arrangements of segments. Discrete Math. **162**(1), 187–197 (1996)

59. Kaneko, A., Kano, M., Yoshimoto, K.: Alternating hamilton cycles with minimum number of crossings in the plane. Int. J. Comput. Geom. Appl. **10**(01), 73–78 (2000)

60. Kynčl, J., Pach, J., Tóth, G.: Long alternating paths in bicolored point sets. Discrete Math. **308**(19), 4315–4321 (2008)

61. Lee, C.W.: The associahedron and triangulations of the n-gon. Eur. J. Comb. **10**(6), 551–560 (1989)

62. Lucas, J.M., Vanbaronaigien, D.R., Ruskey, F.: On rotations and the generation of binary trees. J. Algorithms **15**(3), 343–366 (1993)

63. Merino, C., Salazar, G., Urrutia, J.: On the length of longest alternating paths for multicoloured point sets in convex position. Discrete Math. **306**(15), 1791–1797 (2006)

64. Oda, Y., Ota, K.: Tight planar packings of two trees. In: Twenty-second European Workshop on Computational Geometry Delphi, Greece March 27–29, 2006, p. 215 (2006)

65. Pach, J., Tóth, G.: Monochromatic empty triangles in two-colored point sets. Discrete Appl. Math. **161**(9), 1259–1261 (2013)

66. Sharir, M., Sheffer, A.: Counting plane graphs: cross-graph charging schemes. Comb. Probab. Comput. **22**(06), 935–954 (2013)

67. Sharir, M., Sheffer, A., Welzl, E.: On degrees in random triangulations of point sets. J. Comb. Theory Ser. A **118**(7), 1979–1999 (2011)

68. Souvaine, D.L., Tóth, C.D., Winslow, A.: Simultaneously flippable edges in triangulations. In: Márquez, A., Ramos, P., Urrutia, J. (eds.) EGC 2011. LNCS, vol. 7579, pp. 138–145. Springer, Heidelberg (2012). doi:10.1007/978-3-642-34191-5_13

69. Tóth, C.D.: Connectivity augmentation in planar straight line graphs. Eur. J. Comb. **33**(3), 408–425 (2012)

70. Urabe, M.: On a partition into convex polygons. Discrete Appl. Math. **64**(2), 179–191 (1996)

71. Urabe, M.: On a partition of point sets into convex polygons. In: Proceedings of the Ninth Canadian Conference on Computational Geometry, August 11–13, Queen's Universiy, Kingston, Ontario, pp. 179–191 (1997)

72. Urrutia, J.: Coloraciones, tetraedralizaciones, y tetraedros vacíos en coloraciones de conjuntos de puntos en $\mathbb{R}^3$. In: Proceedings of the X Encuentros de Geometría Computacional: Sevilla, Junio 16–17, 2003, pp. 95–100 (2003)

Erratum to: Discrete and Computational Geometry and Graphs

Jin Akiyama[1](✉), Hiro Ito[2], Toshinori Sakai[3], and Yushi Uno[4]

[1] Tokyo University of Science, Tokyo, Japan
ja@jin-akiyama.com
[2] The University of Electro-Communications, Tokyo, Japan
[3] Tokai University, Tokyo, Japan
[4] Osaka Prefecture University, Sakai, Japan

Erratum to:
J. Akiyama et al. (Eds.):
Discrete and Computational Geometry and Graphs, LNCS,
DOI: 10.1007/978-3-319-48532-4

The sequence of the editor names was incorrect. Yushi Uno was not listed as volume editor. The updated online version of the cover and the front matter pages III–V can be found at DOI: 10.1007/978-3-319-48532-4

The updated original online version for this book can be found at DOI: 10.1007/978-3-319-48532-4

© Springer International Publishing AG 2016
J. Akiyama et al. (Eds.): JCDCGG 2015, LNCS 9943, p. E1, 2016.
DOI: 10.1007/978-3-319-48532-4_26

Author Index

Printed in the United States
By Bookmasters